Toxic Hazard Assessment
of Chemicals

Toxic Hazard Assessment of Chemicals

Edited by

Mervyn Richardson, BSc., C.Chem., F.R.S.C.,
C.Biol., M.I.Biol., M.I.W.E.S.
Environment Group, The Royal Society of Chemistry

The Royal Society of Chemistry
Burlington House, London W1V 0BN

7346 - 711X

CHEMISTRY

British Library Cataloguing in Publication Data

Toxic hazard assessment of chemicals
  1. Toxicity testing
  I. Richardson, Mervyn   II. Royal Society
  of Chemistry
  615.9'07   RA1199

  ISBN 0-85186-897-5

Printed in Great Britain by
Whitstable Litho Ltd., Whitstable, Kent

Foreword

The ultimate objective of those working on assessing toxic hazards associated with chemicals, which must be to provide a sound body of knowledge underpinned by validated theory, is still some way off. Nevertheless, the scientist responsible for recommending acceptable or safe levels of exposure to particular substances must use to best advantage present knowledge, which, although it may be incomplete, is still considerable. This book gives information and advice from many of those concerned with research, application and legislation relevant to toxic hazards and should materially assist in the formulation of such recommendations.

It is important to note that assessing the probability that a chemical will be toxic to man depends on several factors. In order to make the perfect assessment in a particular instance, one would need to be an expert in many areas, such as methods of chemical analysis for tiny concentrations, the physical chemistry of the relevant environment and of the pharmacology and toxicology of the substance. Such a combination of specialisms must be very rare, but a study of the methods, and particularly of the traps and pitfalls to be avoided, which are described here by the specialist authors should help the reader to make the best possible assessment. It should also help those working in the field to decide on the best direction for future research.

Many of the chapters are based on presentations made at several scientific meetings (given in detail in the Preface). The most important conclusions arising from these presentations appear to be, first, that good quality data are vital - many problems in assessing toxic risk have sprung from poor quality data. Secondly, that the uncertain step of extrapolating experiments with mammals to the human situation can only be made more certain by determining the chemical-biological mechanisms, metabolic routes, pharmocokinetics, and tissue dose of the proximate toxin responsible for toxicological effects on animal tissues. Only by doing this and by finding ways to monitor effects in humans at the early stage before these effects become irreversible would the ultimate objective be realized of making sound predictions in particular instances.

JOHN M. WARD, CHAIRMAN,
PUBLICATIONS AND INFORMATION BOARD,
ROYAL SOCIETY OF CHEMISTRY

Preface

Mankind, other animals, and the environment are continually exposed to potential toxicants. Such exposure can arise from chemicals and materials handled at home, at work, or from food, water, or air.

Hazards can occur from exposure to both man-made chemicals and natural products. In recent times, man-made chemicals have become essential to man's well-being. Certain risks are therefore inevitable, but when problems occur the greatest use should be made of the information available to minimize any recurrence and understand the mechanisms involved.

For the purpose of this book, hazard is defined as: the set of inherent properties of a chemical substance or mixture which makes it capable of causing adverse effects on man or the environment, when a particular degree of exposure occurs; and risk is defined as: the predicted or actual frequency of occurrence of an adverse effect of a chemical substance or mixture from a given exposure to humans or the environment. These definitions are consistent with those recommended by the Organization for Economic Co-operation and Development and the Royal Society.

The aim of the book is to provide basic guidance on means of retrieving, validating, and interpreting data in order to make a toxicological hazard assessment upon a chemical; it is not a dictionary of hazards associated with any specific chemical.

Over the past few years, the Royal Society of Chemistry has organized a number of meetings and workshops dealing with hazard assessment of chemicals; this book brings together some of the chemistry and other sciences which are necessary in the multi-disciplinary approach required in toxic hazard assessment.

Much of the content of the book has previously been discussed at the following Royal Society of Chemistry sponsored meetings:

Workshop on Information Retrieval of Environmental Chemicals, London, 29th February 1984,
Environmental Chemicals, Health and Behaviour, London, 23rd October 1984,
Physico-Chemical Properties and Their Role in Environmental Health Assessment, Canterbury, 1st-3rd July 1985,
Chemical Pollution of Water Supplies - Hazard Assessment and Emergency Action, York, 18th-20th September 1985,
Hazard Assessment: Chemicals, part of the Royal Society of Chemistry Annual Chemical Congress, University of Warwick, 8-11th April 1986.

The Editorial Board has attempted to minimize overlap between chapters; however, in some cases, and in particular in dealing with such important topics as hazard, risk, and the 6th Amendment Directive, some duplication between chapters has been inevitable. Such repetition should enhance the content of the book in view of the various and diverse opinions expressed by authors from the United Nations, EEC, UK Government Departments, multi-national industry, and research associations.

Section I, Introduction, outlines the work of the Health, Safety, and Environment Committee of the Royal Society of Chemistry.

Section II deals with retrieval of data and includes chapters on both computer and manual methods of data retrieval, including an indication of how to obtain information from various international and national government sources. A chapter on data selection and quality connects this section with the next.

Section III reviews the problems of selecting information and common sources of error in the verification of data.

Interpretation of data is dealt with in some depth in Section IV, with introductory chapters on the interpretation of both toxicity and ecotoxicity data. This section will be useful to those scientists and others who are concerned with the notification of 'new' chemicals. Other chapters cover carcinogenic risk of chemicals to humans and the interpretation of toxicity data in relation to colorants. The concluding chapter deals with the interpretation of epidemiological data.

Section V is concerned with risk assessment and case histories. The introductory chapter outlines how the developed world has seen some dramatic changes in hazards faced by its inhabitants over the past 100 years; the concern of how to reduce the rate of premature death still further, coupled with the potential hazards in the environment, is stressed. Concepts on how to assess and manage risks arising from hazards associated with chemicals are explained with reference to case studies on di-(2-ethylhexyl) phthalate, asbestos and lead. Matters of absolute safety, acceptable and accepted risk, perception of risk, voluntary and involuntary risk, and the offsetting of benefits are outlined. Chapters in this section also deal with risk to work forces, and risks to non-target organisms. The concluding chapter outlines the differences in risk assessment between Great Britain and the United States.

Section VI deals with chemically related legislation including a general background of EEC Directives, an industrialist's view on worldwide legislation and a concluding chapter on the important topic of labelling.

Finally, the book includes a Glossary of Terms, a Compendium of Useful Addresses, and a detailed Index.

As Editor and a member of the Royal Society of Chemistry's Environment Group, Chemical Information Group, and Water Chemistry Forum Committees, I am particularly fortunate in being able to draw on the advice and expertise of members of these committees. In particular, I have been well supported by members of my Editorial Board, D.L. Miles, R. Purchase, D.M. Sanderson, and D.

Taylor. Thanks are also due to friends and colleagues who have acted as referees; and their comments have enhanced the quality of this book. They include F.S.H. Abram, G.M. Belsman, G.J. Dickes, P.A. Gilbert, H.P.A. Illing, D.L. Miles, T.G. Onions, H.A. Painter, R. Purchase, D.M. Sanderson, D. Taylor, W.G. Town, J. Vosser, and E.A. Walker.

The creation of this book would not have been possible without the secretarial assistance of Pauline A. Sim of the Gascoigne Secretarial Services who not only retyped all the manuscripts but carried out the typing of the considerable correspondence which the preparation of this book required. Similarly, I am indebted to A. Cubitt of the Society's Book Department for his unstinted assistance in the desk editing of this book. Finally, I thank my wife Beryl who so well tolerated my working on this book for the past year.

MERVYN RICHARDSON, EDITOR
ENVIRONMENT GROUP AND CHEMICAL INFORMATION GROUP,
ROYAL SOCIETY OF CHEMISTRY

Contents

Section 1: Introduction

Chapter 1   3
By S.G. Luxon
  1. Background   3
  2. The Role of the Royal Society of Chemistry   4
  3. Statistics   5
  4. The Role of the Health, Safety and Environment
    Committee   9
  5. The Way Forward   10

Section 2: Retrieval of Data

Chapter 2: Methods of Data Retrieval - Computer   15
By D.M. Sanderson
  1. Introduction   15
  2. Databanks   16
  3. Databases   17
  4. Selection of Sources   19
  5. Abstracts   20
  6. Filling the Gaps   20
  7. Re-use of Searches   21
  8. Training and Experience   21
  9. References   22

Chapter 3: Methods of Data Retrieval - Manual   24
By D.M. Sanderson
  1. Introduction   24
  2. Choice of Books   24
  3. Abstract Journals   25
  4. Other Sources   26
  5. Value of Older Literature   26
  6. Other Uses of Library Facilities   27
  7. References   28

Chapter 4: International and National Government
       Information Services   29
By J.A. Deschamps
  1. Introduction   29
  2. United Nations Environment Programme Systems   29
  3. European Communities Systems   33
  4. Organisation for Economic Co-operation and
    Development (OECD), Chemicals Information
    Switchboard   35
  5. United Kingdom Publications and Services   35
  6. Conclusions   38
  7. ·Final Word   39
  8. References   39

Chapter 5: Data Selection and Data Quality in IRPTC   41
By M. Gilbert
    1. Some Concepts Regarding Information Systems on
        Chemicals   41
    2. IRPTC: A Databank   41
    3. Ambitions and Limitations of the IRPTC Databank   42
    4. Sources of Information and Reliability of Data
        in IRPTC   44
    5. Data Validation during Data Profiles Preparation   45
    6. Selection Rules for Toxicity Data in IRPTC   45
    7. Conclusions   47
    8. References   48

              Section 3: Verification of Data

Chapter 6: Separating the Wheat from the Chaff - The
           Selection of Appropriate Toxicological
           Data from the World Literature   51
By D. Taylor
    1. Introduction   51
    2. Why Is Validation Necessary?   51
    3. Accessibility   53
    4. Relevance   54
    5. Quality   55
    6. Reliability   56
    7. Scientific Merit   57
    8. Consistency   60
    9. Conclusions   63
   10. References   63

Chapter 7: Some Common Sources of Error in Data
           Derived from Toxicity Tests on Aquatic
           Organisms   65
By R. Lloyd
    1. Introduction   65
    2. Concentration of Test Substance   66
    3. Response   68
    4. Exposure Time   69
    5. Needs for Critical Assessment of Experimental Data   70
    6. Conclusions   71
    7. References   71

Chapter 8: The Application, Generation, and Validation
           of Human Exposure Data   72
By P. Leinster and M.J. Evans
    1. Introduction   72
    2. Occupational Hygiene   72
    3. Occupational Exposure Limits   73
    4. Objectives of Data Collection   74
    5. The Application of Published Data   76
    6. Sources and Nature of Chemical Exposure   76
    7. Measurement of Personal Exposure   77
    8. Sampling Strategy   78
    9. Calibration   79
   10. Sampling and Analytical Procedures   79
   11. Record Keeping   81
   12. The Interpretation and Validation of Exposure Data   81
   13. References   82
        Appendix: Abstracting Journals and Other
        Information Services   82

Section 4: Interpretation of Data

Chapter 9: Interpretation of Data   87
By A.D. Dayan
    1.   Introduction   87
    2.   Toxicity   87
    3.   Biochemistry and Toxicological Data   90
    4.   Hazard and Risk   91
    5.   How Does the Toxicologist Operate?   92
    6.   The Responsibilities of the Toxicologist   96
    7.   Reading List   96

Chapter 10: Interpretation of Ecotoxicity   98
By J.H. Duffus
    1.   Introduction   98
    2.   Ecotoxicity Testing   99
    3.   Prediction of Environmental Distribution   103
    4.   Biodegradation   107
    5.   Bioaccumulation, Bioconcentration, and
         Biomagnification   108
    6.   Conclusions   109
    7.   References   109

Chapter 11: The European Community Chemicals
            Notification Scheme and Environmental
            Hazard Assessment   117
By J.L. Vosser
    1.   Introduction   117
    2.   Competent Authorities   117
    3.   Action Arising from Assessment of Notified Data   118
    4.   Definition of a New Substance   118
    5.   Types of Notification   118
    6.   Hazard Assessment Using Notified Data   120
    7.   Estimation of Exposure   120
    8.   Use of Notified Data in Exposure Assessment   120
    9.   Additional Information on Flow in Rivers,
         Sewage Treatment, etc.   121
   10.   Application of the Procedure   123
   11.   Predicting Environmental Concentrations -
         Conclusions   124
   12.   Environmental Effects Data   125
   13.   Actions Arising from Hazard Assessment   126
   14.   Summary   126
   15.   Acknowledgement   126
   16.   References   126
         Appendices   127

Chapter 12: The European Community Chemicals
            Notification Scheme: Human Health
            Aspects   133
By H.P.A. Illing
    1.   Introduction   133
    2.   What Testing and When?   134
    3.   Assessing the Result: What Next?   136
    4.   What Should Appear on the Label?   136
    5.   Interpreting Individual Tests   137
    6.   Problem Areas   142
    7.   Final Word   146
    8.   Acknowledgements   146
    9.   References   146
         Appendix: List of Risk and Safety Phrases   147

Chapter 13: Identification of the Carcinogenic Risk of
             Chemicals to Humans   150
By H. Vainio and J. Wilbourn
     1. Introduction   150
     2. The IARC Monographs Programme   151
     3. Working Procedures   152
     4. Criteria for the Evaluation of Data   154
     5. Supplements to the IARC Monographs   158
     6. Results to Date   159
     7. Conclusions   164
     8. References   164

Chapter 14: Organic Colorants - Interpretation of
             Mammalian-, Geno-, and Eco-toxicity Data
             in Terms of Potential Risks   166
By R. Anliker
     1. Introduction   166
     2. Classification and Application of Colorants   167
     3. Mammalian and Genotoxicity Database   170
     4. Environmental Data   177
     5. Conclusions   183
     6. References   184

Chapter 15: Interpretation of Epidemiological Data -
             Pitfalls and Abuses   188
By P.C. Elwood
     1   Introduction   188
     2. The Design of Epidemiological Studies   188
     3. Pitfalls: Errors in Design   192
     4. Conclusions   201
     5. References   201

Section 5: Risk Assessment and Case Histories

Chapter 16: Risk Assessment - General Principles   207
By D.P. Lovell
     1. Introduction   207
     2. What Is Risk?   208
     3. How Are Risks Measured? - Estimation   210
     4. How Do We Evaluate Risk?   214
     5. Managing Risks   217
     6. Risk Management in Practice   219
     7. Concluding Remarks   220
     8. References   221

Chapter 17: Risk to Work Forces   223
By H. Holden
     1. Introduction   223
     2. Hazard Identification   225
     3. Risk Assessment   227
     4. Exposure Risk   228
     5. Risk Monitoring   230
     6. Risk Estimation   231
     7. Conclusions   231
     8. References   232

Chapter 18: Use of Toxicity Data - A Case Study of
             Di-(2-ethylhexyl) Phthalate   233
By J.W. Bridges
     1. Approaches to the Assessment of Human Hazard
         from a Toxic Chemical   233
     2. Chemical and Biological Properties of DEHP   234

3. Information from Toxicity Studies   234
4. Information from Metabolism and
    Pharmacokinetics Studies   236
5. Information from Mode of Action (Mechanism)
    Studies   238
6. Overall Assessment of Human Hazard from DEHP
    Exposure   242
7. Future Prospects   243
8. References   243

Chapter 19: Effects of Pesticides on Non-target
            Organisms   247
By D. Osborn
1. Introduction   247
2. General Points   248
3. Types of Living Organisms Affected by Pesticides   249
4. Birds   250
5. Effects on Mammals   255
6. Effects on Invertebrates   255
7. Effects on Plants   256
8. Conclusion   257
9. References   257

Chapter 20: Comparative Risk Assessment - The
            Lessons of Cultural Variation   259
By S. Jasanoff
1. Introduction   259
2. The Causes of Divergence   260
3. Risk Assessment in Operation   266
4. Asbestos   266
5. Lead in Petrol   272
6. Science and Risk Assessment   276
7. Conclusion   278
8. References   278

Section 6: Legislation on Chemicals

Chapter 21: European Community Regulation of 'New'
            and 'Old' Chemicals   285
By C. Whitehead
1. Introduction   285
2. The European Community's Approach to
    Chemicals Control   286
3. The Regulation of 'New' Chemicals   286
4. The Regulation of 'Old' Chemicals   290
5. Future Developments   292
6. References   294

Chapter 22: Regulatory Affairs - A European View
            Relating to Legislation, Hazard/Risk,
            and Chemicals   296
By B. Broecker
1. Introduction   296
2. Definition of the Legislative Problems   297
3. Possible Concepts for Legal Rulings on Hazard
    Assessment of Chemicals   299
4. Discussion of the EC Concept   301
5. Problems Still to be Resolved   304
6. References   308
    Appendices   309

Chapter 23: Labelling   314
By S.G. Luxon
    1. Historical   314
    2. User Labelling Systems   315
    3. The Packaging and Labelling Regulations 1978
        and Subsequent Amending Regulations   319
    4. Preparations and Mixtures   325
    5. Solvents   326
    6. Paints, Varnishes, Printing Inks, Adhesives,
        and Similar Products   327
    7. Pesticides   327
    8. Environmental Hazards   327
    9. Summary   329
   10. References   329

Appendix A: Glossary of Terms   331

Appendix B: Useful Addresses   343

Subject Index   347

Contributors

R. Anliker, Ecological and Toxicological Association of the
Dyestuffs Manufacturing Industry, Basle, Switzerland

J.W. Bridges, University of Surrey, Guildford, Surrey, UK

B. Broecker, Hoechst Aktiengesellschaft, Frankfurt am Main,
West Germany

A.D. Dayan, St. Bartholomew's Hospital Medical College, London, UK

J.A. Deschamps, Department of the Environment, London, UK

J.H. Duffus, Heriot-Watt University, Edinburgh, UK

P.C. Elwood, MRC Epidemiology Unit, Cardiff, UK

M.J. Evans, British Petroleum plc, Sunbury-on-Thames, Middlesex, UK

M. Gilbert, IRPTC, Geneva, Switzerland

H. Holden, BICC plc, Prescot, Merseyside, UK

H.P.A. Illing, Health and Safety Executive, Bootle, Merseyside, UK

S. Jasanoff, Cornell University, Ithaca, New York, USA

P. Leinster, FBC Limited, Hauxton, Cambridge, UK

R. Lloyd, Ministry of Agriculture, Fisheries and Food, Burnham-on-
Crouch, Essex, UK

D.P. Lovell, British Industrial Biological Research Association,
Carshalton, Surrey, UK

S. Luxon, Health and Safety Consultant, Bourne End,
Buckinghamshire, UK

D. Osborn, Monks Wood Experimental Station, Huntingdon,
Cambridgeshire, UK

D.M. Sanderson, FBC Limited, Saffron Walden, Essex, UK

D. Taylor, ICI Brixham Laboratory, Brixham, Devon, UK

H. Vainio, World Health Organization, Lyon, France

J.L. Vosser, Department of the Environment, London, UK

C. Whitehead, Commission of the European Community, Brussels,
Belgium

J. Wilbourn, World Health Organization, Lyon, France

Editorial Board

'This is as strange a maze, as e'r men trod,
and there is in this business, more than nature
Was ever conduct of: some oracle
Must rectify our knowledge'

(The Tempest)

Section 1: Introduction

1
Introduction
By S.G. Luxon

MAIDENSGROVE HOUSE, RIVERDALE, BOURNE END, BUCKS. SL8 5EB, UK

1. Background

From the earliest times, danger has been linked with the use of chemicals but it was not until the middle of the nineteenth century that properly documented evidence setting out a link between specific chemicals and health hazards began to appear. As a consequence, the legislation of this period contained the first requirements to control such hazards. Simultaneously fire and explosion risks arising during chemical processes were causing increasing concern. Contemporary writers of the day, including Charles Dickens, sought to highlight these problems and arouse public anxieties. The resulting pressures ensured that by the beginning of the twentieth century the first real steps were being taken to investigate the hazards in depth and lay down requirements to control particular industrial poisons. During this period surveys were carried out to determine the extent of respiratory disease in the pottery industry, of mercury poisoning in the manufacture of hats, and the bodily disfigurement caused by phossy jaw from exposure to yellow phosphorus used in the manufacture of matches, which incidentally led to it being the first substance to be prohibited in UK legislation. Also in this period the first notifications to central government were required for certain types of poisoning, namely lead, phosphorus and arsenic; and diseases such as anthrax. As a result the first health statistics were born. Gradually the list was extended until the compilation of comprehensive accident and disease statistics became a legislative requirement in the UK.

After the 1939-45 War, changes in public attitude were occurring which were to influence the course of events. A number of disastrous accidents involving not only workers, but also the general public, had occurred and aroused public alarm. As in the previous century the underlying causes lay in the scaling-up of plant and the associated transport facilities together with the historic location of the industry in areas of high population density. Public awareness was increased by the wide media coverage and pressure built up for more comprehensive controls. Concurrent with these developments and probably promoted by increased prosperity and job security was an awakened worker-interest in, and an awareness of, the dangers which could arise coupled with the underlying fear of possible long-term effects, particularly in the area of carcinogenicity about which disturbing information was now becoming public knowledge. This in turn created a need for reassurance and information on hazards previously unknown or under-estimated. These new pressures resulted in a demand for more detailed information and greater consultation on the adequacy of the precautionary measures. As a consequence, the development

of systems for the labelling of hazardous substances came under
active consideration. Unfortunately, while a need for a universal
system of labelling was acknowledged and indeed, proved possible
within the European Community, it has thus far proved impossible
to reconcile the systems in Europe with those in the United
States. While of course the information is similar, such
differences highlight the further international co-operation which
will be necessary in this area.

Much of what has been said applied to workers in the plants
using hazardous chemicals and it must be emphasized that
experience shows that the greatest danger often arises in those
plants where chemical substances are used as an adjunct to the
main process *e.g.* trichloroethylene baths in engineering works.
Public concern at the possibility of a catastrophic accident has
already been mentioned, but in the post-war period there also
developed an increasing awareness of the effect of chemicals on
the environment. Links, some very tenuous, began to be
demonstrated between the presence of certain chemical substances
in the environment and adverse effects. The demand by the media
for information and consultation reached unprecedented levels and
public anxieties were accordingly heightened. Because many of the
questions were highly complex, they were were difficult to explain
in simple language and hence the public found itself unable to
evaluate the problems. Perhaps because of this, professional
views were increasingly questioned and not infrequently
mistrusted. Clearly professional bodies had an increasing role to
play in protecting their members' interests, both directly in
respect of possible injury and indirectly, in upholding
professional standards. At a time when the general public and the
media disbelieved the views of central government and industry on
such complex technical issues, professional bodies have a
responsibility for providing impartial advice which is seen to be
free from any pecuniary or political bias.

2. The Role of the Royal Society of Chemistry

A Royal Charter was granted in 1848 to the Chemical Society.
Following amalgamation with the Royal Institute of Chemistry and
other bodies representing chemists, the Society in 1980 became the
Royal Society of Chemistry. At that time a new Royal Charter was
granted stating that the object with which the Society is
constituted is the general advancement of chemical science, and
for that purpose it is:

i)   To foster and encourage the growth and application of such
             science by the dissemination of chemical knowledge.

ii)   To establish, uphold and advance standards of
             qualification, competence, and conduct of those who
             practise chemistry as a profession.

iii)   To serve the public interest by acting in an advisory,
             consultative, or representative capacity in matters
             relating to the science and practice of chemistry

iv)   To advance the aims and objectives of members of the
             Society so far as they relate to the advancement of the
             science or practice of chemistry.

In respect of the matters set out in the preceeding section, these objectives clearly place upon the Society a requirement:

i)   To safeguard the health and safety of its members and advise them in regard to their responsibilities to others.

ii)   To maintain a professional overview in respect of the public interest in health, safety, and the environment and to represent such interest and advice as necessary.

iii)   To help in furthering the knowledge of such hazards in so far as they are chemical in nature, and to give such advice as is necessary to those in authority.

iv)   To liaise with such other bodies as have an interest in these objectives.

In order to ensure that these requirements are fully and continuously kept under review the Society's Professional Affairs Board has set up a Health, Safety and Environment Committee (HSEC).

3. Statistics

In order to set the problem into a proper perspective and to see what action the Society should take, it is first necessary to look at the currently available national statistics in the field of health and safety and those due to exposure to chemicals in particular. Unfortunately, much of the available information is incomplete, because until 1975 Health and Safety legislation and hence notification procedures have not been applicable to chemical laboratories where much exposure to chemicals has occurred. In order better to understand the statistics it is necessary to review briefly the historical reasons for this incompleteness.

Early legislation in the past century concentrated first on the exploitation of women and young persons and then on mechanical safety in the major areas of industrial expansion at that time, *i.e.* textiles and engineering. Subsequent legislation followed this same piecemeal pattern until, by the middle of the present century, coverage had been extended to all persons employed in the manufacturing industry. Unfortunately, the premises defined in the legislation included only certain chemical laboratories operated solely for plant control purposes. The great majority of chemical laboratories, therefore, remained outside the scope of the legislation until the passing of the Health and Safety at Work Act in 1974. Even then, this Act provided only enabling powers, and it was not until specific regulations were made in 1981 that any general requirement to notify accidents and dangerous occurrences in all laboratories came into effect. Given that there is a two to three year delay in the publication of government statistics, we can only use as a guide those statistics available before 1983. Before doing this however, we must have some knowledge of the way in which these statistics have been compiled so as to gain a better appreciation of their significance.

For many years, the following classes of accident in manufacturing industry have been reported to central government:

i)   Fatal and serious accidents involving fractures or
            hospitalization for twenty-four hours or more.

ii)   Accidents involving an absence for more than three days.

    In addition, it has been a requirement to report certain
classes of dangerous occurrences which provide information on
serious malfunctions of equipment or systems of work which have a
high potential for injury, but which occur only rarely. The
current regulations require the notification of seventeen such
classes of dangerous occurrences. Of these, three are of
particular interest to chemists. In recent years they have
accounted for no less than 70% of the dangerous occurrences
reported in the chemical industry. They are:

i)   The uncontrolled or accidental release or the escape of
            any substance or pathogen which, having regard to their
            nature and the extent and location of the release or
            escape might have been liable to cause the death, injury
            or damage to health of any person. The key words here
            are 'uncontrolled release' and 'liable to cause injury'.
            Much will depend on the circumstances of each case, but
            careful consideration must be given as to whether or not
            such an escape could, given customary work patterns, have
            caused injury whether or not that injury did in fact
            occur.

ii)   The sudden uncontrolled release of one tonne or more of
            highly flammable liquids or gases from any system, plant,
            or pipeline.

iii)   Any process explosion or fire resulting in a stoppage or
            suspension of work for more than twenty-four hours.

    As distinct from the actual number of incidents occurring in a
period of time, the frequency rate is considered to be a better
indicator of safety performance as it takes account of the hours
worked. If we look at the general trend in respect of accidents
over the past decade, we see that the frequency rate shows a
reduction of some 30% over the period. Fatal accidents, the
statistics for which are more reliable, show a similar trend,
because they will all be reported. However, it must be remembered
that changes in the economy during this period may have affected
figures as much as improved accident prevention techniques. The
situation in the chemical industry is perhaps worth comparing
with that in other sectors. It is shown in Table 1.

Table 1

All reported accidents

Incident rates per 100,000 at risk - yearly averages

	-or 3 days absence	*Fatal*
Chemicals	3500	2.5
Food and Drink	4500	3.0
Metal manufacture	5000	7.0
Engineering	3500	3.5
Vehicle manufacture	3000	2.0
All industry	3000	2.5

Table 1 shows that incident rates for all reported accidents in the chemical industry are about the average for all manufacturing industry. This may be considered to be a good performance in view of the inherent dangers present in the industry. Looking in more detail at the differences in the industry between the various product groups we see from Table 2 that drugs and fine chemicals have the lowest frequency rate. Chemical laboratories, for which separate figures are not available, are probably most closely related to this group.

Table 2

Chemical industry product groups

$$Frequency\ rate = \frac{total\ accidents\ X\ 100,000}{total\ hours\ worked}$$

Heavy Chemicals	1.4
Drugs and Fine Chemicals	1.0
Plastics	1.2
Fertilizers	1.0

Table 3 shows the effect of the size of the undertaking on accident performance.

Table 3

Effect of size of individual undertaking in chemical industry

Number of employees	Frequency rate	
	Heavy chemicals	Other
Less than 100	1.6	1.5
100 - 500	1.5	1.6
500 -1000	1.4	1.2
Over 1000	0.9	0.9

With increasing size of organization there is a lower accident rate in spite of what is probably a higher reporting standard. This may be due to many factors, but of course it must be appreciated that all the accidents in these Tables for the chemical industry do not arise from the use of chemicals and it is important to consider causation, as shown in Table 4.

Table 4

Causation as a percentage of total accidents

Chemicals	2%
Fires and Explosions	0.5%
Machinery	26%
Lifting Articles	24%
Building and works surfaces (falls)	14%
Lifting containers (muscular)	11%
Transport (striking against)	6%
Working environment (falls)	2%
Other	14.5%

It will be seen that 62% of all accidents involve working practices which are dependent on operator awareness and in respect

of which physical precautions play only a small part. Those
arising directly from the manipulation and use of chemicals form
only 2% of the total. Very recent figures suggest that this
general pattern also applies to research and development
laboratories.

The other national statistic of interest is that relating to
the numbers of gassing accidents and industrial diseases. Tables
5 and 6 show the averages over a ten year period.

Table 5

Gassing accidents: yearly averages

	Yearly average
Chlorine	43
Carbon Monoxide	16
Ammonia	10
Sulphur Oxide	15
Hydrogen Sulphide	9
Hydrogen Chloride	7
Nitrous Fumes	5
Phosgene	4
Trichloroethylene	10
Dichloroethylene	5
Formaldehyde	5
Other Chlorinated Hydrocarbons	6

Table 6

Industrial diseases qualifying for benefit: yearly average

Lead	25
Benzene	2
Cadmium	6
Nitro/amino chloro aromatics	1
Mercury	2
Phosphorus	2
Other	5

These statistics, while not specifically related to chemical
hazards, indicate certain general guidelines:

i)   More can be achieved by the education and co-operation of
        everyone involved than by any other single factor.

ii)   Falls, striking objects, and muscular injury account for
        more than half of all accidents and these can only be
        reduced by improved work practices, better supervision,
        and an awareness of the danger by all concerned.

iii)   Chemical accidents account for only a small part of the
        total but their severity may be greater.

iv)   Many long-term effects of chemicals are probably not
recognized and hence are not included in the statistics.
Where there is a delay between the exposure and the onset
of symptoms this is even more likely to be the case,
particularly where the employee has retired. More work
on the toxicology of many chemicals is clearly needed.

In respect of safety all the available statistics demonstrate
that training coupled with clearly understood systems of work and
a personal involvement on the part of all concerned will achieve
more than money spent on physical precautions. That is not to say
that the latter should be totally neglected. Rather that the
effort to push forward in the area of management should have
priority.

In respect of health risks there is a need for more reliable
information and it is clear that in the immediate future, value
judgements will be required in the evaluation of risk.

The Royal Society of Chemistry's morbidity study will, it is
hoped, provide a greater amount of statistical information on the
hazards to which members of the Society are exposed.

4. The Role of the Health, Safety and Environment Committee

The Society and its predecessor, the Royal Institute of Chemistry,
has always been concerned with the health and safety of
professional chemists. After the war it widened its activities to
include environmental matters and since the granting of its Royal
Charter in 1980 it has also been responsible for keeping a
professional overview of the public interest in such matters.

Out of such considerations, the Health, Safety and
Environmental Committee was formed in 1976. The Committee is
responsible to the Professional Affairs Board for all health and
safety matters of concern to the Society. It consists of six
members and a chairman drawn from as wide a background as
possible, with experience in both sides of industry, some with
medical and toxicological interests and some being independent
consultants. Thus it can fairly be said to be unbiased and
independent, and is uniquely placed to offer a non-partisan
professional view on chemical matters.

The Committee has also set up specialist sub-committees and
working groups to research on specific matters of interest. Such
sub-groups draw on the expertise of the Society's membership more
generally as necessary. Close liaison is also maintained with
other groups in the Society having similar interests, *e.g.* those
on environmental and toxicological matters. Normally, the
chairmen of such groups sit on the committee as observers. Its
activities are normally publicized via the Professional Bulletin,
which, amongst other matters, carries invitations for experts to
sit on the relevant groups.

The work programme of the Committee falls into two main areas,
reactive and prospective. Reactive work in the health, safety,
and environment field at the present time may be summarized as
follows:

i)   Response to members' enquiries and requests.

ii)   The Society's response to national and international
            initiatives.

iii)   Consideration of government papers on health, safety, and
            environmental matters.

iv)   Servicing of contracts with the Commission of the
            European Communities and preparing publications in
            connection with such contracts.

v)   Liaison with other bodies and professional societies.

vi)   Representing the Society at national and international
            meetings.

vii)   Reacting rapidly to major incidents or problems which
            have far reaching effects on members or the public, and
            responding to media enquiries.

viii)   Informing members of the Society's views on major issues.

Prospective work includes:

i)   The preparation of fact sheets and briefing papers.

ii)   Briefing Members of Parliament.

iii)   Discussions on policy with government departments and
            agencies and with international bodies.

iv)   Preparation of Codes of Practice giving up to date
            guidance on health and safety matters.

5. The Way Forward

As has already been indicated, we are witnessing at the present
time an explosion in the demand for information and consultation
by the public which is mirrored by the media. There has been, as
a result, increased legislative activity on many fronts.
Unfortunately, much recent legislation has necessarily been highly
complex, owing partly to more detailed consultation on proposals,
which has brought forward many amendments designed to modify and
improve the original text but which necessarily increase its
complexity. The result of all this has been to produce regulatory
requirements that are incomprehensible to many of those they are
designed to protect, and more importantly, to those charged with
their implementation. This must be a matter of concern and
perhaps now is the time to stand back and seek a more practical
and economic approach to control. On a practical basis it is
desirable that the legislative requirements leave sufficient room
for manoeuvre at local level to meet the multiplicity of specific
problems which in many cases could not be covered in sufficient
detail in the legislation as drafted. Such flexibility is
important, not only in encouraging consideration in depth of the
problems at local level, but also to enable a more cost effective
approach to be adopted, both by central government and by industry
at a time of financial stringency.

Central to any control of hazardous substances is the question of how standards can be formulated and promulgated. Given that it is now possible to determine nanogram quantities of many toxic substances with ease and monitor individual workers by personal samplers with reasonable confidence, standards have become of paramount importance. Again, in the environmental field, widespread monitoring of chemical hazards is now becoming commonplace but the preparation of norms for the control of such substances has proved extremely elusive. Proposals have been made for control levels, action levels, short-term levels, and other refinements, which unfortunately all invariably beg the question: what level is acceptably safe? 'Acceptably' here must surely mean acceptable to those who may be affected, *i.e.* those coming into contact with the substances under consideration. Safety committees and workers' representatives have become a feature of industrial safety and health and at a local level can play a part in such deliberations. In some cases, particularly in the radiation field, local liaison committees have been set up to involve the public in such considerations. On the scientific front the investigation of the toxicological hazards of substances has proceeded apace. New methods have been developed to determine the long-term effects of many hazardous chemicals. Unfortunately, the results of such investigations are often ambiguous and therefore need careful interpretation if they are to gain general acceptance. To develop meaningful standards, a dose-response relationship must be established from such information. At a time when there is public mistrust of both government and industry, professional societies clearly have an important role to play in such evaluations. This book sets out to provide an unbiased view of these problems.

In the development of regulatory standards account must be taken of acceptable risk, which will involve political considerations. Perhaps, the way forward has been indicated over the past thirty years by bodies such as the American Conference of Governmental Industrial Hygienists who have made a very great contribution by publishing annually recommended standards for several hundred toxic substances. These standards have been used extensively with minor modifications by regulatory bodies in many countries. Recently the Society *via* the HSEC has itself contributed to this work under a contract to the Commission of the European Communities.

Given that standards can and must be adopted, then the future of control may not lie in an extension of the present detailed regulatory requirements. Rather, there could be a move towards providing a simpler, more flexible framework for the control of the risks at local level, always providing that the local working rules adopted are acceptable to those who may be affected, coupled with an over-riding requirement that the mandatory exposure standard must be met. Perhaps the best way of achieving this objective might be for the legislation to require the employer to appoint a competent person, *i.e.* one competent in the eyes of his peers in the profession, as an advisor, to whose appointment the employees would also have to agree. In this way the advisor could assume a more independent professional middle role and would hence be better able to arbitrate in such matters. This course has been canvassed, albeit somewhat tentatively, in recent regulatory proposals.

The principal duty of such an advisor could be to ensure that the processes were operated in such a way that hazards to the workforce, the general public, and the environment were minimized as far as practicable. In this way legislation could be simplified to provide a minimum administrative framework and lay down the appropriate standards to be observed. All the detailed anticipatory precautionary measures could be worked out at local level and laid down in working procedures and rules, thus providing the necessary flexibility, but always subject to the over-riding proviso that the mandatory exposure standard must be met at all times, and hence the workforce and the public adequately protected. The enforcement authority's primary task would then be to ensure the professional competency and integrity of the advisor, thus greatly simplifying their task and reducing the need for direct intervention by central government. More importantly, the general procedures would be brought into line with other related professions, *e.g.* medical, dental, pharmaceutical, and nursing, where skill, integrity, and judgement are relied on to meet general legislative requirements. It pre-supposes, of course, that a professional society such as our own has the determination and ability to ensure that high ethical and professional standards are maintained.

If the Society can meet this challenge and gain the confidence of the public, government, unions, and industry, then the way may well be clear for a greater independent professional involvement in this field, thus advancing the status of our members, an objective to which the Health, Safety and Environmental Committee has in the forefront of its long-term planning.

Section 2: Retrieval of Data

2
Methods of Data Retrieval - Computer

By D.M. Sanderson

FBC LTD., CHESTERFORD PARK RESEARCH CENTRE, SAFFRON WALDEN, ESSEX CB10 1XL, UK

---

1. Introduction

This chapter and the following one will review sources and methods for retrieval of medico-toxicological and safety information on chemicals. The information sought may be on adverse effects to man, laboratory animals, farm-stock, wildlife, or the environment in general, and may be for a variety of purposes often involving interpretation for non-toxicologists.[1] The information retrieved may make it unnecessary to carry out practical toxicity tests, thus saving time, cost, and the use of experimental animals.

In addition to the traditional library and bookshelf resources (see Chapter 3), an increasing number of external, commercial, or governmental online computer-based sources of scientific information are becoming available. This is as true of medico-toxicological and safety information on chemicals as for other fields, and many organizations are now using a variety of computer-based information sources as a major component of information retrieval. The history of the development of some of these online computer-based systems has recently been reviewed.[2] A recent study in the occupational health field suggested that retrieval of toxicological information presented particular problems, and that online computer-based searching is especially useful in this field.[3]

In very simple terms, the system used for consulting computer-based sources consists of a typewriter-type keyboard for the operator, connected to a printer and perhaps also to a visual display unit. Through an electronic interface, this equipment is then connected *via* a standard telephone line to a telephone network that communicates with the chosen distant computer which stores and searches the information. These distant computers are operated commercially by 'host' organizations, and can provide the retrieved information either directly online or by post as an offline print-out.

The situation is complicated by the different command languages and access structures when computer-based information files are available from different hosts. In some areas, the situation is becoming so complex that pressure is growing for some international standardization or convertibility of command languages.[4]

This standardization was attempted in the European Common Command Language, but few hosts in the medico-toxicological field, other than the German medical host DIMDI, use it.

One result of this growing complexity is the evolution of the 'information finder', who acts as an intermediary between the subject specialist or enquirer and the computer.[5] Such an intermediary is likely to have considerable knowledge of the computer-based information sources and their use, but less of the technical subject areas in which searches are requested. Difficulties may arise if the intermediary and subject specialist are unable to work closely together and understand one another; this aspect will be referred to again later.

2. Databanks

Online computer-based information sources are of two main types: databanks and databases.

Databanks contain pre-chosen factual information in summary form, with a sophisticated search system to enable the sought information to be located; they may thus be considered as computerized textbooks. Examples in the chemical toxicology field are the Registry of Toxic Effects of Chemical Substances (RTECS) and the Toxicology Data Bank. In Great Britain, RTECS is available from several hosts including BLAISE-LINK and DIMDI, and the Toxicology Data Bank from DIMDI and from BLAISE-LINK *via* TOXNET. Of these two databanks, RTECS covers a larger number of chemicals (currently about 80,000), but the information on the Toxicology Data Bank is much fuller for those fewer chemicals (currently just over 4000) which it does cover. A recent comparison of search techniques on RTECS[6] showed that computer searching has advantages over searching of a hard copy format. Another databank, first developed by the EEC, is ECDIN[7,8] (see also Chapter 4). This is available from Datacentralen in Copenhagen and includes some data on about 65,000 compounds, though toxicological coverage is incomplete on many,[9] is partly copied from RTECS, and is not always up to date, though the environmental information may be valuable. A further databank recently made available from BLAISE-LINK *via* TOXNET is the Hazardous Substances Data Bank; this also contains some toxicological, medical, and environmental information on many of the over 4000 chemicals covered, and the Toxicology Data Bank is now a sub-set of it which includes the medical and toxicological information on all the same chemicals. In the pharmaceutical field, MARTINDALE ONLINE is particularly useful, and is available from Data-star.

In general, databanks are often useful for rapid location of key information on toxicological properties of a chemical or product. However, it is important to make sure that the information in the databank has been objectively evaluated when first mounted, has been entered correctly with the right units, and has been kept up to date. If this is not done, it is very easy to be misled and reach wrong conclusions.[10] Information on RTECS is not evaluated, but that on the Toxicology Data Bank and ECDIN is. Both RTECS and the Toxicology Data Bank indicate when a chemical entry was last updated. It is important to be aware of the policy of the databank on selection and indexing of information for inclusion;[6] for example, it appears that RTECS sometimes includes the most adverse information found in a given field rather than the most representative or most reliable, and this may be a possible cause of unnecessary alarm. Also, RTECS does not contain negative findings.

For most enquiries in the field of chemical toxicology, databanks should not be used alone, but as a supplement to databases (see below), and, where appropriate, hard-copy sources; where thought necessary, information should be checked in the original references to eliminate transcription errors.

3. Databases

Online computer-based bibliographic databases are of particular value for obtaining detailed information on the medico-toxicological and safety properties of chemicals from the more recent literature. They are equally valuable for the two main types of search carried out, firstly where the greatest possible completeness of retrieval is required and it is important that no key information is missed, and secondly where only limited information is needed or retrieval of recent reviews will be adequate.

These databases are searched using specific command languages, which vary between hosts and individual databases, and which generally incorporate the ability to combine individual search statements to make up the search strategy using Boolean logic. This uses the three Boolean operators, AND, OR, and NOT to combine the search statements, as indicated in Figure 1.

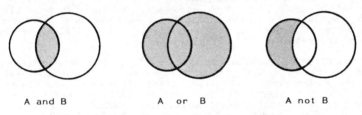

A and B   A or B   A not B

Figure 1 Boolean Operators

For example, a search for 'acetone and Ames' would only retrieve those records on acetone which also include mention of Ames tests. A search for 'acetone or chloroform' would retrieve all records covering either of these topics. A search for 'acetone not Ames' would retrieve all records on acetone except those including mention of Ames tests. Various combinations of these operators can be used to define every specific search. It must be remembered that the NOT operator has limitations, in that it will remove all records which include the search term it qualifies, even if they cover several topics. For example, if, as above, one wished to retrieve all information on the carcinogenicity of acetone, but exclude Ames test reports, use of the NOT operator would eliminate records mentioning both carcinogenicity and an Ames test, so that some required carcinogenicity references would be lost. The search statements used to compose the search strategy in this way may include free text terms from title or abstract, the Chemical Abstracts Service registry number, specific keywords (on some databases), authors, publication years, languages, *etc.*, giving great searching flexibility.

In addition, unlike hard-copy sources, a computer-mounted

database or databank can be used as an 'inverted file', so that for example it can be searched not only for the effects of a chemical, but also for what chemicals are reported as causing particular effects.

Some large databases have had to be divided into segments according to the entry date of the references, giving a current file, together with one or more backfiles containing the older references, which may need to be searched separately because of system capacity limitations.

In the field of chemical toxicology, TOXLINE and CHEMICAL ABSTRACTS, with, of course, the relevant backfiles as well as the current files, are probably the most useful databases to search first.

TOXLINE is a database which needs considerable familiarity before it can be used to best effect. It is administered by the National Library of Medicine in Washington, USA, and is available in Great Britain *via* BLAISE-LINK and DIMDI, with different command structures. It is a free text database, meaning that, for good recall, all relevant synonyms of search terms must be thought of and included in a complete search. They may be fed in individually, or pre-assembled as a stored group of search terms or 'hedge' intended for combination with others specified at the time of the search.[10] With its backfiles (see above), all of which are available online, TOXLINE covers literature back to 1965 with a small number of earlier items, and contains nearly 2 million references.

TOXLINE has a chemical dictionary file, CHEMLINE, which it is often useful to consult first to ensure that the chemical is adequately described with its alternative names and registry number, or to identify the chemical from a trade name; it currently lists about 700,000 chemicals. CHEMLINE also indicates which of the current or backfiles of TOXLINE, RTECS, or the Toxicology Data Bank have information on the particular chemical.

CHEMICAL ABSTRACTS is a computerized version of the large hard-copy source found in many scientific libraries. It is available from a number of different hosts,[11] and is easy to search by chemical using registry numbers. However, non-chemical terms are sometimes not very specific; the Index Guide is often useful as an aide memoire to choosing such terms, and their development has recently been reviewed.[12] On most hosts, coverage of CHEMICAL ABSTRACTS is back to 1967, though on STN International it is being extended back to 1962, and possibly further in due course; only STN International has full abstracts online; other hosts do not have this facility. References from some sections of CHEMICAL ABSTRACTS covering much of the toxicological field also appear in TOXLINE, with abstracts.

Further information on some chemicals, particularly on their effects on humans, may be obtained from MEDLINE and its backfiles. MEDLINE is not primarily a free text database, but relies on controlled index terms, *i.e.* medical subject (MeSH) headings with a hierarchical 'tree' structure which relates narrow and broad terms; it is available from several suppliers.

Other supplementary toxicological and medical information may be obtainable from EXCERPTA MEDICA online, BIOSIS, and CAB

ABSTRACTS/ANIMAL from the Commonwealth Agricultural Bureaux; whilst the latter two are strong on veterinary and wildlife aspects, the first one is not.

For some enquiries, it may be very valuable to search some appropriate smaller more specialized databases, whose names are usually self-explanatory, such as CANCERLIT (particularly useful on carcinogenicity and mutagenicity), ENVIROLINE, AQUALINE and ASFA (aquatic and fish), POLLUTION ABSTRACTS, AGRIS (agriculture and environment), CIS-ILO (occupational medicine and hygiene), HSE-Line (safety), *etc.*; the list is growing all the time. Some of these specialized sources will be more useful for common chemicals than for unusual ones. Most can be searched by keywords or free text, but not all have abstracts.

Conference proceedings, which may give useful information but are not well covered by many of the above sources, are covered by the database ISI/ISTP & B, available from DIMDI.

4. Selection of Sources

There are a number of considerations to bear in mind when deciding to which computer-based sources to subscribe, particularly since some databases and databanks are available from more than one host or supplier.

The first consideration must be which sources best cover the field of interest, bearing in mind their degrees of overlap.

Secondly, where a chosen source has more than one host, it may be advisable to choose that host which you need for other sources, because there may be facilities to transfer search strategies between sources without re-entry (as on DIMDI). Moreover, the fewer different command languages which need to be learnt and used, the more efficient searching is likely to be. In addition, subscription costs may be reduced by choosing certain hosts.

Thirdly, the speed with which the information is needed is important. If the search results are often needed immediately online, one should choose a host which has all the backfiles always available; alternatively, if offline prints are often required, postal delays may be shorter from some sources than others.

Fourthly, costs and their calculation basis vary considerably between sources and hosts.

A recent collaborative evaluation of information sources in chemical toxicology[10,13] analysed the findings of a number of searches in response to a series of test questions. The evaluation confirmed that, whilst it is necessary to use a variety of sources for comprehensive retrieval, it is possible to choose a limited number of sources which, with judicious use, will miss little key information. The exercise confirmed that CHEMICAL ABSTRACTS and TOXLINE were generally the two most valuable computer-based sources in the field of chemical toxicology, with additional information coming from others selected according to the nature of the query. For example, in a search for data on the toxicity of the insecticide propoxur to fish, out of 18 relevant primary references, 14 were retrieved from TOXLINE, 13 from CHEMICAL ABSTRACTS, 3 each from BIOSIS and EXCERPTA MEDICA, and 2 from CAB ABSTRACTS. All but one of the references could have been found from a combination of TOXLINE and CHEMICAL ABSTRACTS, and all if CAB ABSTRACTS were also included.

These general conclusions are supported by an evaluation by King.[14] A more recent comparative exercise,[15] which concentrated on occupational and environmental health and did not use CHEMICAL ABSTRACTS, found TOXLINE particularly valuable together with EXCERPTA MEDICA and BIOSIS, with MEDLINE contributing rather less information, and ENVIROLINE only strong on environmental quality aspects.

Obviously, if an enquiry is only for a few references rather than one needing completeness of retrieval, only one or two sources may be needed.

5. Abstracts

With some of the bibliographic databases, the computer record contains abstracts which can be included in the print-out. Abstracts are provided for most entries on TOXLINE, CANCERLIT, EXCERPTA MEDICA ONLINE, BIOSIS, and a number of others, and for many of the references on MEDLINE. However, until 1983, abstracts were not provided online for CHEMICAL ABSTRACTS and the hard copy had to be consulted for these. CHEMICAL ABSTRACTS with abstracts available online was launched by STN International in 1983.

The abstracts may often be sufficient to give the information required, but there will be many occasions when it is necessary to consult the original paper for more detailed information, secondary references, or both. Access to library facilities is clearly necessary for this. Again, the references may well be to papers in foreign languages, and translation facilities may be needed for these unless a previous translation can be located, for example through the ASLIB system. It is for these foreign language papers that a good quality abstract is particularly valuable, either to provide the required information or to give an indication of whether it is worthwhile to obtain and translate the paper. It may sometimes be useful to seek an abstract of a retrieved reference from another hard copy or computer source, if the abstract on the database is poor or absent; again, a well-chosen library may save much time (see Chapter 3). Awaiting and translating original papers are often the most time-consuming parts of a literature search.

6. Filling the Gaps

A properly conducted full literature search will identify not only available data, but also data gaps. It may then be possible to fill some of the gaps by further more specific or broader seaching.

It may be that one is unable to trace any useful toxicological information on a particular less-common or novel chemical. It may then be worthwhile seeking toxicity data on chemically related compounds. One way of seeking related compounds on which toxicity data are available is by substructure searching in CAS-ONLINE (a graphical computer dictionary file of structures in CHEMICAL ABSTRACTS) and then automatically transferring the registry numbers obtained to CHEMICAL ABSTRACTS online for subject searching. This has the minor disadvantage that the majority of the compounds first retrieved from CAS-ONLINE will have no published toxicological data, but these are mostly eliminated by introducing CHEMICAL ABSTRACTS subject search terms or subject sections. Substructure searching can also be done on DARC (a French equivalent of CAS-ONLINE). Two alternatives which can be very useful with appropriate experience are name fragment searching on CHEMLINE or RTECS, or searching of truncated Wiswesser Line Notation on RTECS.[16] In addition, Chemical Information Systems Inc. provide a substructure searching system (SANSS) which can be used with RTECS.[6] Once the related known compounds are identified, toxicological information on these can be retrieved in the usual way, and then used to make some prediction of possible toxicological properties of the original chemical.

It has been suggested that, as a last resort, toxicity can be predicted from computer analysis of structural fragments, against stored background data, by quantitative structure-activity relationships. While such approaches, with or without computer aid, have long been feasible and useful within a chemically-related series of compounds of similar mode of toxic action (for example organophosphorus anticholinesterases),[17] many workers feel that much more development is needed before the approach can be used, with any reliance, for predicting toxicity of a completely novel compound of unknown mode of action.[18]

7. Re-use of Searches

It is well worth considering storing completed search print-outs and documents, with suitable indexing, so that they can be retrieved and, if necessary, updated or extended, to prevent repetition when answering subsequent enquiries. Such a storage system would make an additional valuable source of information, and reduce expenditure and time.

8. Training and Experience

Inexperienced use of computer-based information sources may not only miss important information, or produce a mass of irrelevant information, but may also incur high searching costs, and possibly even a liability claim for failing to find key adverse information on a chemical. For these reasons, the person doing the search must have adequate experience and training in the use of the relevant databases and databanks. In addition chemical knowledge is often needed to identify synonyms of the chemical sought, and particularly for substructure or name fragment searching. However, it is also vital that retrieved specialized information is evaluated for scientific validity and relevance, using expert subject knowledge, during both the search and its review. A searcher with adequate subject knowledge can then act quickly and efficiently to modify or add to the search strategy as gaps or deficiencies in the original search or retrieved information emerge. This subject knowledge and experience is also important if the best way of narrowing a search is by retrieval and online display of a batch of reference titles, from which references are then selected for full display.

It follows from this that there are considerable advantages in training subject experts to do their own computer-based searching, rather than relying on intermediaries who may be highly proficient at use of online systems, but lack detailed subject knowledge.

There is now a good range of short training courses and reference books on online systems available.[19-22] If staff with appropriate backgrounds are chosen, it is clearly easier to train an experienced toxicologist with chemical knowledge to use computer-based information sources in his field than it is to train an information scientist or librarian to be a competent toxicologist. Such online training, and use of the experience gained in information retrieval, should be encouraged, so that more subject specialists are able to do their own computer-based literature searching, and then verify and evaluate the information retrieved.

9. References

1. H. M. Kissmann and P. Wexler, in 'Annual Review of Information Science and Technology', ed. M. E. Williams, Knowledge Industry Publications Inc., USA, 1983, Vol. 18, Chapter 7, p.185.

2. H. M. Kissmann and P. Wexler, *J. Chem. Inf. Comput. Sci.*, 1985, **25**, 212.

3. F. E. Wood, *J. Inf. Sci.*, 1985, **9**, 141.

4. C. A. Wells, *Online Rev.*, 1983, **7** (1), 45.

5. P. Duckitt, *ASLIB Proc.*, 1984, **36** (2), 79.

6. A. Costigan, F. E. Wood and D. Bawden, *J. Inf. Sci.*, 1985, **10** (2), 79.

7. N. H. Pedersen, in 'Workshop on Information Retrieval of Environmental Chemicals', Proceedings, ed. M. Boni and M. L. Richardson, Directorate-General for Science, Research and Development Joint Research Centre, Ispra, Italy, 1984, p.8.

8. M. Boni, O. Norager, W. Penning, R. Roi, and W. G. Town, in 'Workshop on Information Retrieval of Environmental Chemicals', Proceedings, ed. M. Boni and M. L. Richardson, Directorate-General for Science, Research and Development Joint Research Centre, Ispra, Italy, 1984, p.12.

9. R. Roi and M. Boni, 'Environment and Quality of Life; Development Criteria and Future Plan for the Occupational Health and Safety Sector of the ECDIN Data Bank', Commission of the European Communities, Ispra/Brussels, Rept. EUR 9098 EN, 1984.

10. D. Bawden and A. M. Brock, *J. Inf. Sci.*, 1982, **5**, 3.

11. Royal Society of Chemistry, 'Education Programme; Toxicity and Hazard Information, Especially in Chemical Abstracts', RSC, Nottingham, 1984, p.30.

12. D. F. Zaye, W. V. Metanomski and A. J. Beach, *J. Chem. Inf. Comput. Sci.*, 1985, **25**, 392.

13. D. Bawden and A. M. Brock, *J. Chem. Inf. Comput. Sci.*, 1985, **25**, 31.

14. J. King, 'Searching International Databases: A Comparative
    Evaluation of their Performance in Toxicology', Lib. Inf. Res.
    Rept. 3, British Library, London, 1983.

15. P. K. Corbett and S. L. Ifshin, *Med. Ref. Serv. Q.*, 1983, **2**
    (3), 25.

16. D. M. Sanderson and F. A. Parkar, *Blaise Newsletter*, 1983,
    No.56, 16.

17. R. D. O'Brien, 'Toxic Phosphorus Esters', Academic Press, New
    York, 1960.

18. T. Höfer-Bosse and R. Kroker, in 'QSAR in Toxicology and
    Xenobiochemistry', ed. M. Tichý, Elsevier, Amsterdam, 1985,
    p.65.

19. 'Going Online', Online Information Centre, London, 1983.

20. 'Medical Databases', Online Information Centre, London, 1983.

21. W. M. Henry, J. A. Leigh, L. A. Tedd and P. W. Williams,
    'Online Searching; an Introduction', Butterworths, London,
    1980.

22. C. H. Fenichel and T. H. Hogan, 'Online Searching; a Primer',
    Learned Information, Marlton, NJ, 1981.

3
Methods of Data Retrieval - Manual

By D.M. Sanderson

FBC LTD., CHESTERFORD PARK RESEARCH CENTRE, SAFFRON WALDEN, ESSEX CBIO IXL, UK

1. Introduction

When reviewing sources and methods for retrieval of medico-toxicological and safety information on chemicals, similar general considerations apply as for many other types of information retrieval. In particular, it is important not to restrict oneself to online computer-based sources, and so neglect the more traditional manual sources, such as the well chosen bookshelf or library.[1] When dealing with the simpler information requests, it may often be quicker, more convenient, and more cost effective to use a few selected reference books or files than to use the computer terminal (Chapter 2). Often, urgently needed simple information should be sought manually in this way, rather than by online consultation of bibliographic databases or databanks which may initially give much unwanted information as well as the specific item needed. On many more occasions, manual sources will be needed or desirable to supplement online searching, for example to add older information or proprietary information, or to provide original documents or general background.

2. Choice of Books

Most information specialists will have their own favourite list of the most useful and reliable text books in their particular subject field, and will be able to keep their bookshelf up to date as new titles or editions appear and are found suitable. Awareness of worthwhile purchases can arise from publishers' announcements, reviews, and, in particular, discussion with others who have similar interests. In the field of chemical toxicology, particularly useful books which should be readily available include:

'Patty's Industrial Hygiene and Toxicology', Wiley-Interscience, New York, 3rd Edn., 1981-82, Vol. 2 A-C.

N. H. Proctor and J. P. Hughes, 'Chemical Hazards of the Workplace', Lippincott, Philadelphia, 1978.

A.C.G.I.H., 'Documentation of the Threshold Limit Values', American Conference of Governmental Industrial Hygienists Inc., Cincinnati, 4th Edn., 1980 with updates.

M. L. Clarke, D. G. Harvey and D. J. Humphreys, 'Veterinary Toxicology', Baillière Tindall, London, 2nd Edn., 1981.
World Health Organization, International Agency for Research on Cancer, 'IARC Monographs on the Evaluation of the

24

Carcinogenic Risk of Chemicals to Humans', I.A.R.C., Lyon, 1972 onwards.

W. J. Hayes, 'Pesticides Studied in Man', Williams & Wilkins, Baltimore, 1982.

'Casarett and Doull's Toxicology; The Basic Science of Poisons', Macmillan, New York, 2nd Edn., 1980.

In addition, other reference books may be needed when dealing with pharmaceuticals; a very good source is:

'Martindale, The Extra Pharmacopoeia', Pharmaceutical Press, London, 28th Edn., 1982.

There are, of course, many other general or more specialized textbooks which may be found useful; the choice of these depends on the particular subject areas to be covered, the size of the budget, and the experience and preference of the user, which will take account of the coverage, reliability, and date of individual books.

3. Abstract Journals

The library could well contain hard-copy secondary journals providing abstracts of recently published papers or proceedings. Abstracts from different sources vary in quality and in the selection of information from the original paper for inclusion. The most familiar of these abstract journals is *Chemical Abstracts*, which is particularly important for searching the older literature. The 'Index Guide to Chemical Abstracts' is particularly useful in finding search terms, and the development of non-chemical search terms has recently been reviewed.[2] Other abstract journals which may also be useful either for searching or for providing abstracts or better abstracts of retrieved references include:

Toxicology Abstracts, Cambridge Scientific Abstracts, Bethesda, USA, from 1978.

Veterinary Bulletin, Commonwealth Agricultural Bureaux, Slough, England, from 1931.

CIS Abstracts, International Labour Office, Geneva, Switzerland, from 1974.

Abstracts on Hygiene and Communicable Diseases, (formerly *Abstracts on Hygiene* and *Bulletin of Hygiene*), Bureau of Hygiene and Tropical Diseases, London, England, from 1926.

Others may also be useful,[3] but it may be thought unnecessary to duplicate too many abstract journals which are also readily available online in computer-readable form (see Chapter 2) for the whole of their time span. Microfilm versions of certain abstracting journals (e.g. *Chemical Abstracts*) may also be available, saving space and cost. In the case of *Chemical Abstracts*, only the microfilm version is now obtainable back to 1907, except at a few specialist libraries such as the Royal Society of Chemistry in London, England.

Such abstracts may be particularly helpful in deciding whether it is worth obtaining a foreign language paper or obscure publication, and if necessary translating it for extracting key information quickly.

4. Other Sources

Another useful source of information may be the manufacturer's data sheet on the chemical in question. These sheets may give proprietary data not available elsewhere, though the format and the quality of the toxicity data included are very variable and could be better standardized.[3] Certain chemical suppliers' catalogues also include brief hazard statements.

Patents form another type of literature which can usefully be searched for some types of enquiry, for example on pharmaceuticals.

In-house information collections are also of value, particularly for finding specialized material such as reports.[4] Such collections could also usefully include indexed original papers obtained for one search or enquiry which may also be useful for future ones, thus reducing delays and costs.

In-house collections can also very usefully include indexed information or print-outs from completed computer or manual searches, so that these can be retrieved and, if necessary, updated or extended, to prevent repetition when answering subsequent enquiries. The search strategy and sources used should be included in the records.

5. Value of Older Literature

It is important not to forget that the computer-based information sources discussed in Chapter 2 cover relatively few years, with hardly any including much information published before the middle 1960s, and many starting much later than this.[5] Obviously, while scientific standards and abilities have improved over the years, there will be many occasions when older information is needed to provide a complete picture or indeed, on occasion, the only information.

For example, recent collaborative evaluation of information sources in chemical toxicology[4,6] analysed the findings of a number of searches in response to a series of test questions. One of these questions sought all information on the toxicity of sodium bismuthate. Of the 21 known primary references, 13 were published during 1930-39, 6 in 1941-45, 1 in 1956, and 1 in 1967. Not surprisingly, those using only computer-based search techniques failed to find most of the information. In fact, 10 of the 21 references were only found by one of the seven searchers. Of the 21 references, 14 could have been obtained from hard-copy *Chemical Abstracts*, and most of the remainder from citations in textbooks or as cited references in the papers already retrieved.

This clearly illustrates the importance of searching manual sources and the older literature when appropriate, and following up secondary references, rather than relying entirely on a computer terminal. Liability claims could arise if failure to find key published adverse information on a chemical results in alleged harm.

6. Other Uses of Library Facilities

One of the most obvious uses of library facilities is in following up an online search carried out as described in Chapter 2. Some databases provide abstracts which can be included in the print-out, while others do not. It may sometimes be useful to find an abstract of a reference from another hard-copy abstracting journal (see above), if the abstract on the database is poor or absent; the author index is usually the best way of doing this. The abstracts may be sufficient to give the information required, but there will be many occasions when it is necessary to consult the original papers for more detailed information. The original papers may also cite important secondary references. Ready access to suitable library facilities is vital for these purposes. The library will be able to borrow or obtain copies of documents it does not hold from other libraries, including the British Library; it can also check for commercially available translations of foreign language papers.

Useful information may be buried in papers or books which appear to be mainly non-toxicological; for example a paper on chemical synthesis of a group of compounds may include a footnote that the one of interest was found to be an unpleasant lachrymator or irritant during its manipulation.

Scanning the latest journal issues may reveal important new information which has not yet been abstracted and included in a bibliographic database.

Another circumstance is when it is difficult to define precise needs, or suitable search terms for them, or when seeking what is new in a broad field of interest. Such 'browsing' is much easier in a library than at a computer terminal.

On other occasions, the library may be useful for providing the background against which new information obtained from a specific search must be evaluated.

There is no need to be put off by apparent difficulty in finding material in a large library. Most have clear subject classifications on the shelves, and have a catalogue of holdings in some form. If one is in doubt, the librarian is always very pleased to help.

The online computer terminal must clearly be regarded as one of a number of tools for information retrieval, rather than the only source, and the more traditional hard-copy sources provided by the library and the personal bookshelf have an important part to play in dealing with many information requests.

After all, libraries have served our information retrieval needs well, without computer systems, at least since the days of the baked clay tablet libraries of Nineveh, in use well before 600 BC, the libraries in Egypt of Khufu (4th dynasty) and Rameses II (1300-1236 BC), and the two large libraries of Alexandria founded by Ptolemy I in 305-283 BC which were among the first to have a bibliographic catalogue.'

7. References

1. D. M. Sanderson, *U.K. Online User Group Newsletter*, 1984, No.32, 10.

2. D. F. Zaye, W. V. Metanomski and A. J. Beach, *J. Chem. Inf. Comput. Sci.*, 1985, 25, 392.

3. F. E. Wood, *J. Inf. Sci.*, 1985, 9, 141.

4. D. Bawden and A. M. Brock, *J. Inf. Sci.*, 1982, 5, 3.

5. D. M. Cipra and C. F. Damron, *Database*, 1985, (June), 23.

6. D. Bawden and A. M. Brock, *J.Chem. Inf. Comput. Sci.*, 1985, 25, 31.

7. *Encyclopaedia Britannica*, 1951, 14, 1.

4
International and National Government Information Services

By J.A. Deschamps

DEPARTMENT OF THE ENVIRONMENT, 2 MARSHAM STREET, LONDON SW1P 3EB, UK

---

1. Introduction

The information services described in this chapter fall into two main groups.

> Those sponsored by international organizations, such as the United Nations Environment Programme and the Commission of the European Communities; and,
> United Kingdom Government and related organizations' publications and services.

The commercial services are covered in Chapters 2 and 3. The aim of this chapter is to provide an introduction to some less widely publicized activities which include information on environmental chemicals.

2. United Nations Environment Programme Systems

In the late 1960s and early 1970s there was a growing awareness in many countries that international co-operation was needed to deal with environmental problems such as oil spills and pollution of rivers, where the effects transcended national boundaries. Various United Nations agencies already dealt with environmental areas, for example forestry and foods, but by 1968 the UN saw that a comprehensive approach was needed. The United Nations General Assembly, therefore, called a world conference on the environment. This was held at Stockholm in 1972. The United Nations Environment Programme (UNEP) was established during the conference. It was given three functions:

> Environmental assessment, including monitoring, research, information exchange, evaluation, and review.
> Environmental management, including law.
> Support, such as education and training. UNEP was not intended to 'take over' the activities of existing agencies, but rather to have a catalytic role.

2.1. The International Register of Potentially Toxic Chemicals (IRPTC).
This system, together with other UNEP services, was largely designed to meet the needs of developing countries, so that they could benefit from the experience of the more advanced nations. Nevertheless it, and the other services described in this section, have much to offer to information seekers in the developed world.

Note: The addresses of the various agencies mentioned are given in Appendix B.

IRPTC became operational in 1976. Its aim is to help reduce the hazards associated with chemicals in the environment by supplying information to those responsible for human health and environmental protection. It provides data for evaluating the hazards associated with particular chemicals, and seeks to identify gaps in existing knowledge on the effects of chemicals. Rules applied by IRPTC to select the most informative data are discussed in Chapter 5. IRPTC also holds information on national, regional, and global policies, regulations, and standards. It operates as follows:

From a central unit at Geneva which holds data on computers; IRPTC will answer questions put to it, although it does not offer an online service.

Through a network of National Correspondents who publicize the service, distrubute the *IRPTC Bulletin*, answer questions forwarded to them by Geneva, and liaise with other bodies (mainly Government Departments) in order to provide input to the system on matters of current concern and to obtain help in answering enquiries.

By building up a network of data suppliers, such as the United States National Institute of Occupational Safety and Health (NIOSH) which provides the RTECS (Registry of Toxic Effects of Chemical Substances) file. International bodies contributing material include the Chemical Group of OECD and the International Group of National Associations of Manufacturers of Agrochemical Products (GIFAP). IRPTC also collaborates with the European Communities ECDIN (Environmental Chemicals Data and Information Network) in the collection and exchange of data. The *IRPTC Legal File*, referred to below, is accessible online *via* ECDIN. Further details of ECDIN are given later in this chapter.

IRPTC has devised a detailed format for inputting data to its computer files. Examples of the categories of information held are toxicity to mammals and man, and effects on non-mammalian organisms and plants.[1] It also disseminates the information it has collected, either in the *Bulletin* or in monographs such as:

IRPTC Legal File (United Nations Publications, Geneva, 2nd Edn., 1986, 2 vols.). The volumes contain legal data profiles (details of regulations on named chemicals). Information is held on 600 substances; 12 countries and 6 international organizations have contributed. *The Legal File* can, for example, be used to trace regulations governing the use of methyl isobutyl ketone as a pesticide ingredient in the USA, or Indian recommendations for packaging and labelling of nitrobenzene. The *IRPTC Legal File* also includes occupational and environmental standards for chemicals (*e.g.* TLVs). For each record full details are given of the related documentation.

Toxicometric Parameters of Industrial Toxic Chemicals under Single Exposure (1982). This publication summarizes data in the Soviet literature. Lethal and threshold doses or concentrations are given for more than 700 industrial chemicals. It was prepared in co-operation with the USSR State Committee for Science and Technology.

Scientific Reviews of the Soviet Literature on Toxicity and Hazards of Chemicals (1982-). The reviews comprise a series of short monographs. About 80 titles had been published at the time of writing. Chemicals which have been reviewed include asbestos, formaldehyde, mercury, cadmium, nitrates, and a large number of pesticides.

IRPTC is a Special Sectoral Source in the INFOTERRA system which is described later in this chapter; that is, it has agreed to provide substantive information and, where appropriate, publications to enquirers, especially those from developing countries. It also works closely with UNEP's Industry and Environment Office (see below). IRPTC disseminates information on behalf of the International Programme on Chemical Safety (IPCS) which is described immediately below.

2.2. The International Programme on Chemical Safety (IPCS). The Programme came about as a result of a resolution passed at the 1977 World Health Assembly, to examine the options for international co-operation in the evaluation of health risks from chemicals. It is a joint endeavour of UNEP, the International Labour Office (ILO), and the World Health Organization (WHO); its objectives are to:

Evaluate the effects of chemicals on human health and on the environment.

Develop guidelines on exposure limits for chemicals in air, food, water, and the working environment.

Develop methodology for toxicity testing, epidemiological and clinical studies, and risk assessment.
Co-ordinate laboratory and epidemiological studies internationally, and promote research on dose-response relationships and on mechanisms of the biological action of chemicals.

Develop information for coping with chemical accidents; promote technical co-operation on the control of toxic substances, and the training and development of manpower.

IRPTC acts as the lead institution within IPCS for the collection, retrieval, and dissemination of information. News of current activities is published in the *IRPTC Bulletin*. IPCS is, of course, not a brand-new activity but rather a strengthening and extension of existing work, such as that carried out by the International Agency for Research on Cancer (IARC) who issue the series of monographs on the evaluation of the carcinogenic risk of chemicals to humans,[2] and publish reviews of research in progress[3] (see also Chapter 13).

IPCS has a central unit at WHO Headquarters which is supported by various committees, and a network of national and other lead institutions. Its publications include the *Environmental Health Criteria* series. Titles in this series aim to summarize, review, and evaluate the available information on the effects of specified chemicals, and groups of chemicals; 63 documents had been produced at the time of writing. Chemicals covered include metals such as mercury, arsenic, and cadmium; air pollutants, for example oxides of nitrogen, sulphur oxides, and suspended particulate matter; and asbestos and other fibrous particles; pesticides such as

organophosphorus compounds, paraquat and diquat, and 2,4-D;
solvents, for example trichloroethylene and toluene; and physical
factors such as noise, and radiofrequency and microwaves.

2.3.   The Industry and Environment Office. This organization is
located in Paris. It was set up, as the Industry Programme, in
1973 to bring government and industry together to co-operate in
reducing the adverse impact of industries on the environment. It
carries out its information role in two ways. It has been
building up since 1978, in collaboration with UNESCO, a
computerized databank, *The Industry and Environment File*. It
obtains information from key organizations in specific industrial
sectors, and enlists their help in evaluating items intended for
the databank. Industries covered so far include pulp and paper[4]
and petroleum.[5] The information contained in the databank
comprises:

> Pollution abatement and control technologies.
> Ambient quality standards.
> Environmental technology costs.

Each file is cross-referenced to IRPTC and to INFOTERRA. The
information is designed for the use of decision-makers and
technical personnel in government, industry, and scientific and
educational institutions. The databank is not available online,
but the staff of the Paris office will access it and other
information sources at their disposal, including individual
experts, to answer questions put to them.

The Office also publishes a quarterly newsletter, *Industry and
Environment*. Each issue focuses on a specific example, such as
'The Chemical Industry', 'Agrochemicals', 'Environmental Impact
Analysis', and 'Industrial Hazardous Waste Management'. The
newsletter is issued free of charge at present.

**2.4.   The International Environmental Information System
(INFOTERRA).** INFOTERRA is a rather different system from those
already described in that:

> It covers the broad environmental subject area.
> Its operation is based on the referral principle, that is, it
> refers enquirers to sources of information or expertise,
> rather than providing substantive information direct to the
> user. However, factual information is often sent to
> enquirers, especially those in developing countries.

Nevertheless, it should be considered a relevant service for
obtaining information on environmental chemicals. Firstly, many
of the sources included in its *International Directory* are
concerned with chemicals and related topics such as pollution and
wastes. Secondly, it has a network of National Focal Points from
well over 100 countries world-wide who can be approached for
information. INFOTERRA is of particular value where multi-
disciplinary topics are concerned. It has two key components:

> The *International Directory* containing descriptions of *ca.*
> 6000 sources such as government departments, research
> organizations, university departments, and private
> consultants. A copy of the Directory is held by each National
> Focal Point and plans are in hand to make it, and sub-sets of
> it, more widely available.

The second, and equally important component is the network of National Focal Points. Each of these has links with, and knowledge of, information services and centres in its own country. Requests for information are made through National Focal Points.

INFOTERRA has also designated a number of Special Sectoral Sources who offer a wider range of services, including the provision of substantive information. Examples of such sources, in addition to IRPTC and the Industry and Environment Office already mentioned, are the Environmental Law Information Service (ELIS) of the International Union for the Conservation of Nature and Natural Resources (IUCN) and the Asian Institute of Technology (which covers the subject areas water supply and sanitation, and renewable energy).

3. European Communities Systems

3.1. Environmental Chemicals Data and Information Network (ECDIN). ECDIN, which has already been mentioned in Chapter 2, was initially conceived as a network-based system.[6] However, progress in this direction has been impeded, although not abandoned, by failure to obtain general agreement to the principle and to a mechanism for data exchange. It is, therefore, at present the online equivalent of a chemical handbook. The files which compose the commercial version, or which will be available shortly, are:

Identification
    Chemical Synonyms
    Chemical Structure Diagrams

Physico-chemical Properties

Production and Use
    Chemical Producers
    Chemical Processes
    Production and Consumption Statistics
    Export Statistics
    Import Statistics
    Uses

Legislation and Rules
    IRPTC Legal File (to be updated early 1986)
    Directive 67/548/EEC

Occupational Health and Safety
    Hazards and Prevention
    Occupational Exposure Limits (taken from IRPTC data)
    Occupational Diseases Prevention
    Symptoms and Therapeutic Treatment
    Occupational Poisoning Reports

Toxicity
    Classical Toxicity
    Carcinogenicity
    Aquatic Toxicity
    Effects on Soil Micro-organisms
    Mutagenicity
    [Teratogenicity (will be available during 1986)]

Concentration and Fate in the Environment
   Concentration in Environmental Matrices (Cost 64b)
   Concentration in Human Media (EPA)
   Concentration in Animal Media
   Metabolism in Soil
   Aquatic Bioaccumulation

Detection Methods
   Analytical Methods
   Odour and Taste Threshold Values

ECDIN is available from the Datacentralen Host Centre which is operating a trial marketing exercise for the European Commission. The databank is accessed using a simple menu-driven system. It contains synonyms and identifying numbers for around 65,000 compounds, together with 50,000 chemical structures. Examples of the number of chemical substances covered by some of the larger files are:

Chemical Producers, about 6000
IRPTC Legal File, about 6000
Physico-chemical Properties, over 4000
Classical Acute Toxicity, over 18,000 (including a large number taken from RTECS); more extensive toxicity information is available on about 1000 - 2000 compounds.
Aquatic Toxicity, about 800.

Many of the more specialized files are considerably smaller, but the total amount of data held has recently been increased considerably.

3.2. Directory of Environmental Research Projects in the European Communities (ENREP). The ENREP record, which dates from 1980, exists both as a title-only printed publication[7] and as on online database which is available on the European Commission's ECHO host. The printed version covers over 20,000 projects from some 4400 organizations. Projects are indexed using a specially developed multi-lingual thesaurus. Subject areas covered which relate to chemical substances include agriculture and forestry, air, environmental chemicals/pollutants, health and occupational safety, solid wastes, and water. Each entry in the database contains some or all of the following: research organization, status of project, start and finish dates, research workers, co-operating organization(s), title of project, abstract, indexing terms, costs, date information supplied, details of associated publications.

Input from the EC Member States is provided by National Focal Points who collect information from organizations in their country. For example, the United Kingdom National Focal Point collects details of research being carried out by over 4000 organizations, together with details of supporting documentation. Information about United Kingdom research projects up to 1978 was published by the Departments of Environment and Transport in its *Register of Research and Surveys* which was issued annually from 1974-1978.

3.3. System for Information on Grey Literature in Europe (SIGLE). SIGLE is an European Communities-wide attempt to deal with the gap in the knowledge and supply of 'grey literature', that is, documents which are issued informally, and which are not readily available through normal bookselling channels. Examples of such literature are scientific and research reports, theses, proceedings, technical notes, internal reports, private communications, and local authority documents.

The database became publicly available in 1984, and contains information dating from 1981 onwards. It is accessible both on the hosts INKA and on BLAISE-LINE. The United Kingdom input is provided by the British Library's database, *British Reports, Translations and Theses* which is described later in this chapter. SIGLE's subject range is wide, covering science and technology, social sciences, and humanities. Relevant areas include agriculture, biology, chemistry, earth sciences, and energy.

A SIGLE record, which can be searched by subject category, as well as by author, title, *etc.*[6], contains details of: ISBN, personal and corporate authors, title, publisher, date of publication, and subject category. British items identified using SIGLE can be ordered online using the British Library Document Supply Centre's Automatic Document Request Service (ADRS).

4. Organisation for Economic Co-operation and Development (OECD), Chemicals Information Switchboard

Although this project is still in its pilot phase, it should be mentioned as a potentially useful information source.

The Switchboard, which was set up by the OECD Chemicals Group, aims to help with the problem of finding information on chemicals which is available only in the 'grey literature' (see section on SIGLE above) or which is unpublished but available on request, for example data intended for future publication or negative results which are often not published.

The scheme operates on the referral principle. For the purposes of the pilot study, National Focal Points (NFPs) and networks of participating organizations have been set up in a small number of selected countries (Japan, Netherlands, Sweden, United Kingdom, United States). Questions received by an NFP are passed to relevant domestic sources, and to the NFPs of other participating countries. Information providers respond direct to the requesters, with whom they will negotiate any fees or special conditions for the release of information which may be commercially sensitive. Organizations making requests will be asked to co-operate in evaluating the information they receive, both during and at the end of the pilot study.

The Switchboard commenced operation in mid-1985, and a decision on its future will be taken between one and two years after that time.

5. United Kingdom Publications and Services

5.1. Current Research in Britain (CRIB).
CRIB is the new title for the expanded version of *Research in British Universities, Polytechnics and Colleges* which, with the exception of 1985, was published annually by the British Library. The first edition of CRIB was published at the end of 1985. It comprises four volumes:

Physical sciences (including chemistry, chemical engineering,

and technology).
Biological sciences (including agriculture, biochemistry, biology, biomedical sciences).
Social sciences (including law, community and social medicine).
The Humanities.

It contains details of 60,000 projects from almost 4000 departments, together with comprehensive keyword and name indexes.

5.2.   British Reports, Translations, and Theses (BRTT).   This monthly journal is also published by the British Library. It is a bibliography of 'grey literature' (semi-published literature, including reports from government organizations, industry, and universities, together with most doctoral theses accepted from 1970 onwards) which can be difficult to identify. It also covers reports and unpublished translations from the Republic of Ireland, and selected British official publications that are not published by Her Majesty's Stationery Office (HMSO). Some relevant subject areas are agriculture, biological and medical sciences, and chemistry.   Keyword, author, and report number indexes are provided.

As has been noted earlier in this chapter, BRTT provides the UK input to the European Communities' grey literature database, SIGLE.

5.3.   Pollution Research and the Research Councils.   This publication is issued annually by the Inter-Research Council Committee on Pollution Research (IRCCOPR). It contains details of projects carried out by, or on behalf of, the Agriculture and Food, Medical, Natural Environment, Science and Engineering, and Economic and Social Research Councils. The information given includes title, sponsor, laboratory carrying out the work, value of contract, and expected finish date. The broad subject areas covered include effect of pollutants on man and other organisms, and on populations and communities, abatement studies, and technological aspects.

5.4.   The Medical Research Directory.   This commercially produced directory was published in 1983.[9]   It gives details of current and recent medical and nursing research conducted in Britain. The Medical Research Council, the Department of Health and Social Security, and the British Library have co-operated in its compilation. Relevant subject areas include cancer, biochemistry, epidemiology, food science, toxicology, and occupational medicine. A short summary of the research is given, together with details of the institution performing it, research workers, funding authority, and tenure of grant. Full name, subject, and techniques indexes are provided. The directory is also available in a magnetic tape version.

5.5.   Laboratory of the Government Chemist.   The Laboratory, which is an industrial research establishment of the Department of Trade and Industry, provides, on a commercial basis, a range of analytical, advisory, and information services. The comprehensive analytical service is described in a brochure available from the Laboratory.   The Biotechnology Unit promotes the application of biotechnology in industry.   Other specialist advisory services are:

Agricultural Materials Analysis Information Service (AMAIS). This provides details of methods of analysis for pesticides residues, fertilizers, and animal foodstuff additives.

Chemical Nomenclature Advisory Serice (CNAS). The Service gives guidance and advice on the correct use of IUPAC nomenclature. It maintains a list of roughly 20,000 chemicals for Her Majesty's Customs and Excise.

Consumer Hazards Analytical Information Service (CHAIS). CHAIS provides advice on consumer safety regulations.

Dangerous Goods Advisory Service (DAGAS). This offers technical advice to subscribers on the chemical hazards associated with the carriage of dangerous goods, including the classification of dangerous substances according to the various international systems.

Microbial Culture Information Service (MiCIS).

5.6.   The Waste Management Information Bureau (WMIB), Harwell Laboratory. The WMIB is one of the services operated by the Harwell Environmental Safety Group to assist industry, national and local government, and others in any way concerned with the production, treatment, disposal, and recycling of non-radioactive wastes. It is supported by the Department of the Environment. The Bureau aims to give access to the information already published in this area. Information is held in a computerized databank which can be searched, for a fee, using a keyword system. Further investigations involving Harwell's consultancy or library services can also be caried out. WMIB produces a bulletin which is issued, as a commercial journal, six times a year.[10]   The bulletin is arranged under subject headings such as recycling, recovery, reclamation, and environmental hazards. Summaries of recent developments and a diary of forthcoming events are also included.

5.7 The Marine Biological Assocation of the United Kingdom (MBA), Plymouth. The Association is a grant-aided institute of the Natural Environment Research Council. Its library is one of the most comprehensive in the world on marine biology, oceanography, fisheries, pollution, and related subjects, and offers a range of services. It includes a Marine Pollution Information Centre which collects documents on marine and estuarine pollution, covering the detection, remote sensing, monitoring, and analysis of pollutants; the levels of pollutants in seawater, sediments, and organisms; the biological effects and fate of pollutants; and the control and removal of pollutants. Some 2500 to 3000 documents are added to the collection each year and these are listed in a monthly information bulletin *Marine Pollution Research Titles*. The library is the United Kingdom Focal Point and main input centre for the United Nations' Aquatic Sciences and Fisheries Information System (ASFIS), and it provides data on toxic chemicals for both ECDIN and IRPTC.

5.8 The Water Research Centre (WRc). WRc is the national centre for water research in the United Kingdom. The Centre is a private research company, financed by a subscribing membership, which also undertakes work on a commercial basis. Its many information activities include a technical enquiry service and input to AQUALINE, a commercially available online database. The Centre

also produces *Aqualine Abstracts* (formerly *WRC Information*) which
is published fortnightly by Pergamon Press.   The research
programme of WRc is published separately.

WRc also provides an information service on toxicity and
biodegradability, INSTAB.   The service supplies information and
advice on the effects of chemicals on biological sewage treatment
processes, both aerobic and anaerobic, and their biological
effects on fish and invertebrates.

Bibliographic details and assigned key words for research
reports, papers, and unpublished work are recorded on a
computerized database.   The relevant documents are stored as hard
copy by the information service.   This service is provided free of
charge to members of WRc.   Non-member organizations may also use
INSTAB but will be asked to pay at current commercial rates for
all work undertaken.

In addition, WRc provides an information service on the
mammalian toxicity of chemicals.   A small, specialist database
covering some 6000 records has been compiled and material is
continually being added.   The service is free to member
organizations but there may be a charge for non-member enquiries
depending on their nature and size.

Finally, WRc has supplied ECDIN with the CICLOPS file which
contains data obtained from the published records of chemicals
found in the aquatic environment world-wide.

6. Conclusions

The characteristics, strengths, and weaknesses of the types of
services which have been described are outlined below in order to
assist potential users to select appropriate sources, and to make
optimum use of them.   They have been sub-divided into four
categories, which are, however, not mutually exclusive.

6.1.   Databanks, *e.g.* IRPTC, UNEP Industry and Environment File,
ECDIN.   These systems are most useful for providing an overview
for a named chemical substance.   They are particularly valuable as
a source of information on legislation, and have good coverage of
data taken from key works produced by various international
agencies.   A limitation is that, in general, in-depth and
comprehensive information is available for a comparatively small
number of chemicals.   Also, data taken from publically available
files (*e.g.* RTECS) are repeated in more than one databank.   The
extent to which data have been evaluated and validated is not
always clear.   IRPTC in particular gives access to a wide range of
information sources through its network of National
Correspondents.

6.2.   Referral Systems (INFOTERRA, OECD Switchboard).   Referral
systems do not hold data, but provide, through their national and
international networks, access to an extensive range of sources.
NFPs are often the focal points for a number of complementary
international systems.   The chief weakness of referral systems is
their inbuilt communications delay, making them less suitable for
very urgent requests.   Their strength is that they can provide
access to publications and data which are not generally available,
and to specialist advice.

6.3. Bibliographic Systems and Registers of Research, *e.g.* **SIGLE, CRIB, ENREP**. These records do not usually include data. The original document will have to be acquired or the research worker contacted in order to obtain the desired information. Coverage between contributing bodies and countries is not uniform, as the quality and quantity of input will depend on the commitment and resources of the various NFPs, *etc*. Some of the services are available on-line, giving rapid access. The most valuable feature of such listings is that they assist in tracing information which is often not published in the open literature.

6.4. Specialized Sources, *e.g.* **MBA, WMIB, WRc**. These sources usually offer a range of servics in a specified field such as the marine environment or waste disposal. They often produce a specialist abstracts journal, provide information on request, and give access to experts. Requests are dealt with by an experienced information officer who will be able to advise on the best approach to a problem. A possible difficulty is that many organizations exist mainly to serve their own members; a fee is often payable for services to non-members. Such sources do not always give instant access to data, although some may, but they can be used to obtain detailed advice from subject specialists.

7. Final Word

From the foregoing, it is clear that a significant number of less obvious sources can be used to locate information on chemical substances. The information-seeking procedures may be rather cumbersome and diffuse, and often the success rate will be low. However, on the occasions when it is essential to explore all possible avenues, such an approach can produce positive results and may well be justified.

8. References

1. 'International Register of Potentially Toxic Chemicals, Part A', United Nations Environment Programme, Geneva, 1985.

2. 'IARC Monographs on the Evaluation of Carcinogenic Risk of Chemicals to Man', International Agency for Research on Cancer, Lyon, 1972→.

3. 'Directory of On-going Research in Cancer Epidemiology', International Agency for Research on Cancer, Lyon/German Cancer Research Centre, Heidelberg, 1977→.

4. 'Environmental Aspects of the Pulp and Paper Industry; an Overview', UNEP Industry Programme, Paris, 1977.

5. 'Environmental Aspects of the Petroleum Industry: an Overview', UNEP Industry and Environment Office, Paris, 1978.

6. J. M. Hushon, J. Powell, and W. G. Town, 'Summary of the History and Status of the Systems Development for the Environmental Chemicals Data and Information Network (ECDIN)', *J. Chem. Inform. Comput. Sci.*, 1983, 23, 38–43.

7. 'ENREP Directory of Environmental Research Projects in the European Communities, Part 1: Research Projects; Part 2: Indexes', Peter Peregrinus Ltd. (for the Communities of the European Commission), Hitchin, Hertfordshire, 1982.

8. J. M. Gibb and M. Maurice, *ASLIB Proc.*, 1982, **34**, (11/12), 493.

9. 'The Medical Research Directory', Wiley, Chichester, 1983.

10. 'Waste Management Information Bulletin', Routledge and Kegan Paul Ltd./The Waste Management Information Bureau, Bi-monthly.

5

Data Selection and Data Quality in IRPTC

By M. Gilbert

INTERNATIONAL REGISTER OF POTENTIALLY TOXIC CHEMICALS, PALAIS DES NATIONS, CH-1211 GENEVA 10, SWITZERLAND

---

1. Some Concepts Regarding Information Systems on Chemicals

Information on chemicals relating to human health and the environment can be obtained from several sources: publications in scientific journals, bibliographical databases, and factual databanks.

The first two are closely linked but differ significantly with respect to the amount of detail supplied. The scientific literature (journals, periodicals, books, and reports) contains research results and experimental evidence reported in such a way that the scientific and technical quality of the study can be assessed. Moreover such 'primary literature', as it is often called, supposedly reports a sufficient number of details so as to permit reproducability a described experiment. The second type of source, computer-managed databases, contain bibliographical references with possibly brief abstracts. In such cases, no validation of the data is possible (without first obtaining the paper from the journal) this should not, however, be considered as a drawback since bibliographical systems should offer comprehensive coverage of the literature, allowing simultaneous retrieval of very limited information on well-defined topics.

A factual databank is another type of information tool with specific characteristics related to the audience addressed, the objective pursued, the completeness, and the reliability of its content. (see also Chapters 2-4).

Some salient features of databanks as compared with databases can be summarized as follows:

1. They contain factual data and not merely bibliographical references.

2. Databanks provide definite answers to specific requests for information, being easily searchable and retrievable systems.

3. The user will save time in:

The selection of data sources pertinent to his concern.

The scanning of the relevant documents.

2. IRPTC: A Databank

One of the objectives assigned to the International Register of Potentially Toxic Chemicals concerns the development of central

files containing sufficient information for an understanding of
health and environmental hazards caused by toxic chemical
substances.

The pertinent information is extracted from scientific and
other relevant documents and presented in the International
Register, as a data collection which should be adequate in itself
to offer a complete 'data profile' of a particular chemical.

Data profiles for chemicals provide the information necessary
to evaluate the potential threat posed by chemicals to man's
health and to his environment, or indicate the absence of such
information in available literature. They are intended mainly for
the use of those who have responsibility for the protection of
human health and the environment from noxious effects of
chemicals.

The IRPTC data profiles contain seventeen information files:

Identifiers, physical and chemical properties
Production and trade
Production processes
Use
Pathways into the environment
Concentrations
Environmental fate tests
Environmental fate
Chemobiokinetics
Mammalian toxicity
Special toxicity studies
Effects on organisms in the environment
Spills
Sampling/preparation/analysis
Treatment and disposal methods for waste chemicals
Treatment of poisoning
National and international recommendations and
Legal mechanisms for the control of chemicals

Each file, divided when appropriate into subfiles, contains
data records. A record in a file represents a complete item of
information and is accompanied by a cited reference.

IRPTC is confident that its system design, as suggested above
in the list of files, is comprehensive regarding the type of data
to be provided to decision-makers in order to substantiate their
best-informed assessment. It is also believed that the data
elements given in the files are sufficiently detailed and numerous
to allow full understanding and validation of the reported
information. Admittedly the laboratory scientist will not find in
the databank all the detailed information that may be wished for
carrying out practical experiments.

3.   Ambitions and Limitations of the IRPTC Databank

Bearing in mind the objective of IRPTC, which is to help
decision-makers in the field of chemicals control, it may be
worthwhile to outline what can and cannot be expected from a
factual databank such as IRPTC.

Prior to discussion of these points, some definition of terms
is appropriate.

<u>Toxicity</u>:-Any harmful effect of a chemical or a drug on a target organism.*

<u>Hazard</u>:-The set of inherent properties of a chemical substance or mixture which makes it capable of causing adverse effects on man or the environment, when a particular degree of exposure occurs.*

<u>Risk</u>:- The predicted or actual frequency of occurrence of an adverse efect of a chemical substance or mixture from a given exposure to humans or the environment.*

The IRPTC factual databank contains information on toxicity, hazards, and pertinent dose-effect relationships. It may report on hazard appraisal when indeed the potential threat posed by a chemical to health and/or the environment has been assessed by national or international expert groups.

In practical situations, the hazard and *a fortiori* the risk will depend on the likelihood and degree of exposure. As stated by Balk and Koeman[1] 'the exposure of a species or ecosystem depends on the likelihood that a pesticide or its conversion products will actually reach the system and its availability for the biotic components. For the assessment of the ecological vulnerability of an area with regard to an intended use of pesticides, factors regarding behaviour and exposure should be considered in relation to the ecological characteristics of the area'. A similar reasoning can be applied to other domains of toxicology.

Data relating to factors which may influence the likelihood and degree of exposure, *e.g.* stability and mobility in the environment or metabolism and rate of excretion in humans, can be found in the databank. However, data on exposure pertinent to specific (sub)populations or ecosystems have to be generated on a case-by-case basis in order to perform hazard and risk assessment. Such information will not be provided in the IRPTC unless very significant resources are specifically mobilized. National information systems may be the solution to such a problem.

The acceptability of risk based on in-depth cost-benefit analysis is also a concept which goes beyond the objectives of an information system such as the Register's factual databank.

In summary, it can be said that the target audience of databanks, and in particular of IRPTC, should be reliably informed about the potential hazards of chemicals; key elements indicating the validity of the data should be provided; care should be taken not to add to the complexity of the issue with too many technical details which may not contribute significantly to the understanding of the hazard.

Such constraints imply quality control and careful selection of the data while the databank is being compiled. These can be based on a choice of the sources where the information can be found; experience shows that, not infrequently, shortcomings in the publications permit useful screening of 'less satisfactory' data; well-defined selection criteria can help scientific scrutiny.

*see also Appendix A

In simple terms, the needs are for reliable data in manageable amounts. The following section suggests some guidelines in that respect.

4. Sources of Information and Reliability of Data in IRPTC

The first step in the preparation of data profiles consists in identifying pertinent sources of information.

When one first attempts to extract data from the literature, multiple decisions concerning the selection of data must be faced. A possible approach would be to include all data which have been published on the chemicals under study. In this case, the factual databank would very quickly become unwieldy and it's development very costly. Moreover, not all data in a scientific study play an important role in hazard assessment of a chemical.

Another possibility would be exclusively to use secondary evaluated documents, such as review documents prepared by various international or national groups of experts. The secondary documents are extremely useful in that they provide a mechanism for dealing with the vast quantity of primary literature available for some chemicals. In order to avoid reviewing all data, particularly for the well-studied chemicals, only those data which have been selected by a panel of experts for publication in a secondary document are considered. It is expected that the use of such critical reviews for data selection will both increase the quality of the data in the Register and reduce the task of literature screening. Moreover, if stringent selection has been used during the preparation of such reviews to quote only the most significant primary sources relevant to the evaluation made, this would be in perfect agreement with IRPTC's policy to present to it's users only the best quality data pertinent to perform risk and hazard assessment.

Data evaluations are being undertaken by the International Programme on Chemical Safety (IPCS). One of the objectives of IPCS is to carry out and disseminate evaluations of the effects of chemicals on human health and on the quality of the environment.

The International Agency for Research on Cancer (IARC) of World Health Organisation (WHO) (see Chapter 13) evaluates carcinogenesis and mutagenesis; the Food and Agriculture Organization (FAO) of the United Nations, in collaboration with WHO, publishes evaluated data for food additives' and pesticides' residues; the Economic Commission for Europe (ECE) and the Inter-Agency Group of Experts on the Scientific Aspects of Marine Pollution (GESAMP) cover the physical, chemical, and biological hazard attributes, as related to transport of chemicals by rail, road, air, and sea; and the International Labour Office (ILO), evaluates the data for the toxicology and workplace standards' attributes. Other organizations of international character which produce reviews are: the Council of Europe; the Council for Mutual Economic Assistance (CMEA); the European Economic Community (EEC); and the Organization for Economic Co-operation and Development (OECD).

The above efforts are augmented by evaluations executed by national agencies, *inter alia*, the United States Environmental Protection Agency (EPA), the United States National Institute for Occupational Safety and Health (NIOSH), the National Research

Council of Canada (NRCC), and the Health and Safety Executive (HSE) of the United Kingdom.

The following list illustrates the hierarchy adopted by IRPTC when searching the scientific literature as well as other literature, for information which best qualifies for entry into the Register.

The sources are used in the following order of priority:

1. International monographs and criteria documents containing evaluated information, *e.g.* those of the International Agency for Research on Cancer and the International Programme on Chemical Safety.

2. National monographs and criteria documents containing evaluated information, *e.g.* those prepared by expert panels convened by national agencies.

3. National monographs, reviews, and reports containing non-evaluated information, *e.g.* individual articles from symposia organized by national agencies.

4. Primary literature, *e.g.* articles in scientific journals, reports made available by industry.

Although the available secondary documents are used by the IRPTC to facilitate literature selection, the data themselves are extracted from the primary literature where the information is more comprehensive. For the majority of the chemicals in the Register, primary literature will often be the only source of published information available. In fact, for a large number of chemicals, data selection may not be an issue as there are very few data available.

5. Data Validation during Data Profiles Preparation

Identification of unreliable data is, more often than expected, facilitated by shortcomings such as:

No data on purity/composition are quoted
No data on exposure levels are given
Additional or mixed exposures have occurred
No Good Laboratory Practice has been followed
No control group is mentioned
Unrealistic dosages have been used
Insufficient detail is provided to validate the data

Data are more likely to be of a high standard:

i) if Good Laboratory Practice and internationally agreed protocols have been followed;

ii) if the data have been peer-reviewed;

iii) if they have been published.

6. Selection Rules for Toxicity Data in IRPTC

Practical problems related to the selection of data have been identified and carefully studied by IRPTC. Though open to improvement, workable solutions have been developed. A general presentation of these 'Selection Rules' as they apply to the data files reporting on toxicity data is outlined below. Detailed guidelines specific to each file of the Register are described in the various sections of the IRPTC Instructions Manual entitled 'International Register of Potentially Toxic Chemicals, Part-A' (1981, 1983, 1985).

When many papers are published on essentially the same topic, IRPTC does not attempt to be fully comprehensive. A basic concept has always been to draw the reader's attention to toxic effects without necessarily covering all the species which may be of concern, or all the experimental conditions which may be systematically modifed. Although giving interesting scientific insights it does not necessarily shed new light on the hazardous properties of a chemical substance.

In such cases, it is thought in IRPTC that a large number of studies related to a particular compound points to a compound 'giving rise to concern'. Therefore, secondary literature is most likely to exist. This probably eliminates unreliable and less reliable studies as well as hard-to-interpret observations; *e.g.* one or two seagulls found dead with 'high' DDT level in carcasses - is DDT the causative agent?

Three main doubts may arise when no 'secondary evaluated document' is available.

(a) One toxic effect is described in one species in many papers, but most of them report on changes in results when the experimental conditions or parameters are modified.

(b) Correlations of various types are reported in many documents.

(c) One toxic effect is described in many species.

Many publications contain results of elaborate investigations on the various aspects of the biochemical mechanism of toxic effects.

An 'always true and applicable' priority ranking of selection rules is not easy to develop and often more importance will have to be given to one point rather than to another on a case-by-case basis.

The main points of the strategy followed by IRPTC in extracting data from scientific documents are outlined below:

1.   All toxic effects are reported.

2.   Data on human beings are always reported.

3.   The lowest dose (or concentration) at which an effect is observed is reported.

4.   The 'No (Observed) Effect Level' is always reported.

5.   Studies where the results have been statistically analysed following currently accepted methods are entered by preference. This implies that a laboratory study with a control group is considered as more indicative for the

Register as compared with a hard-to-interpret so-called field study based on a few case observations of the same effect.

6. Epidemiological studies with a matched control group and statistical analysis of the results are always entered. When there are no such 'field' studies, case reports can be entered provided the causal relationship between the chemical and the effect is established.

7. As regards changes in experimental conditions modifying the experimental observations and various correlations resulting in a large number of papers:

(i)   If the results are significantly changed, as stated by the authors, the smallest modification of experimental conditions responsible is reported in the Register.

(ii)   Changes due to age of population and to variations in fat, protein, and other essential nutrients contained in feeding materials are always reported with the limitations as in *(i)* with respect to the significance of the results.

(iii)   Changes observed under other modified experimental conditions (temperature, pH, relative humidity), are only included if the latter are in a 'reasonable' range not too remote from normal (ecological/physiological) conditions.

8. As regards the same toxic effect described in many species:

(i)   In addition to human data, toxicological data on a rodent and a non-rodent species are reported whenever possible.

(ii)   Some fields of special interest to human beings can be useful as criteria of choice, *e.g.* species used for human consumption, species of economic importance.

(iii)   When a food-chain or a food-web is described in a document (or a group of documents) three species belonging to each trophic level are chosen. A range in sensitivity can be a useful criterion of choice: one highly, one moderately, and one poorly sensitive species.

7. Conclusions

Admittedly, the users of the IRPTC databank on chemicals are better served, being provided with a careful selection of the most pertinent data rather than with an exhaustive coverage of the relevant scientific literature.

That selection can be based on criteria and procedure, some aspects of which have been described in this chapter. They are not perfect and can be improved: IRPTC is open to any discussion and suggestion from readers. However, it is paramount to underline that such criteria are of no value without scientific judgement supported by scientific competence and integrity. Experience has also shown that common sense is, more than often, a very useful tool.

8.   References

1.   F. Balk and J. H. Koeman, *The Environmentalist*, 1984, 4, Supplement No.6.

Section 3: Verification of Data

6

Separating the Wheat from the Chaff - The Selection of Appropriate Toxicological Data from the World Literature

By D. Taylor

ICI PLC, BRIXHAM LABORATORY, OVERGANG, BRIXHAM, DEVON TQ5 8BA, UK

1. Introduction

In recent years, the implementation of a number of UK laws relating to health, safety, and environmental affairs has led to a situation where scientists, and in particular chemists, who may have no toxicological training or experience, are nevertheless expected to evaluate, interpret, and apply toxicological data as part of their regular activities. For example, the Health and Safety at Work Act 1974[1] places a duty on the employer to ensure the safety of his workforce. This will involve, among other things, the need to evaluate the potential risk of any of the chemicals which the employees will handle in the course of their job. Similarly, the Control of Pollution Act 1974[2] requires the producer of any waste to decide, mainly on the basis of its potential effects on human health, whether the material should be classified as 'special wastes' requiring disposal in accordance with the Section 17 regulations.[3] In addition, legislation resulting from European Community initiatives, in areas such as transport of chemicals, packaging and labelling of dangerous substances, and notification of new materials, requires the application of some toxicological data. In situations where no toxicologist or occupational physician is available, it is usually the chemist who is called upon for advice.

In the majority of cases, there is no shortage of available toxicological data, either in the form of reference works or as the original literature (see Chapters 2-4). The difficulty arises in selecting the valid data from the mass of sometimes conflicting information. Although this can be a fairly straightforward task for a toxicologist, it can be a daunting task for a non-specialist in the subject. However, considerable progress can be made by the non-specialist in the selection of relevant, reputable, and reliable data which can then be examined for both basic scientific merit and consistency. In the majority of cases this will enable a rational and reasonably accurate assessment of the data to be made. This chapter is intended to provide some signposts to enable the non-specialist to separate the wheat from the chaff.

2. Why Is Validation Necessary?

The simple answer to this question is that some of the data in the literature are incorrect or inapplicable. At first sight, this statement may seem surprising, particularly since there is a tendency to treat the written word, especially if generated by a computer, as being inherently accurate. However, there are a number of reasons why this should not be so, and these can be divided into three broad areas:

51

2.1. Misprints. Despite the combined efforts of authors, publishers, and printers, misprints are a fact of life. Computerized databanks can usually be modified to remove identified errors, although unidentified ones will remain. This is not possible with conventional publications where incorrect data will persist as long as the document survives. Fortunately, the majority of misprints are obvious typographical errors. However, some are undetectable to the reader and these can create difficulties. For example, in a recent proposal from the European Commission for a directive relating to the discharge of dangerous substances,[4] one of the quality objectives relating to DDT was inadvertently printed as 10 mg l^{-1} instead of 10 ng l^{-1} resulting in a published proposal approximately one million times less stringent than had been intended.

2.2. Advancing Knowledge. Toxicology is still a relatively young discipline in which advances in both the conduct and interpretation of toxicological studies are being continually made. The inevitable result of this is that information, which may have appeared to be correct at the time of its publication, can be shown subsequently to be incorrect or is invalidated by further scientific advances. Nevertheless, the results of older toxicological tests may often be of value if their limitations are borne in mind.

A classic example of this effect is given by the case of *N*-phthalylglutamic acid imide. A search of the literature in 1960 would have revealed that this material was an extremely safe sedative, with no addictive effects and such a low acute toxicity that it was impossible to use the drug to commit suicide. A similar literature search today would reveal the devastating effects caused by Thalidomide when taken between the twentieth and fiftieth day of pregnancy. Another example of this effect is shown in the Figure, which displays the values reported in the literature between 1942 and 1976 for the copper content of seawater.[5] Each value was considered to be correct at the time of publication. However, the reduction in copper concentration by three orders of magnitude has been brought about by improved techniques of analysis and sample collection.

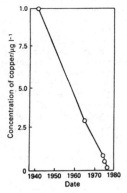

Figure   *Reported baseline values for copper in open ocean waters (1942–1976)*

2.3.   Incompetence.   It is inevitable that some toxicological studies will produce incorrect data.   There are many reasons for this, ranging from the inadvertent use of an impure test substance through the misapplication of test procedures to the use of inappropriate statistics to evaluate the data.   Some of these incorrect data eventually appear in the published literature owing to the incompetence or inattention of a small minority of authors, referees, editors, and publishers.   A number of the more common sources of error are examined in greater detail both in the remainder of this chapter and in the subsequent one.

The importance of data validation depends to some extent on the reason behind the request for the data.   Such requests can be broadly divided into the following categories in which data validation becomes increasingly important.

Firstly, a substantial number of requests for data fall into the so-called 'fire brigade' type.   In other words, a limited amount of information is needed rapidly, usually within minutes, to deal with an unforeseen set of circumstances.   A typical example of this would be the case of a road accident resulting in a chemical spillage.   The emergency services would require immediate data, on both human toxicology and ecotoxicology, in order to decide on the appropriate remedial action.   In such circumstances, time constraints prevent a comprehensive literature search, but in any case the information required is relatively crude and an order of magnitude estimate of acute effects will usually suffice.

The second area involves hazard evaluations.   Here, more time is available to determine the risks involved from the use of the chemicals in question, and thus satisfactory literature review can be undertaken and realistic assessments made.   However, the eventual assessment will be a simple one, usually resulting in advice on working practices, the protective clothing required, or the suitability of a material for discharge to sewer.

The final category is where the data are to be used to set regulatory criteria.   In such instances, scrupulous validation of all the data used must be undertaken, especially in the area of reported no-observable-effect concentrations (NOEC) or doses since the consequences of a mistake in the decision process could be severe.   For example, if the criterion is set at too lax a value, irredeemable damage might ensue.   Alternatively, erring too far on the side of safety by selecting an unnecessarily restrictive standard might result in heavy penalties for society either in terms of an overpriced product, caused by the cost of installing unnecessary treatment plant, or ultimately in the non-availability of desirable materials and loss of jobs.

3. Accessibility

This is only a problem for the more comprehensive searches, since the 'fire brigade' activity by its very nature has to rely on readily available information.   However, any large-scale literature search is likely to produce a number of citations that are not readily available to the enquirer.   This can be because they relate to obscure journals which are not contained either in the enquirer's own library nor in one of the national collections, or because the data are contained in an unpublished document, *i.e.* in the 'grey' literature (see Chapter 4).   Given sufficient time

and patience, most documents of this type can eventually be
obtained, but in the majority of cases, unless there is an almost
complete lack of information on the substance or problem in
question, these sources may only be essential in the case of a
fully comprehensive search.

The other major inhibition on access is that of language. The
majority of European languages can be translated relatively easily
and economically; nevertheless, there is a sizeable and rapidly
growing collection of toxicological data that is only available in
either Japanese or Russian and it is becoming apparent that a
significant body of Chinese literature also exists on this
subject. These data cannot be ignored in a comprehensive search
without running the risk of invalidating the conclusions.
Substantial progress can often be made by using the abstracts
provided by some of the major database suppliers, such as *Chemical
Abstracts*, but costly translation, although sometimes only of
small parts of the total article, may be essential. In some cases
a published translation may be available, *e.g.* ASLIB.

4. Relevance

The initial step in the validation process is to make sure that
all the data selected are in fact relevant to the question that is
being answered. Firstly, do the data relate to the correct
substance? This may seem a very elementary precaution, but it is
an area where substantial effort may be wasted and where
potentially serious errors can occur. The chemical nomenclature
is a minefield in which even professional chemists move with
caution and where considerable confusion can be caused by the
existence of different conventions; *e.g.* benzene hexachloride and
hexachlorobenzene are totally different materials. The former is
more correctly called hexachlorocyclohexane, but the early name is
often found in the literature, even in modern publications.
Similarly, the understandable tendency of authors to use acronyms
for lengthy chemical names is a further source of ambiguity.
Sometimes the resulting confusion is simply tiresome, as in the
use of PCB for both polychlorinated biphenyls and printed circuit
boards. On other occasions the confusion could lead to serious
error: ODCB for example has been used as an acronym for both
ortho-dichlorobenzene, a relatively innocuous solvent, and *ortho*-
dichlorobenzidine, a suspect human carcinogen. Wherever possible,
the *Chemical Abstracts* registry number (CAS Reg. No.) should be
used to identify the substance of interest. This number is unique
to each individual substance and is coming into common use in the
more recent publications and databases and databanks on
toxicology.

A similar warning applies to subject areas, and users of
conventional databases in particular should always be alert for
homophones; *e.g.* an article dealing with the effect of cyclohexane
on seals may be concerned with the toxic effects of the material
on a Cetacean species, but might equally be concerned with the
engineering aspects of corrosion in pipeline flanges.

Secondly, having established that all the data actually relate
to the substance in question, the next stage is to ensure that the
data concern the appropriate form of the material in both physical
and chemical terms. It is of little relevance, for example, to
know the effects caused by the exposure to mercury vapour in an
assessment of the toxicological significance of the consumption of

grain coated with a methylmercury seed dressing. The chemical form of the material, especially in the case of heavy metals, is an extremely significant factor in determining the toxic effect. All cadmium compounds do not have the same impact; for example, cadmium sulphide has a totally different toxic effect from cadmium sulphate. Arsenic compounds can be extremely poisonous to human beings, arsenic trioxide being favoured by many Victorian murderers. However, the consumption of shellfish containing extremely high concentrations of arsenic is without significant effect on the human body since in these circumstances the arsenic is strongly bound in an arsenobetaine complex which passes through the human gut without being broken down.

Finally, the route and mode of administration of the substance are vitally important. Subcutaneous injection of a substance, for example, may provide useful data on the metabolism of the material within the body, but may be less than helpful in identifying the impact of exposure to the substance in air or water. The results can be particularly confusing when tissue damage at the injection site leads to the development of a tumour, which might have been caused by the substance directly or only in conjunction with the damage caused by the mode of introduction.

Care must also be taken to discriminate between information relating to the major routes of occupational or environmental exposure, since substantial differences in effect may occur dependent on whether the exposure to the material was by inhalation, ingestion, or exposure to the skin.

Thus, it is an essential first step in any data appraisal to ensure that the data being studied relate to the substance of concern, its appropriate physical and chemical form, and the exposure route of interest.

5. Quality

The enquirer now has a list of citations, extracted from the scientific literature, that are relevant to the subject in hand. However, the literature itself is a heterogeneous collection of work of widely varying quality, ranging from papers of prime importance to irrelevant and sometimes incorrect observations. An initial guide to the likely validity of the information can sometimes be obtained by reference to either the organization which published the data or the individuals who carried out the work.

Firstly, is the information published in the conventional sense, and if so where has it appeared? There are many ways of publishing data: as a paper in a scientific journal, in the published proceedings of a conference either as a formal paper or as a poster presentation, as a book or section of a book which may or may not have been commissioned. In addition, there is a growing 'grey' literature consisting mainly of internal documents and reports produced by industrial, government, and independent bodies and made generally available. Under normal circumstances, more weight should be attached to the conventional than to the 'grey' literature and, furthermore, substantially more weight should be attached to a publication that has been subjected to a peer review process by an expert group of referees. The peer review process itself is not infallible, but does ensure that a substantial quantity of incorrect information does not see the

light of day. Most reputable international journals have
rejection rates of the order of 50-60%. It should be borne in
mind that some scientific conferences have excellent refereeing
systems, whereas scientific journals do not always require papers
to be submitted to a refereeing process. Similarly, the 'grey'
literature contains some extremely valuable data that should not
be ignored. However, the problem with these data is that they
have not been subjected to independent review or exposure in the
public domain. The 'grey' literature must therefore be seen as
inherently less reliable than conventionally published
information.

Secondly, who carried out the work? The presence of a Fellow
of the Royal Society, Chartered Chemist, head of a distinguished
academic department, respected industrial hygienist, *etc.*, in the
list of authors of the paper lends considerable weight to the data
which it contains, especially if the list of authors is limited to
two or three names. It can be assumed that such people would not
risk their reputations lightly and are thus unlikely to permit
their names to be associated with any works of dubious quality.

Thirdly, who sponsored the research? Unfortunately, a
substantial amount of quasi-libellous material has appeared in the
less responsible sections of the media accusing the sponsored
organization of bias in favour of the sponsor, usually an
industry. This bias is said to lead to the slanted reporting of
data in order to win favour with the client and thus generate more
funded work. The irony of this situation is that virtually all
scientific research in toxicology is funded by third parties,
involving governments, local authorities, and independent research
foundations in addition to industry. Thus, accusations of bias in
the sponsor's favour could be aimed at any research worker; *e.g.*
academic research teams might have an interest, especially when
academic research funding is difficult to obtain, in promoting a
particular issue simply in order to retain their grant support.
Unsubstantiated accusations are unworthy of any scientist and
reflect more on the character of the accuser than the accused.

Data should always be examined on their merits, regardless of
the source. The reputation of the researcher and his sponsor
should be used with caution in data validation and only in a
positive sense, in other words, to add weight to results.
Reputation alone should never be used to reject data as being
invalid. However, reputation can be a useful guide where
suspicion about the data has arisen for other reasons.

6. Reliability

The concept of reliability in scientific research is a relatively
recent one and can be summarized in the following question: are
the data presented in the publication an accurate reflection of
the work that was actually carried out? The first indication that
reliability might be a significant problem in toxicological
research appeared in the mid-1970s when the United States Food and
Drug Administration (FDA) discovered that work undertaken by a
contract toxicology organization was seriously flawed. The data
produced in final reports could not be substantiated by any
written documentation, records of scientific measurements were
absent or incomplete, the quality of some of the staff was felt to
be inadequate to carry out the work in a satisfactory manner, and
it was discovered that some of the data had been completely

fabricated.[6] This incident led directly to the establishment by FDA of a compulsory code of practice enshrined in the Good Laboratory Practice (GLP) Regulations. These regulations have to be strictly enforced by any organization wishing to submit non-clinical test data to FDA in support of registration petitions. This GLP concept has now spread to ecotoxicology and has broadened to include the majority of industrial countries, most of whom have implemented their own regulations, based on OECD GLP guidelines.[7]

The major principles of GLP relate to the control of test substances, the conduct of studies, the recording and retention of data, and the provision of an independent quality assurance unit. Briefly, the organization must uniquely identify and subsequently maintain adequate security of all test substances so that all the material used can be accounted for on completion of the work. A written scheme of work must be prepared prior to any study being undertaken, and the responsibility for all aspects of the study must be clearly defined. All the original data derived from the study, including any corrected errors, must be dated and signed by the operating staff, who themselves must have comprehensive training records indicating their competence in their specific tasks. The original data must subsequently be consigned to a secure archive for permanent retention, and indexed in such a way that a subsequent future inspection would be able to reconstruct totally the conduct of the study solely from the written documentation. Finally, the organization must appoint at least one person, who reports directly to the senior management of the laboratory and does not take part in any of the toxicological studies, to monitor the operation of the regulations within the organization. All laboratories undertaking GLP-controlled studies are subject to regular intensive inspection by government appointed officials to make sure that the laboratory facilities comply with the regulations and that studies are being carried out and recorded in the prescribed manner. Considerable expenditure of time and resources is necessary both to implement and to maintain these regulations. A statement that work has been carried out in accordance with GLP can be taken to imply that the work in question was carried out in the manner described by appropriately trained staff in a competent manner, and is an excellent indicator of reliability.

It is, however, important to realize that a statement that a study was carried out in accordance with GLP regulations shows only that the data presented are an accurate reflection of the study which was carried out. It does not indicate that the work in question was relevant to the problem or, indeed, had any scientific merit whatsoever. Good Laboratory Practice is about 'how' rather than 'what' work is undertaken.

7. Scientific Merit

Having now established that the data being examined are relevant, reputable, and reliable, the next step in the validation process is to examine the scientific merits of the information. It is difficult for a person without toxicological training or experience to make judgements in this area. Nevertheless, there are still a number of points which will be detectable even by the non-specialist and these are outlined below. Additional information can be found in the works of Brown[8] and Ludewig.[9]

7.1. Materials. The first questions relate to the test substance itself. Was it properly characterized and is it adequately described? The material may exist as a number of different isomers or in a variety of physical forms. More importantly it may not be a pure material owing to either the presence of impurities or to the addition of formulating and stabilizing substances. The presence of the latter materials can have a major impact on the toxicity of the test substance, especially since some of the surface active formulating agents themselves can have a very significant toxicity. Secondly, was the concentration of test substance in the dose measured or estimated? If it was measured were the analytical techniques used adequate? The report should contain sufficient information to make an assessment of the precision, accuracy, and specificity of the analytical techniques used. In addition, there should be an indication of the frequency of such measurements together with information on the checks for recovery and precision undertaken during the study. If the dose was only estimated, did this detract from the validity of the study? Some test substances are highly volatile and thus readily lost to atmosphere; others are strongly adsorbed onto surfaces and, at low concentration, can be effectively eliminated from solution by this process. In addition, some substances may be unstable under the test conditions used and thus liable to decompose during the study. In these cases, dose estimation is inadequate and any test result will be open to question.

The use of solvent vehicles in toxicology is of particular concern. The carefully controlled use of appropriate solvents to assist in the preparation of test solutions of sparingly soluble substances is a sensible and acceptable technique. However, studies involving the use of solvents such as dimethyl sulphoxide (DMSO) which materially assist the transfer of the substance across cell membranes should be viewed with caution, as should any data derived from experiments where a solvent has been used to maintain the concentration of the test substance at a level above its true solubility. Under such conditions the information produced will relate to the effect of the test substance/solvent association which may be very different from the effects of the test substance alone.

The next question relates to the test animals used. In ecotoxicology, all species will have some relevance, since they all live in the natural environment. However, the environment is highly variable, and data on saltwater crustacea may not be very relevant to the impact of the material on a freshwater fish in a soft water environment. In the mammalian area, where the object is usually to extrapolate the data to man, there is a very real problem. Where all the species tested give a similar result we can assume that man will show the same susceptibility. However, where there is a wide discrepancy in effect between species, the mechanism of the toxic action needs to be more carefully assessed whilst in the interim assuming that man is as sensitive as the worst affected species. Whatever animal is used, there must be adequate control population data to support the toxicity information. The control population should not exhibit any abnormal behaviour and obviously mortality or morbidity levels should be minimal. Control populations are particularly important in those areas where a vehicle is necessary to effect dosing of a material, *e.g.* the use of solvents to prepare solutions of the test substance. In such cases, there must be a population separate from the conventional control group that receives the vehicle at the highest concentration or dose used in the test, but not the test substance. It should not be assumed that the vehicle will have no effect on the eventual result.

7.2. Procedure. The exposure route tested should be a practical one for the material and its use in question. For example, in cumulative oral toxicity tests gavage and dietary administration may produce different responses, since in the former the material is administered in pulses whereas in the latter it is effectively a continuous dosing. Similarly, exposure of complete animals to a gaseous test substance will lead to dermal absorption, and possibly ingestion, as the animal preens itself, in addition to the anticipated inhalation route.[10]

The duration of the test must also be sufficient to allow the toxic effects, if any, to appear. For example, any negative tests for carcinogenic responses which do not continue for the lifecycle of the animal in question will be open to doubt. The number of animals used must also be adequate to allow the effects to be distinguished from chance occurrences at an appropriate level of significance. One random death in a group of five animals represents a 20% mortality, whereas the same random death in a group of 20 animals is only a 5% mortality.

Finally, does the procedure enable a dose-response curve to be constructed? Many toxicological studies reported in the literature consist of single observations with only one concentration or dose of the test substance. Such studies do not allow the relationship between dose and effect to be examined and are consequently of only limited value, since the dose-response curve may not be linear, may be discontinuous, or may have upper or lower threshold values; however, such evidence of absence of effect at a particular dose or concentration may still be of value.

7.3. Interpretation. Although the scientific study may have been both valid and reliable, the final conclusions may still be invalidated by mistakes in data analysis. The interpretation of quantitative toxicological data relies heavily on the use of statistics and, as in other scientific disciplines, these are often misapplied, abused, and misinterpreted. The proper use of statistics is essential in toxicology and must be used for example to determine whether differences between exposed and control populations are likely to be due to the effect of the substance being tested or to pure chance. Despite this, statistics should never be used as a substitute for biological knowledge and toxicological experience.[11] One of the commonest errors in this area is the attribution of causality to correlation. This is particularly prevalent in epidemiological and biological surveillance studies where it is common for authors to indicate that, because the incidence of a particular disease or biological effect is highly correlated to the concentration of a specific substance, therefore the effect is caused by the substance. Such a correlation is indeed indicative that further studies are desirable but is not on its own proof of causality any more than the positive correlation between the crime rate and the incidence of church membership indicates that one is the cause of the other.[12] However, even the correct use of statistics can be misleading and this has been succinctly put by Weil[13] who stated that a statistical significance may be of little or no biological importance and that, conversely, important biological trends should be examined further even in the absence of statistical significance.

A major problem in the interpretation of data is the assessment of the importance of the reported effects. Mortality is obviously an extremely important response, but other effects are more difficult to assess. Some biochemical and physiological responses, for example, are merely a reflection of the animal adapting to its changed environment and are of little or no importance to its general well-being. For example, the human skin responds rapidly to small changes in external temperature by either sweating or forming goose pimples. These effects are readily observable but neither is detrimental and no one would suggest that man was at risk in situations where either response was observed. However, this area is one in which disagreements amongst toxicologists are common; for example, the significance of the enchanced B_2-microglobulin levels in the blood of people exposed to cadmium at low concentrations is still the subject of considerable controversy.[14]

Finally, it is worth highlighting two common fallacies. Firstly, mutagenicity and carcinogenicity are not synonymous. A wide range of mutagenicity tests is now available and since these are rapid and inexpensive there is a tendency to use them as a substitute for the extremely time-consuming and expensive carcinogenicity tests. However, even in the determination of mutagenicity itself, the results of a number of different relevant tests should be examined before reaching any conclusions, and subsequent extrapolations from mutagenicity to carcinogenicity should never be made without further information.

Secondly, LC_{50} and LD_{50} results should be used with caution since they indicate only mortality and do not give by themselves any indication of morbidity. Thus, a material which causes major biological effects, *e.g.* producing almost complete paralysis at low concentrations, might still have a very high LC_{50} value provided that death itself did not occur during the short period of the experiment.

8. Consistency

The last stage in the validation process, especially in the areas of hazard evaluation and criteria setting, is to examine the data for internal consistency. Do all the data agree with each other or are there some data which appear to lie well outside the majority of values?

It must first be recognized that toxicological data are inherently more variable than data from other scientific disciplines, and experiments on living animals produce data that are imprecise and often inconsistent. This may be difficult to appreciate by scientists more used to the accuracy and precision of chemistry, but the recognition of the inherent variability of biological processes is an essential prerequisite for the effective evaluation of toxicological data. For example, Sharrat points out[11] that the growth rate of a normal laboratory rat can be 70-80% of the growth of a similar individual even in a well controlled experiment using genetically similar animals. The effects of biological variability are unfortunately often clouded by toxicologists themselves, some of whom still insist on expressing their findings, especially LC_{50}s and LD_{50}s and NOECs as numbers with two or three significant figures. This may be mathematically correct, but is less than helpful in toxicology and lends a spurious degree of accuracy and reliability to the data. There are also major differences in sensitivity between different species and strains and thus tests for consistency should in general be related to individual species.[15] A topical example is provided by 2,3,7,8-tetrachlorodibenzo-*p*-dioxin (TCDD) where the guinea pig with an LD_{50} of 1 μg kg^{-1} is three orders of magnitude more sensitive to the material than the hamster with an LD_{50} of 5000 μg kg^{-1}.[16]

After taking into account the variability noted above it is possible that some unexplained outliers will remain in the data. All such values should be re-examined, paying particular attention to those which indicate that effects occur at much lower concentrations than would be expected from the other data. Particular attention should be paid to the units in which the data are expressed. The units are usually, although not necessarily, consistent within one publication, but different publications even in the same compilation and by the same author can differ significantly. Some of the commonest pitfalls with units are outlined below.

The concentration or dose of the test substance can be reported variously in terms of the active ingredient, the formulation, or, in the case of residues, as the principal metabolic product. In studies with trace metals, for example, the metal concentration may be expressed as the metal or as its salt. A factor of four exists between copper concentrations expressed as Cu as opposed to $CuSO_4.5H_2O$.

The use of shorthand expressions of concentration such as p.p.m. can be extremely misleading; *e.g.* parts per billion (p.p.b.) usually refers to one in 10^9 but can also mean one in 10^{12}. The recent popularity of the term p.p.t. for one in 10^{12} (*i.e.* parts per trillion) is potentially more confusing, first since one in 10^{12} can be also expressed by p.p.b., if one uses the word billion in its English meaning, but more importantly since marine scientists have used p.p.t. for many years to mean one in 10^3 (*i.e.* parts per thousand). Salinity measurements are usually quoted in this unit. In addition, a major disadvantage of the use of terms such as p.p.m., p.p.t., *etc.* is their lack of precision since they give no indication of the basis for the ratio. Thus p.p.m. could mean either mg kg^{-1} (weight/weight), mg l^{-1} (weight/volume) or μl l^{-1} (volume/volume). This leads to a situation where a 10 p.p.m. solution of carbon tetrachloride in water could be either 10 mg l^{-1} or 16 mg l^{-1} depending on whether it was prepared on a weight/volume or volume/volume basis.

A further source of error relates to the use of different conventions for the basis of the calculation of concentration in solid material. Residue levels in animal tissue are regularly reported in terms of ash weight, dry weight, flesh weight, wet weight, or lipid weight. In most cases the report does not contain any information by which the data can be converted into a common format, *e.g.* dry weight. Furthermore, there may be no clear definition of the terms themselves; *e.g.* what was the drying temperature?, does wet weight include or exclude body fluids?, *etc.* Confusion may also arise over the way in which the concentration of a substance in one phase of a two-phase system is expressed; *e.g.* the concentration of a substance in an airborne particulate may be expressed or as a concentration in the particulate itself or as a concentration in the air volume from which the particulates were collected. This confusion will be exacerbated if the concentrations themselves are quoted in terms of p.p.m.

The data may also relate to a preselected part of the total sample, or a post analysis normalization technique may have been used to transform the original data. This is common practice in many exposure measurements. In the analysis of soil and sediments for example, it is rare to see data which relate to the total concentration in the matrix; the usual practice is to analyse an extract of the original sample. The extraction techniques used vary widely in their efficiency, and results using different techniques are impossible to compare satisfactorily.

Finally, there are two terms LD_{50} and EC_{50} which regularly cause difficulties. LD_{50} is the dose level which causes half of the exposed individuals to die. It is not a concentration but a load that is related to the weight of the animal concerned. Consequently, it is normally quoted in terms of mg kg^{-1} body weight. Unfortunately, LD_{50} values are often quoted simply as mg kg^{-1} on the mistaken assumption that everyone will appreciate the significance of the kg as being the weight of the animal concerned. Secondly the term EC_{50} has two distinct uses. The normally accepted definition is the concentration of material at which half the exposed animals are affected, *i.e.* a similar definition to LC_{50} and LD_{50}. However, it is also occasionally used to indicate the exposure concentration at which the measured variable has changed by 50%; *e.g.* a 50% reduction in photo-synthetic activity of an algal population may also be quoted as an EC_{50}.

The fact that a data point lies outside the normally accepted range of values does not of course signify that it is incorrect and can therefore be ignored. The case of the environmental background concentration of lead provides a salutary warning in this area. Until the early 1960s, there was general agreement in the literature about the background concentration of lead in biological tissue or natural waters. Then in 1963 Patterson in the USA published results[17] indicating that these values were totally incorrect and represented *ca.* 1000 times the true value. This publication was intitially treated with considerable scepticism by the scientific establishment but has subsequently been shown to be correct, previously reported values being almost entirely due to sample contamination. Nevertheless, all examples of outliers should be carefully viewed before being used as an indicator of a lower general threshold for toxic effects. This is particularly true for unconfirmed data. It is almost unknown for a scientific paper or observation to be withdrawn and thus incorrect data have a tendency to persist in the literature. On the other hand, if an observation is correct it is likely that further observations will be published confirming the initial data. Thus an outlier that has been published for 10 years and is unconfirmed by subsequent data is likely to be of little concern.

Although lack of consistency in the data may cause problems in the interpretation of outliers, too much consistency should also be treated with caution. As was mentioned earlier, biological data are inherently variable and thus one should expect to find some variation in reported toxicological data. However, there can often be a marked agreement on the toxic effects of a substance which may extend over a long time scale. The enquirer should be on his guard in such circumstances so that he does not attach too much importance to apparent consistency which may be only superficial. There is a deplorable tendency towards multiple publication of data.[18] In theory, the majority of scientific journals only publish original data, *i.e.* material not previously published. In practice this rule is often ignored, or circumvented. There are innumerable instances of identical manuscripts published in different journals, similar manuscripts describing the same piece of work published with the authors' names in a different sequence, and the same work being presented at two different conferences in revised form but with no advances in knowledge. Thus, what appears at first sight to be general agreement on the toxicity of a particular material may be the result of the same piece of work being published in a large number of different places. In addition, there are now a growing number of compilations, reference works, and databases which contain information selected from the original literature. In cases where the original literature is sparse, the same data are likely to appear regularly, thus giving a spurious authenticity to the result. This is of greater seriousness when the original data were either incorrect or inaccurately transcribed by the first abstracter and thus always subsequently appear in this incorrect form.

9. Conclusions

The literature contains a large number of articles dealing with the toxicity of chemicals. These are a heterogeneous collection of predominantly good, but occasionally bad, and indifferent work, sometimes inaccurately reported or printed. The validation of such data is an essential prerequisite to any assessment of the actual impact of the chemical in the real world. For the non-toxicologist such a procedure is difficult, but considerable progress can be made even by the non-specialist in the selection of relevant, reputable, and reliable data which can then be examined for both basic scientific merit and consistency. In the majority of cases this will enable a rational and reasonably accurate assessment of the data to be made.

10. References

1. Health & Safety at Work Act 1974, Ch 37, HMSO, London, 1974.

2. Control of Pollution Act 1984, Ch 40, HMSO, London, 1974.

3. Control of Pollution (Special Wastes) Regulations, SI 1980, No 1709, HMSO, London 1980.

4. Anon, *Official Journal of the European Communities*, 1985, C70, 15.

5. D. Taylor, *Anal. Proc.*, 1982, 19, 561.

6. W. Broad and N. Wade, 'Betrayers of the Truth – Fraud and Deceit in the Halls of Science', Century, London, England, 1982.

7. Final Report of the Expert Group on Good Laboratory Practice in the Testing of Chemicals, OECD, Geneva, Switzerland, 1982.

8. V. K. Brown, 'Acute Toxicity in Theory and Practice', Wiley, Chichester, England, 1980.

9. R. Ludewig, in 'Topics in Forensic and Analytical Toxicology', ed. R.A.A. Maes, Elsevier, Amsterdam, The Netherlands, 1984, 20, 69.

10. D. Sanderson, Proceedings of the Eleventh Annual Conference of the Association of Information Officers in the Pharmaceutical Industry, 1984, p.65.

11. M. Sharratt, 'Evaluation of Toxicological Data for the Protection of Public Health', Pergamon, Oxford, England, 1977, p.105.

12. M. J. Moroney, 'Facts from Figures', Penguin, London, England, 1951, p.303.

13. C. S. Weil, *Toxicol. App. Pharmacol.*, 1972, 21, 194.

14. H. C. Gonick, Proceedings of the Fourth International Cadmium Conference (Munich), Cadmium Association, London, England, 1983, p.179.

15. J. Doull, Proceedings of the Fifth International Congress on Pesticide Chemistry, 1983, 3, 433.

16. R. L. Rawis, *Chem. Eng. News*, 1983, 61, 37.

17. M. Tatsumoto and C. C. Patterson, *Nature (London)*, 1963, 199, 350.

18. D. Taylor, *Limnol. Oceanog.*, 1978, 23, 382.

7

Some Common Sources of Error in Data Derived from Toxicity Tests on Aquatic Organisms

By R. Lloyd

MINISTRY OF AGRICULTURE, FISHERIES AND FOOD, FISHERIES LABORATORIES, REMEMBRANCE AVENUE, BURNHAM-ON-CROUCH, ESSEX CMO 8HA, UK

1. Introduction

During the decade 1950 to 1960, there was a rapid awakening of the public to the dangers of environmental pollution, and the consequent political initiatives led to a considerable increase in the funding of research into the effects of chemicals on a wide range of organisms, especially those which live in rivers and lakes. Subsequently, the discovery of global contamination by persistent organochlorine compounds focused attention on the marine environment, the concern here arising from fears expressed by a scientific community rather than by the public.[1] The number of publications on concentration-effect relationships between chemicals and aquatic organisms rapidly expanded. However, very few of the scientists involved were trained in toxicology or pharmacology. Many were ecologists or physiologists who were studying aquatic organisms or communities, and who exposed these systems to solutions of chemicals and measured the resulting perturbations. It was inevitable that at times deficiencies in the experimental methods have led to erroneous conclusions being drawn from the results. Because of the limited number of competent reviewers available to the editors of scientific journals, these conclusions reached the open literature and were incorporated into uncritical reviews and compilations of data.

This summary of events is, of course, a sweeping generalization. However, although the science of aquatic toxicology has gradually improved throughout this period, as shown by a recent text book,[2] and guidelines exist for many experimental procedures, there is still scope for errors. These errors fall into two main categories: deficiencies in test methods which affect the results obtained, and errors in the interpretation of the ecotoxicological significance of the data which form the conclusion. Clearly, the scope for such errors is open-ended; the examples chosen in this paper represent the more important or common ones which have been revealed by a scrutiny of data submitted by manufacturers for the notifications of pesticides and other chemicals, by critical reviews of the literature to provide a database from which water quality standards can be derived for aquatic life, and by refereeing manuscripts submitted for publication in scientific journals.

The three major components of an aquatic toxicity test are the *concentration* of the test substance to which the organism is exposed, the recorded *response* of the organism, and the *exposure period* of the organism to the substance. The data obtained can be used, *inter alia*, for crude hazard *rankings* or for water quality *standards*.

2. Concentration of Test Substance

Results of toxicity tests with aquatic organisms are commonly expressed as EC_{50}s or LC_{50}s, that is, the concentration of a test substance which causes an effect on, or mortality to 50% of a test population within a given exposure period. Unlike the LD_{50}, where the normal practice is to give a test organism a single dose of a test substance, an EC_{50} or LC_{50} test requires that the test organism is exposed to a constant concentration of the test substance throughout the entire exposure period. Furthermore, the test substance should be distributed homogeneously throughout the aqueous medium. This is an area where major sources of error can occur. Also, the chemical characteristics of the dilution water in the test vessels can markedly affect the toxicity of some substances. The following sections describe some of these factors.

2.1. Solubility.

This is a common source of error. Although stock solutions may be prepared with solvents, acids, or alkalis in order to facilitate the addition and distribution of the test substance to the test vessels, the solubility of the test substance may be exceeded in the test dilutions and a precipitate may be formed. This is particularly the case with heavy metals, especially in sea water, where the precipitate may become visible only after several hours. There have been some suggestions that the precipitates of some substances, such as carbonates of heavy metals, may be just as toxic as if they were in solution, but in general the precipitates are less toxic and if they settle to the bottom of the test aquaria they do not come into contact with test organisms such as fish which swim in mid-water.

Other examples are those pesticides for which solubility appears to be lower in the test solutions than would be expected from the determined aqueous solubility in distilled water; under the current provisions of Good Laboratory Practice (GLP) it is likely that the presence of a precipitate in the test solutions will be recorded, but with older data this may not be the case. Claims that the 96 h. LC_{50} of a substance is greater than 100 or 1000 mg 1^{-1} are evidently wrong if precipitates form in concentrations greater than 10 mg 1^{-1}, which is then the highest effective concentration to which the test organisms have been exposed. Also, claims that such substances are 'non-toxic' on the basis of short-term tests with two or three species are also optimistic; sub-lethal effects may occur if those species are exposed to the substance for a longer period, or if there are other species or life cycle stages which have a greater sensitivity.

Where a solvent is used in the preparation of the stock solution, a 'solvent control' should be included in the series of test solutions in which the concentration of solvent is equal to that in the highest concentration of the test substance to which the organisms are exposed. However, even if the solvent alone does not exert an effect, it is still possible that the bioavailability of the test substance may be either enhanced, or reduced, in the presence of solvents or solubilizers, because of their action on the surface epithelium of the organisms or the partitioning of the substance between the organism and the aqueous phase. This is an area which has been little explored, and it is prudent, therefore, to treat data from all such tests with caution, if only because solvents are not part of the natural environment.

2.2. Oils and Volatile Substances. Considerable problems arise where the test substance is a fluid and not miscible with water; oils are a prime example. The preparation of water-soluble fractions of oils for toxicity testing is beset with difficulties and, in the absence of a recognized standard procedure, the results obtained depend to a large extent on the amount of agitation given to the oil/water mixture, the oil:water ratio, the duration of agitation, and the subsequent settling period. Some hydrocarbons are more toxic in emulsion than in true solution. If aquatic organisms are to be exposed to oil dispersed in water, then the oil droplet size may have an effect on the toxicity, and the characteristics of the droplets will change during the exposure period. In both cases, losses of volatile components will have to be avoided. This problem is very important for all volatile substances, for which maintaining test concentrations at the desired level can present considerable technical problems.

2.3. Other Causes of Loss of Test Substances. Some substances, particularly those pesticides where very low concentrations are tested because of their high toxicity, can be adsorbed onto the walls of the test apparatus and are then not available to the test organism. This may be accentuated if the surfaces become coated with a microbial film as a result of the use of solvents or other substances which form a substrate for such a growth. In some cases it is possible that the growth may also biodegrade the test substance, and suspended micro-organisms can also exert this effect. Photolysis and hydrolysis can also cause reductions in test concentrations of organic substances.

2.4. Other Factors which Modify Toxicity. There is an abundant literature on a wide variety of physical and chemical factors which can modify the toxicity of certain substances.[2] An extreme example is that of ammonia, which dissociates into the un-ionized and ionized forms in water, and where the un-ionized form contributes most to the toxicity. The degree of ionization depends on the pH, temperature, and salinity of the water; the toxicity also depends on the bicarbonate and free carbon dioxide concentrations, as well as temperature. Most aquatic organisms are more sensitive to substances at the extremes of their tolerated temperature and salinity ranges, and when the dissolved oxygen content of the water falls below a critical level. Water hardness has a considerable effect on the toxicity of heavy metals and some other substances to aquatic organisms, and some organic substances can chelate metals such as copper and reduce their toxicity. Distilled water can be toxic to freshwater organisms.

2.5. Conclusions. Experimenters should have had full regard to all the likely factors which will reduce the effective concentration and should have designed their experiments to eliminate these sources of error as far as possible. Test data should include an analysis of the dilution water, especially for temperature, dissolved oxygen, pH, hardness, and alkalinity, as well as trace contaminants such as metals and chlorine. Lack of such information can lead to errors in the interpretation of the data. Also, the concentration of test substance should be measured at intervals during the test, or some explanation given for the assumption that the test concentrations remained close to the nominal value during the exposure period. In the absence of such information, quoted concentrations must be regarded as suspect; experience has shown that in many cases it would be reasonable to assume that the results from such experiments are an underestimate of the true toxicity of the substances.

3. Response

For all chemicals, aquatic toxicity tests carried out initially are short-term and the respsonse measured in the test animal is mortality. For those chemicals which occur in concentrations giving rise to environmental concern, further tests may be carried out which measure sub-lethal or non-lethal responses. There are several sources of error which may occur in the interpretation of the data obtained.

3.1. Mortality.
Notwithstanding the observations made in the previous section, it is a general rule that it is easier to cause the death of a test organism in the laboratory than to keep it alive. Therefore, test mortalities among the controls in clean water, and in the sub-lethal exposure concentrations, should be recorded. Tables showing the number of mortalities in each test concentration at fixed observation times (6, 12, 24, 48 h., *etc.*) can indicate whether there are apparent anomalies in the times of death, which may also be reflected in the graphical presentation of the log (concentration)-log (survival time) relationship. If such anomalies are present, consideration has to be given to mortality other than that caused by the test substance, such as rapid change in temperature, drop in dissolved oxygen concentration, disease, or starvation. Provision of a single 48 or 96 h. LC_{50} value prevents validation of the data.

3.2. Sub-lethal Effects.
After tests which measure mortality, those which measure growth and reproduction normally form the next phase of investigation. There should be good evidence of a graded response to the test concentrations; full details of the feeding regime should be given. Initially, a chemical may cause a temporary check in the growth rate, followed by an acceleration towards that achieved by the controls, and therefore the measured effect may be strongly influenced by the exposure period. Similarly, it is possible that reproduction may be influenced by factors other than the test substance.

Because of the pressures outlined in the introduction, many scientists search for 'the most sensitive effect'. Where such effects are at the cellular or biochemical (enzyme) level of organization, it has to be shown that these are deleterious to the organism (and not merely an adaptive response) and that the extent of the effect is of biological significance at the individual and the population level, as well as of statistical significance. It is not uncommon for information on a concentration of a chemical causing a small and biologically insignificant effect in an organism to be incorporated into databanks, and then used in the context of concentrations having a harmful effect.

3.3. Dubious Results.
On several occasions, manuscripts refereed for publication have contained data which are so unusual in some respects as to raise doubts about their veracity. For example, the experimental results might display a regularity, uniformity, or precision which is alien to biological or even chemical data. Such a case might be where the median periods of survival or organisms exposed to a range of very closely spaced concentrations

are provided in tabular form; when these are displayed graphically, all the data lie exactly on a smooth curve or straight line. It is possible that the experiments were indeed carried out but the results were felt to be too irregular for publication; instead, a line may have been fitted through the scatter of points and the reported survival time read from the curve. In such cases there is no certainty that the experiment was carried out.

Another questionable result may be a case where the standard deviation of the average enzyme content of tissues supposedly derived from analyses of ten separate animals exposed to a chemical was so exceptionally small as to have been remarkable even if the tissues were in fact bulked and the analysis replicated ten times. Statistical analysis based on the assertion that the analyses were carried out on individual animals is therefore likely to be invalid. Such a device might be used to demonstrate the existence of a significant effect of a treatment (even though it may be very small) and thus support some hypothesis of toxic action, in cases where the real experimental evidence was inconclusive or non-existent.

The accuracy and precision of the chemical analytical methods used should be determined under realistic conditions. Analysts are prone to claim spurious accuracy by assuming that the precision of results obtained in distilled water is applicable also to the test conditions. Internal standardization is a neglected technique in this context, particularly where chromatographic methods are used on minute quantities of determinand; similarly, expression of concentrations to four or more significant figures may imply a spurious accuracy.

These are obvious examples which can be readily identified by a competent referee. A request for the raw data is usually met by silence and the manuscript should then be rejected. However, it is not impossible for such papers to slip through the net and be subsequently published, and again the results can become incorporated, *via* uncritical abstracts, into a variety of databases. A heavy responsibility is placed on editors and referees of scientific journals to prevent such an occurrence.

There remains a problem of unknown dimensions where experimental data have been adjusted or fabricated in a way that cannot be readily identified by a referee. If the experimental results are supported by other published data, it may be that they are not too far from the truth and no damage is done in the context of deriving wrong conclusions leading to inappropriate action or controls. The implications of dubious or aberrant data are discussed below.

4. Exposure Time

Some aspects of exposure time as an essential factor in data assessment have been touched on above. The time taken for a substance to produce an effort in an organism, or for that organism to adapt to the insult, varies widely between substances and between species, and with environmental factors such as temperature. There are two major implications of this variability.

4.1. Acute Toxicity Tests. The normal duration of acute toxicity tests is 48 or 96 h. These exposure periods have been chosen for the convenience of research workers or technicians who work a 5 day week. It is true that, for the great majority of chemicals, results obtained from tests of this duration are adequate for the purposes of a hazard assessment, but there are exceptions. There are, for example, some oils which produce mortalities in the brown shrimp only after 3-4 days exposure and there may be other cases where mortality was not produced within 96 h. but would have been with an extended exposure period. For some chemicals, short-term exposure to low concentrations may produce irreversible effects which lead to a delayed mortality. Such cases are rarities, but the chances that they may apply to a chemical must always be kept in mind; small but significant metabolites in a range of concentrations towards the end of the exposure period may be an indication of such a delayed effect.

4.2. Sub-lethal Effects. As noted above, such effects may be transient and pass through stages of inhibition or stimulation, depending on the exposure period. The possibility of such variability should be recognized, particularly where different research workers appear to have reached different conclusions for the effect of specific chemicals.

5. Needs for Critical Assessment of Experimental Data

Where ecotoxicological data are required for a general hazard ranking, such as those used to regulate the carriage of chemicals and to advise users and others of their potential harmfulness in the event of an accidental spillage, the need for accuracy of information is not great; a general indication of potential hazard for the aquatic environment may be all that is required.

However, where the environmental concentrations of a chemical as a result of normal usage approach a level where damage may be caused to aquatic living resources, then the need for accurate information is paramount in order to formulate the appropriate controls. Under these circumstances, a critical assessment has to be made of all the appropriate data available for that chemical, in order to identify a core of information which is mutually supportive and coherent. Such a literature review may reveal the influence of modifying factors which need to be taken into account when evaluating data from other experiments, thus removing possible errors in interpretation. Other data may then have to be excluded from the review if insufficient detail is given of the experimental conditions.

A critical review will also identify data which do not fit into the general pattern of results. It is these experiments which have to be given especially close scrutiny to assess whether there are any anomalies or errors which invalidate the results. Any supporting data, no matter how fragile, have to be taken into account for such an assessment. Where the data are for sub-lethal effects, the biological significance of the results, not only at the level of the organism but at the population and community level also, has to be assessed. In recent years there has been a tendency for uncritical databanks to be compiled for specific chemicals, from which the lowest conentration which it is claimed has produced an 'effect' on an organism is identified and then for that concentration to be divided by 10 in order to produce a water quality standard. This can lead to a wholly unnecessary stringency in control measures if that specific experiment was flawed for one or more of the reasons outlined above. A more critical approach to literature reviews was adopted by the European Inland Fisheries Advisory Commission's Working Party on Water Quality Criteria for European Freshwater Fish,[4] these reviews on the efects of common pollutants on fish were made by scientists who had relevant practical experience of the individual chemicals and who knew of the specific problems and potential sources of error. The tentative water quality criteria proposed by the Commission were based on defensible scientific data; they have stood the test of time, and recent published information generally supports the criteria rather than causes them to be significantly modified. There are a number of other reviews undertaken to derive water quality criteria which have been based on similar critical appraisals of the literature, but there are others which do not meet this standard.

When assembling the relevant information for specific chemicals, established databanks are a useful source of references to published information, but they have their limitations. Unfortunately the summary information on effects derived from the published results and incorporated into the databanks cannot be used for the preparation of water quality standards; they can be used for hazard assessments for ranking or emergencies, but the generally limited nature of the information precludes their use for more detailed reviews.

6. Conclusions

The published data on effects of chemicals on aquatic organisms are of a very variable quality, because of the large number of errors which can occur within the experimental methods and in the interpretation of the results. Only a few of the major sources of error have been outlined here. There is a very strong case for all such data to be subjected to expert critical review before they are incorporated into databanks or other compilations of information. Ideally, such experts should have personal experience of carrying out experiments with the chemicals under review. For the reasons given in the preceding sections, the conclusions drawn by the author of a research paper or technical report should not be accepted without question.

7. References

1. R. Lloyd, 'Marine Ecotoxicological Testing in Great Britain', in Ecotoxicological Testing for the Marine Environment, ed. G. Persoone, E. Jaspers, and C. Claus, State University; Ghent and Institute of Marine Scientific Research, Bredene, Belgium, 1984, Vol. 1, 39–55.

2. 'Fundamentals of Aquatic Toxicology: Methods and Applications', ed. G. M. Rand, and S. R. Petrocelli, Hemisphere Publishing Corporation, Washington, 1985.

3. R. F. Christman, 'Ethical Consideration in Scientific Publication', *Environ. Toxicol. Chem.*, 1985, **4**, 583–586.

4. J. S. Alabaster, and R. Lloyd, 'Water Quality Criteria for Freshwater Fish', 2nd Edn., Butterworths, London, 1982.

8

The Application, Generation, and Validation of Human Exposure Data

By P. Leinster

FBC LTD, HAUXTON, CAMBRIDGE CB2 5HU, UK

and M.J. Evans

BRITISH PETROLEUM PLC, CHERTSEY ROAD, SUNBURY, MIDDLESEX TW16 7LN, UK

1. Introduction

The potential for exposure to chemical agents exists in many work situations; if not in connection with the main tasks or processes, then during repair, maintenance, cleaning, or other associated activities. There are many factors which must be considered when evaluating the risk associated with the use of a chemical including the toxicity (the ability of the material to produce an unwanted effect), the route of entry to the body, the chemical and physical properties (*e.g.* volatility, physical form), and the level, duration, and frequency of exposure. However, the major consideration is the way in which the chemical is used/handled as this determines whether exposure actually occurs. For example, if a process is totally enclosed then highly toxic chemicals may be handled safely during normal operations, although there may be a risk during maintenance procedures. On the other hand unacceptable levels of exposure can occur with materials of low toxicity if inadequate control measures are associated with a particular operation. The exact nature of the process is also important; the risk may be modified by concurrent exposure to other chemical and physical agents - synergistic, additive, or antagonistic effects. In addition the control measures applied to a given process vary between plants. Therefore, although the toxicity of the materials handled in a particular process in a number of plants may be the same, because of variations in the effectiveness of control measures and in the standard of maintenance, the actual exposures and hence the risk can be considerably different.

2. Occupational Hygiene

Occupational hygiene can be defined as; 'the applied science concerned with the recognition, evaluation, and control of chemical, physical, and biological factors arising in or from the workplace which may affect the health or well-being of those at work or in the community'.

2.1. Recognition. This requires a detailed knowledge of the process involved, the chemicals present, the operations which may give rise to exposure, *e.g.* by inhalation of gas, vapour, fume, mist or dust, or skin contact with liquids or solids, and the effects of the various chemical, physical, and biological hazards.

2.2. Evaluation. This should take account of exposure by all routes and involves observation and suitable measurement, and assessment in terms of occupational exposure limits. Chemicals may enter the body by inhalation *via* the respiratory tract, by absorption through the skin, or by ingestion *via* the gut. The first two are the most important routes to be considered when evaluating the work environment. Data may be acquired, for example, by the measurement of airborne concentrations of chemicals, by analysing biological samples such as hair, nails, blood, urine, and expired air for the compounds of interest or metabolites, or by determining changes in levels of normally occurring biochemical materials, *e.g.* enzymes in blood.

2.3. Control. If an evaluation indicates that a health hazard exists, it is then necessary to introduce control measures such as substitution with less toxic materials, enclosure, segregation, extract ventilation, personal protection, *etc.* It should be noted that a considerable amount of expense can be spared if control measures are introduced at the design stage of projects.

3. Occupational Exposure Limits

Exposure to chemicals should be reduced to as low a level as is reasonably practicable. However, chemicals exhibit an enormous range of toxicities and therefore guidance is required concerning acceptable exposures. The largest group of occupational exposure limits in the Western World is the Threshold Limit Values (TLVs) set by the American Conference of Governmental Industrial Hygienists (ACGIH).[1] TLVs refer to airborne concentrations of substances and represent conditions to which it is believed that nearly all workers may be repeatedly exposed day after day without adverse effect. The ACGIH sets 8 hour time-weighted average TLVs (TLV-TWA) and 15 minute time-weighted average TLVs (TLV-STEL) known as short-term exposure limits. For some compounds ceiling levels are set (TLV-C) which represent the exposure concentration that should not be exceeded during any part of the working day. TLVs have been established for about 700 materials and documentation is published giving the basis for the limits.[2] Approximately 140 of the chemicals with TLVs have been assigned 'Skin Notations' which indicate that skin absorption may occur and therefore exposure *via* this route should be considered.

For many years the United Kingdom Health & Safety Executive (HSE) reprinted the ACGIH TLV list as a Guidance Note (EH15), printed for the last time in 1980. In 1984 a United Kingdom system of exposure limits was introduced and published in Guidance Note EH40.[3] Both long-term exposure limits (8 hour time-weighted average) and short-term exposure limits (10 minute time-weighted averages) are set. Some limits are designated Control Limits; these are contained in Regulations, Approved Codes of Practice, or European Community Directives or have been adopted by the Health & Safety Commission (HSC). They are limits which should not normally be exceeded and are used by the Health & Safety Executive (HSE) in determining whether relevant legislation is being observed. The remainder of the limits (the majority) are termed Recommended Limits. These are recommended by the Health & Safety Executive on the advice of the Advisory Committee on Toxic Substances (ACTS) and are considered to represent good practice and realistic criteria for the control of exposure, plant design, *etc.*

In the USSR limits, known as Maximum Allowable Concentrations (MAC) are set. These are defined as those concentrations which, in the case of daily exposures at work for 8 hours per day throughout the entire working life, will not cause any diseases or deviations from a normal state of health detectable by current methods of investigation. The threshold of harmful action of a substance is defined as the minimum concentration in the environment, which (under defined conditions of entry of substances into the body) gives rise to changes in the person exposed, going beyond the limits of physiological adaptive responses or a latent (temporarily compensated) pathological condition. As such changes are often sub-clinical, and therefore more sensitive than the criteria used in Western countries, many MACs are set at levels considerably lower than the equivalent Western occupational exposure limits.

Exposure standards are set on the basis of the following information:

1.   Human research and experience.

2.   Animal experimentation.

3.   Analogy with other materials.

The first occupational exposure limits (TLVs were introduced in the late 1940s) were based mainly on observations of exposed human populations and were generally set to protect against acute effects. Considerable use was also made of chemical analogy, *i.e.* assuming that similar chemicals produce similar biological responses. However, as time has passed more use has been made of human epidemiology and animal experimentation.

There are certain general rules concerning the use of occupational exposure limits:

1.   Values do not represent a clearly defined demarcation between safe and unsafe working conditions.

2.   Exposure limits have been set for only a small proportion of the chemicals in use in industry today. If the chemical does not have an exposure limit it does not mean that the substance is safe, merely that insufficient information is available on which to base an exposure limits.

3.   Exposure limits are based on the best available information. As further data become available on the toxic effects of chemicals the limits change. Good industrial hygiene practice is to keep exposure to airborne contaminants to as is reasonably practicable.

4.   Objectives of Data Collection

Chemical exposure data may be required for a number of reasons. The techniques and the strategy used depend upon the reason for the data collection and therefore information collected for one purpose may not be applicable to others.

4.1. Exposure Evaluation. The most common reason for measuring worker exposure to a toxic chemical is to evaluate the health significance of that exposure by comparison with recognized occupational exposure limits. Where there is exposure to a substance for which there is no occupational exposure limit, it is necessary to develop a local limit for use in a particular plant or company. Exposure evaluation is normally conducted for groups of workers doing essentially the same job ('job function'), *e.g.* process, maintenance, supervisory, laboratory, *etc.* The purpose of the exercise is to determine whether the exposure experienced by a particular group of workers is acceptable.

The data required to determine the risk from exposure to a chemical may be the time-weighted average airborne concentration over a full-shift period, or over shorter periods depending on the type of the operations and the nature of the toxic effects of the chemicals. Personal monitoring is the preferred technique for assessing exposures. It may also be necessary to carry out measurements to determine the physical characteristics of the airborne contaminant, *e.g.* particle size, vapour/particulate ratio, in order to assist in estimating the dose which may be absorbed.

4.2. Compliance with Legal Requirements. When conducting monitoring to determine compliance with occupational exposure limits associated directly with legislation, the measurements must be undertaken in the manner defined in the documentation associated with the standard. In addition a statistically developed sampling strategy should be applied aimed at indicating whether the worker, or group of workers conducting similar tasks, are in compliance with the limit. Where hazardous materials are used on a regular basis periodic monitoring should be undertaken, the frequency dependent on the level of exposure in comparison with the occupational exposure limit.

4.3. Epidemiological Studies. When conducting epidemiological studies on industrial populations it is important to have detailed information on the chemicals to which the workforce were exposed together with an indication of the range of exposures for each identified job function. The data normally required for epidemiological studies are the personal full-shift exposures and the total number of years of exposure to the chemical of interest. There is an argument for collecting and recording exposure data as if these were needed for epidemiological purposes.

4.4. Evaluation of Control Measures and Plant Integrity. Fixed position or direct reading instruments can be used to monitor the efficiency and establish the effects of changes in control measures. The information gained often has little direct relevance to personal exposure measurements. Monitoring may also be undertaken to detect leaks and to investigate other situations where control has been lost.

4.5. Investigation of Complaints of Ill Health. When a case of occupational disease is suspected, or an employee has suffered symptoms that may be due to over-exposure to a particular material, it may be necessary to conduct monitoring to determine the cause and extent of the exposure. The results obtained are unlikely to represent typical exposures and are more likely to be due to the failure of control measures or to poor working practices.

4.6. Toxicology. It may be necessary to characterize workplace exposure for the purpose of designing inhalation toxicity studies.

5. The Application of Published Data

The collection of occupational hygiene data of the required quality is costly and time consuming. Therefore in recent years collaborative studies have been carried out by industry groups, trade associations, and professional organizations; the pooling of effort and resources enables the data to be collected in a relatively short time span. The advantages of such studies are that standardized sampling and analytical protocols can be adopted allowing data obtained from many sources to be compared.

Although it is possible to obtain information from various sources related to the hazards associated with many industrial processes it is clear that, within different plants, control measures, production rates, operating techniques, age of equipment, standard of maintenance, and individual work practices can all vary. Published data refer to specific situations and may not be applicable to assessing the risks associated with other circumstances. Therefore, the major emphasis for the occupational hygienist is the generation of data for each particular situation encountered rather than the extrapolation of published data. However, the information that is available in the literature can assist in the setting of priorities for occupational health assessments.

Information on many aspects of occupational hygiene, for example results of exposure measurements, sampling and analytical procedures, sampling strategies, evaluation of protective clothing devices, engineering control measures, can be obtained by the data retrieval techniques described in Chapters 2 and 3 of this book. For general searches the following on-line data bases are useful: CHEMICAL ABSTRACTS, MEDLINE, TOXLINE, BIOSIS, HSE-LINE, CIS-ILO, as are the abstract journals listed in the Appendix.

6. Sources and Nature of Chemical Exposure

If a meaningful assessment of exposure to chemicals is to be obtained then a structured approach is essential. As a first step basic information relating to the workplace has to be collected and evaluated.

6.1. Process Description. The equipment layout and sequence of operations should be determined and the flow of materials and products ascertained. The physical and chemical principles of the process should also be documented.

6.2. Chemicals Involved. A complete inventory of all the materials involved should be established and maintained and should include raw materials, intermediates, finished products, by-products, waste materials, and any significant contaminants. Depending on the nature of the material, all or some of the following data should be obtained for each of the identified materials:

Name, synonyms, supplier, quantities handled.

Composition, physical and chemical properties.

Toxicological data, occupational exposure limits, first aid procedures.

Fire and explosion data, incompatibility with other chemicals.

Spillage and emergency procedures, waste disposal.

Personal protection requirements.

Information can be obtained from standard sources, from a literature search as described in Chapters 2, 3, and 4 and from in-house tests on novel chemicals. In addition data about raw materials should be available from suppliers.

6.3. Personnel Involved. Job/task descriptions are required for all those involved including process, maintenance, laboratory, cleaning, and supervisory personnel. Information is required regarding maintenance activities and other operations which may cause exposure to chemicals, *e.g.* sampling, charging. The hours worked and the nature of the shift system should also be noted. If available, sickness absence, accident frequency and severity rates, and labour turnover should be reviewed. The Occupational Health Department may be able to indicate problem areas from treatment and medical records.

6.4. Initial Survey. Once all the basic information has been gathered a walk-through survey should be undertaken to determine the potential for exposure and to define monitoring requirements. If possible the walk-through survey should be carried out in conjunction with personnel with a detailed knowledge of the process. It may be necessary to carry out some monitoring to identify the chemical composition and physical form of the airborne contaminants. The following information should then be available:

The chemicals present.

The airborne contaminants present and their physical form – dust, aerosol, fume, gas, vapour.

Relevant hygiene standards.

Likelihood of exposure by inhalation, ingestion, or skin absorption.

Where, when, and for how long exposure occurs.

Protective equipment worn.

Any other relevant exposure data.

7. Measurement of Personal Exposure

When the intention is to obtain data relating to the exposure of an individual then personal sampling is the preferred technique. The collection device (filter, sorbent tube) should be located so that the sample collected is representative of the air being inhaled. It should be remembered that such monitoring indicates an airborne concentration; it is not a measure of body dose. The actual exposure is a function of the measured level together with a consideration of factors such as work rate and rate of absorption. Personal monitoring equipment should be convenient to wear, and capable of collecting a representative sample over an extended time period. In the majority of techniques the contaminants once collected are analysed in the laboratory, although monitors that provide an alarm and direct reading instruments, some with data logging facilities, are now available.

It is necessary to ensure that the monitoring technique adopted allows for the evaluation of either acute or chronic effect as appropriate. To protect against acute effects continuous monitors capable of giving a rapid indication when preset levels are exceeded are preferred; such instruments are only available for a limited number of chemicals. For chronic hazards time-weighted average measurements over extended periods are appropriate. In addition, when evaluating hazards arising from airborne particulate matter, the particle size fraction that is required to be sampled should also be considered.

Many attempts have been made to estimate exposure by the use of fixed-position samplers in conjunction with time and motion studies. However, the results obtained by personal samplers are invariably higher than those estimated from fixed-position samplers.

An estimation of skin contamination by non-volatile chemicals can be made by the analysis of small pads positioned on the body or of complete suits after they have been worn during particular operations; contamination with radioactive chemicals may be determined by means of direct-reading radiation monitors. In an industrial situation the evaluation of skin absorption usually relies on the observation of work practices, including an assessment of the use of protective clothing, the state of personal hygiene, *etc.* and/or by biological monitoring. A complicating factor in assessing the contribution from skin absorption is that protective clothing does not provide absolute protection; chemicals can permeate or penetrate protective clothing. Therefore some account must be taken of this when evaluating the potential for skin absorption. In addition smoking and eating habits should be recorded so that an assessment can be made regarding the potential for ingestion.

In some situations biological monitoring has advantages over air monitoring as absorption by all routes of eating is taken into account. An analysis of biological samples such as blood, urine, exhaled breath, hair, nails, *etc.*, either for the chemical in question or for metabolites, can provide an indication of the body burden.

8. Sampling Strategy

When assessing exposure to airborne contaminants the measurements must be conducted according to a well developed protocol to ensure that the objective of the monitoring is achieved. Consideration must be given to the type, duration, and number of samples; they should represent typical exposures for a particular task or job function. Sufficient samples must be collected to ensure that a good approximation of the usual exposure situation is obtained. The National Institute for Occupational Safety and Health (NIOSH) has developed a statistically based sampling strategy to establish the number of samples required to determine compliance with exposure limits.[4] A Guidance Note has also been issued by the Health & Safety Executive on monitoring strategies for toxic substances.[5]

The results of time-weighted average exposure monitoring for a particular job function may show considerable variability due to differences in environmental and operational conditions. Data of this kind have been found generally to approximate to a log normal, rather than a normal, distribution. The log normal distribution is described by the geometric mean (GM) and the geometric standard deviation (GSD). Exposure data often have a GSD of 2.0-2.5. This means for example that the 90th percentile value is approximately three times the mean value; it is essential to know the extent of the variability if the majority are to be protected.

If the results of exposure monitoring are very low or very high in relation to the occupational exposure limit only a few samples are required before a decision can be made. However, if the levels are in the region of the occupational exposure limit many more samples are necessary to indicate the distribution of the exposure results particularly in relation to the limit.

9. Calibration

Whatever technique is adopted to monitor exposure to chemicals, calibration is required with each element of the sampling/analytical procedure being considered.

Pumps used for obtaining time-weighted average samples should be calibrated before and after the sample is taken. It is necessary for the pump to draw a known constant flow rate over the whole of the sampling period if a representative sample is to be collected. The flow stability requirement is very important when size-selective sampling systems are used since their performance is very dependent on flow rate. Pumps in direct-reading instruments and continuous-monitoring systems are often neglected but these should also be the subject of routine calibration procedures. The calibration of sampling systems should be carried out with the collection device in the sampling train.

Direct-reading instruments and continuous-monitoring systems should be calibrated on a regular basis using the chemical of interest. The whole system should be tested, including sampling ports, transfer lines, *etc.*, not just the detectors. If a secondary calibration technique is employed such as a gas blend in a cylinder then some procedure for checking the integrity of the standard should also be adopted as these blends can deteriorate with time owing to absorption of the compound of interest onto the walls of the cylinder.

10. Sampling and Analytical Procedures

Governmental agencies in Europe and the USA publish recommended sampling and analytical procedures.[6,7] In addition methods are also published in analytical and occupational hygiene journals. Care should be taken to study manufacturers' literature to determine the limitations of individual pieces of equipment. Whenever possible proven and established sampling/analytical techniques should be used, particularly by inexperienced persons. Whatever technique is used it should have been subjected to a rigorous validation programme and it should satisfy the following criteria:

It should have a known, high, reproducible, collection efficiency. If information relating to the collection efficiency is not available in the literature, from those who developed the method, or if it has not been determined by the intended user, then the method should be treated with caution. Indeed claims made by manufacturers or in the literature should be investigated and supporting evidence sought. Where necesary a limited amount of work may be required to verify the statements.

The breakthrough volume of a sorbent trap should be determined, that is the sampled air volume that can be drawn through the trap before collected material is lost. This value is dependent on factors such as temperature, humidity, and sampling rate.

Sufficient material should be collected to enable the analysis to be carried out and to provide the required limit of detection.

The collection technique should be compatible with the subsequent analysis procedure. The desorption efficiency should be known and should be high.

The presence of other airborne contaminants should not interfere in the collection or subsequent analysis; for example the presence of other airborne dust can interfere in the determination of low concentrations of asbestos fibre. Any interferences should be noted and due care taken in the interpretation of results.

The possibility of chemicals undergoing reactions during sampling and analysis should be recognized and the potential minimized.

The sample should be sufficiently stable to allow transfer to the laboratory. The permissible storage time and conditions should be determined.

Losses in sample transfer lines due to reactions and absorption of chemicals onto surfaces should be considered.

Sampling and analytical procedures should be kept as simple as possible and should not present a hazard to the wearer/user/environment.

Matrix effects should be overcome by matching calibration standard and sample matrices.

The precision and accuracy of the overall technique should be determined. Methods which consistently read high or low or yield a constant answer regardless of the true result are not unknown.

Procedures should be calibrated over the full range of concentrations of interest by means of suitable blanks and standards. The purity of all reagents used should be checked and possible sources of contamination considered and eliminated.

The monitoring procedures should not be adversely affected by changes in temperature, pressure, or relative humidity.

10.1. Interlaboratory Testing. Interlaboratory testing programmes in which the relative abilities of participants in carrying out standard procedures are assessed are an invaluable aid in identifying problem areas and providing reassurance that results being produced are acceptable.

11. Record Keeping

Comprehensive records are important in determining retrospectively levels to which particular job categories were exposed. Such records typically have not been maintained in many industrial operations, and as a result it is often difficult to estimate past exposures. During the collection of data, information pertinent to the result should be documented, *e.g.* tasks undertaken, duration, use of protective clothing, use of engineering control measures, environmental conditions. The sample should be clearly identified as being associated with a particular location/individual, and information relating to the flow rate and sampling duration should also be recorded.

All samples taken or received for analysis should be logged and identified with respect to type and source. The sample identity should then be carefully and clearly maintained throughout the analysis/reporting procedure. Care must be taken to ensure that calculation and transcription errors are kept to a minimum.

12. The Interpretation and Validation of Exposure Data

The principles outlined in the previous sections on the generation of valid data should be applied to exposure measurements retrieved from the literature to determine whether they have been obtained in a manner which allows them to be used with confidence. For example care must be taken to determine whether the data were obtained using personal sampling techniques or estimated from fixed location samplers. If estimated, an assessment has to be made regarding the validity of the method used to predict personal exposures. If the results were from personal monitoring, was account taken of the use of pesonal protective devices and if so were the actual exposures predicted from published protection factors for the protective devices or by measurement? In addition care must be taken to ensure that the protective devices worn were appropriate.

In statements made about the contribution of skin absorption to total body burden care has to be taken in interpreting data to assess whether the discussion concerns actual or potential exposure. Estimates are usually made by analysing contamination on patches or on complete protective clothing assemblies. This indicates the potential for exposure whereas only biological monitoring can demonstrate whether actual contamination and subsequent absorption has occurred.

The primary concern of the occupational hygienist is the recognition, evaluation, and control of the risks arising from the hazards associated with a particular environment. Evaluating the hazards and consequent risks requires several skills and should be carried out only by trained, competent people. As published exposure data relate to specific circumstances it is not usually applicable to the assessment of risk in another situation. Therefore, the occupational hygienist typically has to generate data for the particular environment encountered.

13. References

1. 'Threshold Limit Values and Biological Exposure Indices for 1985-86', American Conference of Governmental Industrial Hygienists, Cincinnati, Ohio.

2. 'Documentation of the Threshold Limit Values', 4th Edn., Cincinnati, Ohio.

3. 'Occupational Exposure Limits 1986', Environmental Hygiene Guidance Note EH 40/86 from the Health & Safety Executive, HMSO, London.

4. N. A. Lendel, K. A. Busch, and J. R. Lynch, 'Occupational Exposure Sampling Strategy Manual', National Institute for Occupational Safety and Health, Cincinnati, Ohio, 1977.

5. 'Monitoring Strategies for Toxic Substances', Environmental Hygiene Guidance Note EH 42 from the Health & Safety Executive, HMSO, London.

6. 'Methods for the Determination of Hazardous Substances, Health & Safety Executive', Bootle, Merseyside.

7. 'Manual of Analytical Methods', National Institute for Occupational Safety and Health, DHEW (NIOSH) Publication No. (77-157), Cincinnati, Ohio.

Appendix: *Abstracting Journals and Other Information Services*

'Abstract on Hygiene and Communicable Diseases', Bureau of Hygiene and Tropical Diseases, London.

'CA Selects: Chemical Hazards, Health and Safety', Chemical Abstracts Services, American Chemical Society, Columbus, Ohio.

'CA Selects: Pollution Monitoring', Chemical Abstracts Service, American Chemical Society, Columbus, Ohio.

'Chemical Hazards in Industry', Royal Society of Chemistry, Nottingham.

'CIS Abstracts', International Labour Office, Geneva.

'Laboratory Hazards Bulletin', Royal Society of Chemistry, Nottingham.

Other useful sources of information, particularly in relation to the assessment of exposure to chemicals are trade and industry association documents and the following publications and journals:

'Documentation of the Threshold Limit Values', American Conference of Governmental Industrial Hygienists, Cincinnati, Ohio, 4th Edn., 1980 with updates.

'Guidance Notes', Health & Safety Executive, HMSO, London.

'Handbook of Occupational Hygiene', Kluwer Publishing, Brentford, Middlesex, 1980 with updates.

'Hazards in the Chemical Laboratory', The Chemical Society, London, 4th Edn., 1986.

'Patty's Industrial Hygiene and Toxicology', Wiley-Interscience, New York, 3rd Edn., 1978-1982, Vol. 1, 2 A-C, 3.

'Toxicity Reviews', Health & Safety Executive, HMSO, London.

Journals

American Industrial Hygiene Association Journal, American Industrial Hygiene Association, Akron, Ohio.

Annals of Occupational Hygiene, Pergamon Press, Oxford.

Archives of Environmental Health, Heldref Publications, Washington DC.

British Journal of Industrial Medicine, British Medical Association, London.

Contact Dermatitis, Munksgaard, Copenhagen.

Journal of Occupational Medicine, Flournoy Publishers Inc., Chicago, Illinois.

Journal of the Society of Occupational Medicine, John Wright and Sons Ltd., Bristol, Avon.

Scandinavian Journal of Work Environment and Health, National Board of Occupational Safety and Health, Helsinki.

Section 4: Interpretation of Data

9
Interpretation of Data

By A.D. Dayan

DHSS DEPARTMENT OF TOXICOLOGY, ST BARTHOLOMEW'S HOSPITAL MEDICAL
COLLEGE, DOMINION HOUSE, 59 BARTHOLOMEW CLOSE, LONDON EC1 7ED, UK

1. Introduction

The interpretation of information about toxicity represents two
main processes in the form of two-dimensional matrix. One
dimension reflects knowledge and understanding about the mechanism
underlying the toxic action and the other is the practical
circumstances in which the activity is likely to be realized. It
may appear strange in scientific work not to separate the
interpretation of data from the events to which the analysis will
be applied, but realization of the conditions in the real world
under which toxicological work is done and decisions are based
upon it will soon show the empirical and pragmatic nature of
applied toxicology. It is a reactive rather than a primarily
investigative discipline.

To perceive how various types of laboratory data are used as
the basis on which decisions about health and safety (see also
Chapters 1, 11, 12, and 23) can be made, it is necessary first to
define the concepts of toxicity, hazard, and risk.

2. Toxicity

This may be taken as any form of adverse effect due to exposure to
a xenobiotic, a foreign chemical. Harm may also follow exposure
to an excess of substances endogenous to the body, e.g. hormones,
but that is an area of disordered physiology that properly belongs
to medicine.

The toxicity of a substance is not an innate constitutive
property of the molecule. It depends on the conditions under
which living organisms are exposed to the substance, i.e. it is a
substantive property of variable expression. As examples consider
penicillin, which is lethal to bacteria and almost harmless to
man, or rat poison, which is intended to kill rats and thereby
benefit farmers. The importance of making this distinction lies
in its clear demonstration of the *biological* nature of toxicity,
and therefore the variable nature of its expressions depending on
the circumstances of exposure *etc.*

The adverse effect on a living organism may be clinical or
structural in type, or a combination of the two. It may be due
directly to the foreign substance or to a metabolite of the
latter, or it may follow from a disorder of body mechanisms
induced by the exposure.

Toxicity may also be manifested at a different level when it
affects aspects of the environment, *i.e.* ecotoxicity, in which one

or more species may be harmed directly or *via* disturbance of the food chain (see also Chapter 10); for example the complex consequences on plant and animal life attributed to acid rain and nitrate eutrophication of surface waters.

2.1. Clinical Toxicity. Under this heading come functional disorders which may not have a structural basis, or which may only have been diagnosed by examination of the patient, as detailed investigations showing their organic basis may not have been completed. Major classes of clinical toxicity are listed in Table 1.

Table 1 *Broad classification of clinical toxicity**

Body system affected	*Type of disorder*	*Example*
Nervous system	Confusion	Vinyl chloride
Central	Ataxia	Haloalkanes
	Dementia	Organic mercurials
Peripheral	Neuropathy	Acrylamide, arsenic
Muscles	Myopathy	Cortisone
Cardiovascular system	Throbbing headache Syncope	Organic nitrates
Respiratory system	Coughing Asthma	Irritant gases, Toluene di-isocyanate
Gastrointestinal system	Constipation	Inorganic mercury
Reproductive system	Loss of libido	Exposure to di-ethylstilboestrol
Skeletal system	Pain in bones and joints	Fluorosis due to fluoride ingestion
Skin	Reddening and weeping due to eczema	Strong detergents, alkalis

*type examples only are given.

The effects range from the serious to the trivial, but they are all important to the sufferer. The nature of the disorders and their pattern in time may give very useful clues to what has caused them and sometimes to the mechanism involved. In historical terms, the 'Monday morning throbbing headaches' of explosives' workers was a good clue to the toxic action of vasodilatation due to organic nitrates, an effect to which man rapidly develops short-term tolerance. Similarly, the complaints of numbness and abnormal sensations in the fingers and toes of workers exposed to organophosphate pesticides, or to methyl n-butyl ketone as a solvent for glues, were the first indications of grave peripheral neuropathies.

Because the clinical detection and assessment of symptoms are subjective and imprecise, it is dangerously easy to disregard clinical findings and to try instead to rely solely on more objective measures. That is not acceptable, because the disordered functions underlying clinical actions may be the most sensitive indicators of a toxic effect, and certainly they are what cause most immediate suffering to the exposed subject.

In considering the quality of toxicity data for assessment, it is important to be certain that appropriate clinical changes have been sought and either detected or reliably excluded, whether in man or animals. The criteria for adequacy of clinical examination are comprehensiveness and detail. The presenting complaint should have been evaluated and then further evidence of dysfunction of that body system sought, as well as general assessment of the clinical state of health.

2.2. Physical Toxicity. There are two sub-types to be considered, functional and structural.

Functional lesions. This group comprises objectively detectable and usually quantifiable evidence of disordered function, corresponding, for example, to clinical signs and abnormal results in laboratory tests.

The former type covers the results of medical examination, such as damage to the finger tips in the osteoacryolysis syndrome of prolonged vinyl chloride intoxication, or the loss of the knee jerk reflex in chronic organophosphate poisoning. The latter type represents the enormous battery of laboratory tests now available for investigation of biochemical and physiological functions of the body. Examples include the demonstration of abnormal nerve conduction velocity and blood pressure control in certain peripheral neuropathies, the measurement of increased protein loss in urine in cadmium poisoning, the stippling of red cells after exposure to lead, abnormal liver function tests caused by inhalation of certain nitroaromatics, *etc.*

Many of these measures overlap with one another and with the methods used to detect other types of toxicity. The redundancy is valuable, because it increases the likelihood of detecting an abnormality, and multiple positives give greater confidence that there is a true abnormality and not just random variation.

Again, toxicity data should be examined to ensure that there has been adequate employment of appropriate methods to detect these types of abnormalities, based on the nature of whatever lesions have been detected.

Structural Lesions. The conventional view of a toxic effect is of a lesion in an organ detectable by the methods of the pathologist, *e.g.* enlargement or atrophy at the macroscopic level, and, say, necrosis or a tumour as seen by microscopy.

The importance of pathological lesions is that they are probably the commonest end point (if death is excluded) in toxicity tests, they are the only way to detect and classify changes in many organs (*e.g.* consider hepatic enlargement due to enzyme induction, to inflammation, or to neoplasia), and they are almost always incontrovertible. They are of vital diagnostic importance in animal experiments and in man, so ascertaining the adequacy of pathological investigation forms an essential part of assessing the quality of any set of toxicological data.

3. Biochemistry and Toxicological Data

It is not essential to understand the biochemical basis of a toxic effect to be able to detect the latter, or even to make a useful extrapolation from it, but it is of the greatest help if the mechanism is understood. Rational extrapolation and sound, reasoned precautions can then be taken, instead of relying on empirical prediction, although the latter has and still does serve us very well in most instances.

Measuring the concentration or the amount of a substance taken into the body, or present at the site of action, is a more secure measure of true exposure than suggestions based only on information about exposure, for example, the average concentration in inspired air, or in a solution applied to the skin, plus the assumed respiratory volume, percutaneous flux, *etc.*

This is an important part of the pharmacokinetics of the substance, *i.e.* measurement of its intake and disposition in the body, and the important process of metabolism to active (toxic) or inactive metabolites and their excretion.

'Toxicokinetics' is a rapidly growing discipline, as it has increasingly been realized that toxicity may be caused by the parent substance, or a metabolite of it formed in the body (*e.g.* carbon tetrachloride), or by the intestinal flora (nitrobenzenes). Ultimately, the formation of metabolites aids excretion of a compound as a harmless reactant.

Toxicokinetics should always be considered in the assessment of toxicity data, because of the importance of cumulation, metabolic activation and detoxification, and the speed of disposition and excretion in determining what, if any, effects will occur. It is also of the greatest importance in extrapolating data from one species to another, as in predicting a safe exposure level for man from results obtained in animals. It is essential always to be aware of possible species differences in metabolism that may determine whether and to what extent a substance will be toxic; for example, a low dose of chloroform kills mice but not rats or man, because of its effects on the heart, whereas chronic exposure to it can be made to produce kidney tumours in the rat but not the mouse, because of a specific metabolic pathway; mammals are far less sensitive than insects to pyrethroid insecticides because of circulating hydrolases; β-naphthylamine is only a bladder carcinogen in man and the dog but not in the rat, because the species affected manifest a particular metabolic activating pathway, *etc.*

4. Hazard and Risk

The most important result of considering toxicity in relation to its practical expression is the distinction between hazard and risk.

Hazard is best defined as the property of a substance to cause toxic effects if a living system, *i.e.* man or an animal (or the environment) is exposed to it. It is a biological property of the material in relation to some practical event, and, as such, is open to experimental investigation.

Risk, on the other hand, is the predicted or actual consequence of exposure of a defined population to a given dose or concentration of the material. It represents the frequency and intensity of the toxic actions in a given group of people, animals, *etc.* under a defined set of circumstances.

Risk is rarely open to direct experimental examination, usually being calculated from knowledge of the hazard and the exposed population. It may be directly assessed by epidemiological surveys of exposed groups, *e.g.* workers in a factory, or animals or people living in a particular area.

Making this distinction permits the toxicologist to define the limit of his experimental expertise and to show where other data are necessary for prediction of the consequences of exposure in the real world as opposed to laboratory conditions.

In practical terms the use of these terms leads to the simple paradigm:

```
Toxicologist            Toxicologist            Dose
Biochemist  → TOXICITY → Clinician    → HAZARD → population → RISK
etc.                     Occupational            and
                         hygienist               exposure
```

From that comes the further development:

```
                                           { = risk    avoidance
            benefit                        {   ban    exposure
RISKS →              }analysis  → DECISION → ─────────────────────
            cost                           {   accept exposure
                                           { = risk management
```

Many social, legal, and economic factors are caught up in the process of decision making, especially if it is agreed that some risk will be accepted, but that it must be minimized by engineering controls in the plant, or restrictions on field use, so as to minimize exposure. In that sort of decision, there is the need to consider separately the risks and benefits to the manufacturer, worker, consumer, and the general public who may only be incidentally exposed, and who may at the most gain only indirect benefit from that risk.

At this point, it will be apparent that although toxicity data are conventional scientific findings from laboratory studies or man, the design and analysis of toxicity and related tests may be very complex, because they involve relativistic assessments about hypothetical exposure conditions and notionally affected populations. The final evaluation of the information and predictions based on it leaves the area of toxicology and becomes a matter of ethics, medicine, law, and economics, in which the risk of using a substance has to be balanced against the costs of not using it.

5. How Does the Toxicologist Operate?

The work of the toxicologist concerns the detection or exclusion of toxic effects under defined circumstances and study of their mechanisms.

The former may be done in man, as the usual target species, but the occurrence of toxic reactions in him represents a serious failure of toxicological advice and industrial practice. Instead, data about the occurrence of effects and the circumstances of dose and duration of exposure required to produce them are obtained in laboratory experiments.

The experiments may be done *in vitro*, if the actions are sufficiently well recognized and understood for that to be reasonable. More commonly, however, they are done in laboratory animals (most often rats and mice), because they are the only models for man that are available.

The animals are dosed by the expected route of exposure and examined in diverse ways to detect various types of toxicity. The procedures involved are usually set out in official guidelines; they are summarized in Table 2.

Great care is required in the design of each type of experiment to ensure that as much reliable positive and negative information as possible is extracted from it. Particular consideration should always be given to:

(*i*)   Purity or confirmed consistency of the composition of the substances

(*ii*)   The route of exposure should usually be the same as that by which man and other species will be affected.

(*iii*)   Selection of several dose levels so that the dose-response relationship can be examined.

(*iv*)   The duration of exposure or treatment should be related to the nature of the effects being explored.

(*v*)   Use of sufficient numbers of animals to permit statistical analysis of the findings and attribution of some confidence to the results.

(*vi*)   Multiplicity of observations to ensure that effects are not missed. This will cover clinical, haematological, biochemical, and extensive pathological examinations.

Regulatory guidelines from various national and international sources define many of these and other factors, but it is always important for the toxicologist as a scientist to be prepared to adapt the experimental circumstances to the particular properties and uses of the material being examined.

Table 2 *Assessment of toxicity*

Type of toxicity test	Species	Type of effect investigated
Acute ('LD_{50}')	mouse/rat	Dose-response relation and acute clinical actions of major overdose
Sub-acute (7 to 90 days)	rat sometimes dog	prolonged dosing leads to clinical or organic lesions
Chronic (3 months to 1 year)	rat sometimes dog	prolonged dosing reveals more slowly evolving lesions
Carcinogenicity (lifespan)	mouse rat	tumour producing potential
Reproduction Phase I	rat	impaired fertility, fetal
II	rat and rabbit	toxicity, and teratogenicity
III	rat	late pre- and post-natal development
Topical	rabbit guinea pig	local irritancy in the eye or skin
Sensitization	guinea pig	cutaneous allergic reactions
Genetic toxicity	Bacteria (Ames test *etc.*) mice and rats	interaction with DNA causing mutation and chromosomal damage

The results of the tests indicate the nature of the toxic effects produced (and those ruled out) at various dose levels. They are the basic data from which extrapolation to another target species becomes possible.

The limitations on extrapolation are:

(i) Change from one species to another. There are many sources of differences between species that may greatly affect their responses to chemicals, sometimes preventing any effect (as when a substance is not metabolized to the ultimate active form, *e.g.* the lack of carcinogenic response in the bladder to aromatic amines by the rat), sometimes permitting an action of considerable sensitivity (*e.g.* the extreme response of death by the guinea pig exposed to nanograms of TCDD), and on occasion leading to a qualitatively different action because of some physiological uniqueness of the species (*e.g.* the production of hepatic tumours in rodents by certain enzyme inducing agents), whereas man, say, given phenobarbital for years, only develops hepatic enlargement without tumour formation.

There is a general quantitative difference between species, due to the fact that smaller animals have higher metabolic rates, because of their proportionately larger surface area (allometric law). In consequence, the dose to produce a given effect will be larger in the rodent than in man. The generally accepted convention states that for the same effect as in man, the dose in the –

        mouse will be   x 14
        rat   x 12
        dog   x 7
        primate   x 4

This is a crude and naive rule of thumb and it takes no account of factors and responses specific to a given species, especially in xenobiotic metabolism, *e.g.* the inability of the dog to acetylate many compounds, with the consequence of slow excretion and greater effects *etc.*

The state of nutrition and the physiological systems of laboratory animals differ from man, who is heterogeneous rather than selected, eats a variable diet instead of receiving a large and even excessive balanced food supply, and may be of any age and health status.

(ii)   The existence of a threshold, *i.e.* the requirement for a certain dose or exposure to be exceeded before a toxic effect becomes apparent. It may be due to physiological mechanisms (*e.g.* the need to overcome repair or adaptive processes) or to pharmacokinetic factors.

(iii)   Numbers. Even the largest experiments do not involve more than perhaps 800-900 rodents, whereas many tens of thousands of workers, or millions of the general population may be exposed.

To some extent the attempt is made to allow for this by using high doses to increase the responses in tests, but the statistical weakness remains.

(iv)   Pattern of exposure. In man this is likely to be very variable, depending on the nature of work, or domestic and leisure habits in the case of materials released elsewhere. It will be far more regular in the case of experimental animals. If anything, the difference is likely to increase the sensitivity of animal and other laboratory studies by maximizing the opportunity for the substance to exert an effect. Intermittent exposure may allow adaptive or repair processes to operate, which may prevent any toxic response from occurring.

Other methods of extrapolation are important, because they are sometimes based on less immediately empirical observations and procedures.

The most soundly based method, at least in the mathematical sense, is to calculate effects from the shape of the dose-response curve. In a few instances that can be done using results from man, in which case the prediction may be reasonably accurate over a small range close to known doses. As one moves further away

from the area explored in practice, the uncertainty grows rapidly.
Examples of this approach may be found in the literature on
cigarette smoking and the sections in this book on actual risks.

A particular weakness of this approach is its total reliance
on assumptions about the shape of the dose-response curve,
especially at its lower end, as extrapolation there will have to
be done from the much higher dose levels at which effects were
detected. Experimental variability means that there is likely to
be considerable dispute about the shape of the lower part of the
curve, which can readily introduce a 10 - 100-fold variation in
the assumed dose for a given effect. The literature on
carcinogenesis contains these disputes in their clearest form.

Mathematical extrapolation from dose-response curves in animal
experiments is subject to the same uncertainties, plus those
associated with use of a different species.

Nevertheless, extrapolation has worked and demonstrably does
so every day, but by imposition of considerable additional
empirical safety factors. A common practice, for example, in the
case of substances that are added to food, or that may
incidentally get into food, is to take the 'No observable effect
level' (NOEL) in an animal experiment (*i.e.* the highest dose not
causing a toxic action), to divide it by 10, as an arbitrary
safety factor to allow for species differences, and then by a
further 10 to allow for sensitivity. In this way $^1/_{100}$ of the
NOEL is taken as the probable 'Acceptable Daily Intake' (ADI). As
it is only 1% of the dose *not* causing an action, it is likely to
be safe, as experience has confirmed. A similar approach is
commonly followed in establishing 'Control Limits' or 'Threshold
Limit Values' for industrial exposure.

Such predicted values may be altered if experience of man
shows that they are not justified, or if further experimentation
reveals new effects not considered in the original calculation.
The effect on which the critical NOEL is based may be expressed as
clinical or organic toxicity, although the latter is often
preferred because it is easier to measure. However, clinical
measures are used and are extremely important in many instances
because of their sensitivity and their occurrence at lower dose
levels, *e.g.* the TLVs for potentially anaesthetic haloalkanes
reflect their clinical actions on the CNS as much as other factors.

Certain types of toxic action are considered so dangerous that
much lower exposure to the responsible material is required on
ethical or legal grounds. This applies particularly to any form
of permanent damage, and especially to any risk of cancer or of
damage to the reproductive process (*e.g.* infertility or
teratogenesis). One practice in such cases is to assume that
there is no threshold, *i.e.* that no dose can be considered even as
'virtually safe', and to work for conditions of exposure which are
as low as possible, if use of the substance is continued.

That requires careful analysis of the probable causal
mechanism of the effect, to ascertain whether it might be due to
any special propensity of the test species; for example,
alkylating mustards are carcinogenic both in man and animals, and
their use can well be regulated on that basis. On the other hand,
there is good evidence that for various physiological and
metabolic reasons man does not respond to formaldehyde or

trichlorethylene in the same way as the rat or mouse, so it would be unreasonable in these instances to make the simplistic prediction that they would cause cancer in man as they do in laboratory rodents.

6.   The Responsibilities of the Toxicologist

The toxicologist is a scientist and carries the same professional duties of honesty and accuracy as any other experimentalist.   He has further responsibilities too, some unique and others shared.

(i)   To design experiments in such a way that a wide range of toxic effects can be detected or positively excluded with some confidence, whilst always remaining alert to the possibility of novel and previously unrealized forms of toxicity.

(ii)   Whenever possible to assess appropriate findings in man to show the validity of the experimental systems, and to use the analysis to modify the methods for greater accuracy, reliability and economy.

(iii)   To be aware of the uncertainties inherent in his laboratory methods and to make extrapolations with due caution.

(iv)   As any experimentalist to improve methods as new techniques become available.   Like other biological scientists this may include changing when possible from *in vivo* to *in vitro* techniques, and incorporating any data from man, but with the overriding ethical and practical requirement that toxicity studies should help to prevent harm to man, so that there must be clear evidence of the value of a new procedure before it is adopted.

(v)   To co-operate with other disciplines in establishing safety limits, because that decision requires equally his interpretation of toxicity data plus understanding of human physiology, industrial working practices, and occupational hygiene.

          The toxicologist's special duties here are to produce and assess approximate data on toxic actions and the doses producing them for extrapolation to be possible.   At the same time, he must warn about the reversible or permanent nature of the toxic effects, as that may influence decisions about safety, and he must also point out what clinical or laboratory factors it would be appropriate to monitor in the exposed population, in order to detect any toxicity as early as possible.

7.   Reading List

1.   C. Brown and J. Noxiol, Statistical aspects of the estimation of human risk from suspected environmental carcinogens, *Soc. Ind. Appl. Math. Rev.*, 1983, **25**, 151-181.

2. 'Toxicological Risk Assessment', Vols. I and II, ed. D. B. Clayson, D. Krewski, and I. Munro, CRC Press, Boca Raton, Fl, 1985.

3. F. J. DiCarlo, Metabolism, pharmacokinetics, and toxicokinetics defined, *Drug Metab. Revs.*, 1982, 13, 1–33.

4. D. W. Gaylor, J. I. Chen, and R. L. Kodell, Experimental design of bioassays for screening and low dose extrapolation, *Risk analysis*, 1985, 5, 9–16.

5. K. S. Khera, Adverse effects in humans and animals of prenatal exposure to selected therapeutic drugs and estimation of embryo-fetal sensitivity of animals for human risk assessment: a review, *Iss. Revs. Teratol.*, 1983, 2.

6. B. L. Oser, The rat as a model for human toxicological evaluation, *J. Toxicol. Environ. Health*, 1981, 8, 521–535.

7. National Academy of Science 1977, Principles and Procedures for Evaluating the Toxicity of Household Substances, NAS, Washington, DC.

8. National Academy of Science, Risk Assessment in the Federal Government: Managing the Process, NAS, Washington, DC, 1983.

9. J. Van Ryzin, Quantitative risk assessment, *J. Occup. Med.*, 1980, 22, 321–326.

10
Interpretation of Ecotoxicity

By J.H. Duffus

DEPARTMENT OF BREWING AND ECOLOGICAL SCIENCES, HERIOT-WATT UNIVERSITY, CHAMBERS STREET, EDINBURGH EHI IHX, UK

1. Introduction

The first problem in interpretation of the probable ecotoxicity of a given chemical is selecting those species and possible effects which are most significant. This requires a profound knowledge of ecosystems at risk together with an ability to assess the likely toxicity of the chemical from its chemical and physical properties and mammalian effect data which are usually available. Having analysed the situation in this way as far as possible, testing may be carried out at various levels. The first level would consist essentially of relatively short-term tests on single species including assessment of important metabolic effects such as those on photosynthesis. It may be possible to identify species which are particularly sensitive and are therefore useful for monitoring purposes or may require special protection. The second level would cover multispecies and microcosm tests for effects that depend on interactions among organisms and populations. Such effects include those on predation, competition, and nutrient cycling. The third level is that of field ecosystem testing. This is necessary to validate laboratory predictions but is very difficult in practice. The designation of testing levels used here has been kept as simple as possible. Other authors' have subdivided them further.

In the natural environment, chemicals are transported to the affected organisms through air, water, or soil and can pass between these media depending upon their physicochemical properties such as volatility, solubility, stability, biodegradability, and their tendency to be adsorbed by environmental components. The fate of a chemical in the natural environment is also profoundly affected by bioaccumulation. Substances which are readily taken up by living organisms and lost only slowly can reach very high concentrations in the exposed organism: if the exposed organism is part of a food web, bioaccumulation may be observed to occur subsequently in predators above it in the web; such food web bioaccumulation may lead to a disproportionately large uptake of xenobiotics by predators (biomagnification) because one predator may consume many prey organisms.

It will be seen from the above considerations that interpretation of ecotoxicity falls into two main subdivisions: ecotoxicity testing and prediction of environmental distribution. These subdivisions will be considered in detail separately.

2. Ecotoxicity Testing

2.1. Single Species Tests.

The minimum premarket tests described by the Organization for Economic Co-operation and Development (OECD)[2-4] and those included in the Base set of tests of the Commission of the European Communities (CEC)[5] and the UK Health and Safety Commission (HSC)[6] are at level 1 as defined above. These tests include measurement of physicochemical properties, mutagenicity tests, acute and 14-28 day oral toxicity tests on rats, tests on *Daphnia* species (for immobilization, lethality and reproductive effects), a 96 hour test on fish (giving an LC_{50}), and a 96 hour test on an alga. (See also Chapters 11 and 12.)

The interpretation of tests on rats is discussed by Dayan and Paine in the preceding chapter but there the emphasis is on extrapolation to predict direct hazards to human beings. It must not be forgotten that other mammals are also at risk when a chemical enters the natural environment. Some of these are important in natural ecosystems; others are directly important to man either as domestic pets or as agricultural animals. Considerable veterinary expertise will be needed to extrapolate from results with rats to predict likely effects on these other mammals.

Tests on *Daphnia* are justified by their importance in a number of food webs. The OECD test is in two parts. An EC_{50} for acute immobilization is determined over 24 hours and a 14 day reproduction test is performed. During the 14 day reproduction test LC_{50} values are obtained and the time for emergence of the first brood is noted along with the number of eggs and live and dead young. Concentration/response curves are prepared for the effects studied. The ratio of the 14 day to the 24 hour LC_{50} is supposed to be an indicator of chronic effects. From the concentration/response curve, a 'no observed effect level' (NOEL) can be determined. Some workers regard the range of concentration between the NOEL and the LC_{50} as the range producing sublethal effects;[7] logically this is untenable though it might be true for the range between the NOEL and the LC_1. However, in the latter case, delayed lethality could make even this interpretation wrong.

2.2. Multispecies and Microcosm Tests.

If there is a possibility of bioaccumulation or if the amount of chemical to be used is large or if it may enter the environment in quantity, further testing will be necessary (level 2). The nature of this testing will depend on an analysis of the available information to determine the components of the natural environment most at risk. However, particular attention must be paid to possible effects on plants.[8] The plants to be studied must be selected carefully on the basis of their ecological or agricultural importance. Effects which may be assessed include changes in pigments, growth, yield, and rates of photosynthesis.

Acute (*i.e.* short-term) multispecies or microcosm tests can never be adequate but long-term tests are all unsatisfactory to some degree. For example, they can never involve more than a few species, and extrapolation to others is always dubious. Whatever chronic tests are being contemplated, it is essential that they include accurate monitoring of chemical concentrations in the test system along with other relevant parameters such as pH, salinity, dissolved oxygen, and water composition where applicable.

Reaction of ecosystems to stress is not simply the summation of the reactions of individual component species.[9,10] Thus, attempts have been made to develop laboratory tests to predict the integrated reactions to chemicals of biological communities and ultimately of the ecosystem as a whole. However, there are doubts about the reproducibility of current tests and their real value for predicting changes in various habitats. The US Environmental Protection Agency[11] has suggested three prime requirements for proper development of these tests:

(1) To determine the limits to which a biological system can be stressed before its recovery to the initial state is no longer possible.

(2) To define measurable properties indicating adverse community and ecosystem effects.

(3) To establish the cross-system generality of the properties selected in (2).

Multispecies laboratory tests have been reviewed by various authors.[10,12-15] Amongst the simpler and more reproducible systems developed, those considering competitive interactions for essential nutrients which are likely to be most useful are algal systems[16,17] and those using pasture grasses and legumes in greenhouses or small field plots.[18,19]

Predator-prey relationships are another important part of the functioning of any ecosystem and attempts have been made to devise tests to assess possible effects on them. For example, in considering aquatic ecosystems, fish may be used as predators with zooplankton, shrimps, or even other fish as the prey. In such a study the predator is separately exposed to the chemical and then placed in the test system. Thereafter the behavioural characteristics of the predator, such as its reactive distance or capture success, may be determined;[20] changes in prey population numbers may also be followed.[21] Tests applicable to terrestrial systems are not so well developed, except perhaps in the case of predatory mite tests.[22,23] These mite test systems are highly sensitive to species choice, physical conditions, host characteristics, and input ratio of predator to prey. The validity of extrapolating results from mite tests to other predator-dominated systems has still to be established.

There have been two main approaches to the design of laboratory ecosystem models. One has been aimed at the production of a system which shows universal ecosystem properties such as primary productivity, organic decomposition, or nutrient cycling. The other approach has been to try to simulate a specific ecosystem. The most realistic models that have been developed are those for aquatic habitats such as ponds,[24] lakes,[25] streams,[26] and coastal marine systems.[27] Some problems have emerged from studies of these systems, which are rather variable and unpredictable. Macroconsumers, when included, often exert an excessive influence on other components. The small size of test systems exaggerates the influence of sediment and restricts the normal behaviour of some micro-invertebrates. The large surface to volume ratio enhances wall effects, most obviously seen in the growth of attached algae. The most well developed terrestrial systems are soil cores used to examine microbial effects such as decomposition and nutrient cycling.[28] Similar cores may be used to study efects on microbial systems in sediments.[29]

Microcosms are multispecies test systems that are subsets of ecosystems. Six recent publications have reviewed their uses;[11-13,30-32] the following is a summary of their conclusions. Microcosms are limited by their small scale, which affects the size dependence of component organisms. Microcosms are unstable with time and any conclusions drawn from their use must be validated by field experiments. Criteria for using results from microcosms in the regulatory process have still to be developed and will require the derivation of appropriate mathematical models. Hence, no microcosm has yet been accepted as a routine test protocol for prediction of environmental hazards of chemicals.

2.3. Field Ecosystem Tests. Tests of the effects of chemicals on natural ecosystems may be the only way to detect certain phenomena. Coles and Ramspott,[33] for instance, showed that laboratory tests would not adequately mimic the migration of radionuclides in groundwater, perhaps because chemical speciation of elements differs under field and laboratory conditions. However, ecosystem tests are difficult to set up, requiring careful choice of experimental design, test site, and chemicals applied. Testing must include monitoring of the chemical and its major derivatives, as well as quantification of changes in population dynamics, community structure, and ecosystem function. Particular attention should be paid to ecosystem processes such as productivity, organic decomposition, and element cycling. The time required for ecosystem studies may range from months to years, particularly if they are concerned with a dominant plant species like trees which have a normally slow growth rate. A few examples of ecosystem testing may usefully be cited. Schindler and colleagues[34] studied whole lake effects of nutrient addition in the Experimental Lakes Area (ELA) of the Canadian Shield and subsequently examined acidification of a poorly buffered lake in the same area.[35] Hall and colleagues[36,37] described experimental acidification of Norris Brook in the Hubbard Brook Experimental Forest. In the same forest, Bormann and Likens[38] set up a terrestrial ecosystem study involving the entire watershed. Geckler and colleagues[39] studied the effect on the fish community of controlled dosing of a small Ohio stream, Shayer Run, with copper; in addition, Winner and colleagues[40,41] studied the effects on the macro-invertebrate community. Ecosystem tests on pesticides and air pollutants have also been carried out.[42-45] From such investigations, some general conclusions may be reached. Microbial processes must be carefully characterized in the ecosystems used. Site-specific properties must be identified and allowed for in interpreting results. Mobile organisms may avoid chemicals and therefore short-term effects may be minimized in such populations. Reproducibility and precision of ecosystem tests are difficult to assess. A general consideration of this final point, though with little reference to chemicals, has been published by Levin and colleagues.[32]

2.4. General Considerations. Before carrying out ecotoxicity testing, the relevant physicochemical properties of the substance to be tested must be known. Different test systems will be necessary for testing volatile or hydrophobic compounds. High intrinsic degradability will not only affect the concentration of chemicals present but may harm organisms indirectly if the degradation consumes oxygen and causes death through anoxia.

Initial ecotoxicity testing is aimed at identifying the nature of effects of the chemical being tested and the approximate concentration at which such effects become apparent. Special care must be taken to check the concentration of the chemical, especially in long-term tests, as the exposure concentration may be reduced by instability of the chemical, by adsorption onto components of the test system or by transfer between environmental media, for example between water and air. There are three approaches to defining chemical concentration in ecotoxicity tests. Firstly there are static tests in which the test chemical is added once to the test system and the system is then left alone. Secondly there are semi-static tests in which the test medium and test chemical are periodically replaced. Finally, there are flow-through (continuous or intermittent flow) tests in which the test medium and chemical are added to the system at such a rate as to maintain a constant concentration.

Interpretation of ecotoxicity depends upon an understanding of the relationship between biological effect and exposure concentration of the chemical under consideration. The biological effects of particular importance in ecotoxicology are those on the function and structure of ecosystems. Trophic relations that affect the energy and nutrient transfers that contribute to natural cycles require special attention in any consideration of the function of ecosystems. Trophic relations involve primary producers, secondary producers, consumers, and decay organisms: environmental damage may often be seen in changes in number of organisms in one or more of these groups or in changes in their biomass. Such changes may take the form of increased mortality, decreased growth, reduced reproductive success, eutrophication, or abnormal behaviour.

In ecotoxicity testing, one is restricted to studies on, at best, subsystems of those ecosystems at risk, and even if a change is observed to occur, it is difficult to determine whether it is likely to be ecologically significant. Living organisms have considerable ability to adapt to environmental stress. Thus, reversible effects are often considered to be tolerable. Irreversible effects are generally considered to be unacceptable. However, a distinction must be made between effects on individual organisms and those on whole populations. The loss of quite a large proportion of a population may be tolerable if the organisms remaining are unaffected and can quickly repopulate their niche. This is perhaps the most obvious difference between ecotoxicology and human toxicology where efects on individuals are of paramount importance.

A consideration of the current state of ecotoxicological testing shows that there is a need for the accumulation of a great deal of fundamental information on the factors that determine the environmental behaviour of chemicals and their effects on natural communities. From such information, it may be possible to generate ecosystem models to evaluate the likely effects of a toxicant on the environment.[46],[47] Finally, there is the question of selection of ecotoxicological tests. This must be essentially different for each chemical under consideration. It is prohibitively expensive and time consuming to carry out every possible test on every possible chemical. Full use must be made of all available knowledge to identify the most probable and most serious hazardous effects of the chemical and then to adopt tests selected to assess these. If this is to be done properly, a group of experts will be required as no one person is likely to possess all the expertise necessary. Bringing together such a group may well be the chief limitation to effective ecotoxicity testing for some time to come.

3. Prediction of Environmental Distribution

Determination of physicochemical data is the first step in predicting the behaviour of a chemical in the environment. For example, compounds with high vapour pressure under normal environmental conditions will certainly enter the atmosphere. However, knowledge of the properties of the environmental media is also essential. Thus, chemicals with low vapour pressure can spread through the atmosphere because of continuous air movement promoting vaporization and because of airborne transport of particulates.

Various computer models which can be used to predict environmental partitioning of chemicals between air, water, sediments, and soil have been published. Neely's model[48] is based essentially on the physicochemical properties of chemicals, and perhaps does not pay enough attention to the environmental media. Mackay and Paterson[49,50] have based their model on the concept of fugacity, defined as the tendency for a substance to escape from one environmental medium to another: this model assumes that an equilibrium exists between environmental compartments. An example of the application of these models is given by Schmidt-Bleek and colleagues.[51] A major complication to be allowed for in application of these models is chemical reactivity. If chemical reactions occur at rates similar to those of the partition processes, distribution of conversion products becomes as important as distribution of the parent compound and must be incorporated into the model.

3.1. Partition Between Soil or Sediment and Water. The first step in getting a complete picture of environmental movement of chemicals is to consider movement between pairs of compartments. This may start with soil or sediment and water. Soils and sediments have characteristic adsorption properties[52] which may be defined by an adsorption coefficient. Adsorption tests should be followed by desorption tests to ascertain whether adsorption is freely reversible or not. Since determination of adsorption usually is based on measurement of unadsorbed chemical left in aqueous solution, loss of chemical from the solution by other processes such as degradation may be misinterpreted as adsorption. Biotic degradation may be prevented by sterilization of the test system but this may cause alterations in adsorbing surfaces, particularly for soils or sediments high in organic matter. Vockel[53] has suggested that biodegradation may be allowed for by measuring CO_2 released from ^{14}C-labelled chemical during the adsorption test. It is possible to relate soil and sediment adsorption to physicochemical parameters[54-60] but such relationships are still not reliable enough to eliminate the need for laboratory and/or field studies.

The migratory behaviour of chemicals in soils, sediments, and water is usually assessed in the laboratory using soil or sediment columns.[61,62] The chemical is applied to the top of the column which is then eluted with water and the eluate analysed at selected time or volume intervals. Alternatively, the chemical

may be applied to the deepest layer of the soil or sediment column
and the column set in a container to which water is added so that
the column is subirrigated: in this way upward movement may be
assessed.   Effects of air movement over soil surface may be
simulated with fans. Finally, the columns can be sliced and the
presence of chemical in different layers determined.[63]

    If the influence of biota on transfer of chemicals between
media is to be estimated, microcosms must be used.   A number of
these have now been described relating to soil/water systems.[64-66]
Where possible, the validity of predictions from results obtained
in laboratory tests should be checked by comparison with field
data obtained by monitoring the presence of the chemical in water,
soils, and sediments where known releases have occurred.   Field
tests have also been described using large boxes, lysimeters, or
open fields.[67-73]   A large box may not always imitate field
conditions very well but it has the advantage that
radio-isotopically labelled substances can be used inside it.   A
lysimeter is a laboratory column of selected representative soil
or a protected monolith of undisturbed field soil with facilities
for sampling and monitoring the movement of water and chemicals.
The cross-sectional area of the lysimeter may vary from a few
square centimetres for a laboratory column to several square
metres for a field lysimeter.   Depth of the lysimeter is
arbitrary: as a minimum, it may be chosen to relate to the depth
limit of potential uptake by plants, whereas at the other extreme
it may approximate to the likely depth of drainage and
contamination of the water table. A lysimeter represents the soil
profile at a particular point, and its value in predicting
chemical behaviour in the field is inversely proportional to the
spatial variability of the soil in the area likely to be
contaminated.   An open field test is often difficult to carry out
and always difficult to interpret.

3.2. Partition between Water and Air.

Transfer of chemicals
between water and air is important as it represents one of the
most important routes whereby chemicals enter the atmosphere.
Again the chemical flux may be deduced to some extent from
physicochemical data, usually by applying the two-film boundary
layer model of Lewis and Whitman[74] (for a discussion of this, see
the review by Scheunert and Klein[75]).   Liquid and gas
mass-transfer coefficients are strongly dependent on environmental
conditions.   Two methods have been proposed for their estimation.
One depends on calculated values for the annual globally averaged
gas and liquid film mass-transfer coefficients based on water and
carbon dioxide fluxes for the oceans.[76,77]   Alternatively, the
mass-transfer coefficients can be calculated from diffusion
coefficients and environmental conditions.[78-81]   Such calculations
still remain to be perfected and do not, for example, allow for
interactions between chemicals in solution or in the atmosphere.
Hence, experimental assessment of transfer will be necessary for
some time to come.

    Quantification of the transfer of chemicals between water and
air in the laboratory involves mainly determining air-water
Henry's law constants, volatility, and the uptake into water from
air. Henry's law constants are determined by simply establishing
an equilibrium between the chemical in solution and a stream of
nitrogen bubbling through the solution and then measuring the
decrease in chemical concentration in solution.[82] This method may
not work well for chemicals of low volatility.

Chemical volatility may also be determined by simple methods. These methods use open beakers with stirred aqueous solutions.[83-85] Large-scale experiments in wind tunnels have also been used, for example with benzene and toluene[86] and dieldrin.[87] Other methods have been described but not so widely used, for example those of Klopffer and colleagues[88] and Kilzer and colleagues.[89] The uptake of chemicals by water from the air has not been so thoroughly studied. One method has been described by Slater and Spedding[87] who measured the uptake of dieldrin from air using a wetted wall column.

Field tests on the volatility of chemicals from water are expensive and affected by many variables. For instance, loss of chemical from solution may be due to adsorption onto sediments, degradation or uptake by organisms as well as volatilization. This aspect has been reviewed by Scheunert and Klein.[75]

3.3. Partition between Soil and Air. Assessment of transfer of chemicals from soil into the air is important both in relation to atmospheric contamination and retention in soil for uptake by plants. For standardized dry soils in the laboratory, the evaporation rate is proportional to vapour pressure and the square root of the molecular weight.[90,91] Unfortunately, the equation derived is not valid under field conditions. Volatilization from moist soil surfaces is much higher than that from dry soils, probably because it takes place mainly from the aqueous phase.[92] Another important factor is the rate of transfer of chemicals from deeper soil layers to the surface, largely by capillary mass flow, although it should be remembered that chemicals can move downwards in the soil water at the same time. Again, this is considered in more detail by Scheunert and Klein.[75]

Volatility of chemicals from soils may be estimated in the laboratory by gas saturation techniques, passing nitrogen through the soil at a very slow rate and establishing conditions close to equilibrium before trapping the volatilized chemical and measuring the amount.[93-95] Assay of concentration decreases in soil may be used for persistent compounds that do not show much in the way of biotic or abiotic change over the time of volatilization.[96] Alternatively, volatilized substances may be trapped and quantified.[97,98] Closed laboratory ecosystems will give a closer approximation to the natural environment than soil on its own as they will incorporate to some extent the influence of plants and animals.[66,99,100] However, care must be taken to ensure that the air flow in these systems corresponds to that under field conditions[100,101] When experiments in open fields are carried out, losses of chemical from the upper soil layer involve not only volatilization but also mineralization and leaching to deeper layers.[102,103] Because a decrease in soil concentration usually leads to a decrease in vapour pressure, residue loss decreases with time. Hence, a half-life determination is often made as a measure of rate of chemical loss from soil[104,105] Large boxes or lysimeters are used for field experiments where radiolabelled substances are to be studied.[106-108]

Higher plants contain much the largest share of the organic carbon in the biosphere and play a major part in the movement of chemicals from the soil. Accumulation of chemicals by plants may make the chemicals directly available to man where the plants are part of the diet or may be a starting point for food web accumulation or biomagnification. For plants with a 'normal' lipid content, the uptake of un-ionized chemicals into the roots from aqueous solution is positively correlated with the log of the n-octanol/water partition coefficient (log K_{OW}).[109] The root concentration factor (concentration in roots divided by concentration in external solution) is usually independent of concentration for dilute external solutions.[110] The efficiency of translocation of a chemical to the shoot from the root is defined by the transpiration stream concentration factor (concentration in transpiration stream divided by concentration in external solution).[111] Very lipophilic molecules are not readily transported in plants, and compounds of intermediate polarity (log K_{OW} = 1.8 approximately) seem to be more effectively translocated than others, apparently by a passive process.[109] It should be noted that uptake of chemicals by plants from solution is not the same as uptake from soil. This is seen experimentally[112] in the fact that the optimum log K_{OW} for uptake from soil is about 0.5 while the optimum log K_{OW} for uptake from solution is about 1.8. These optima do not apply to plants with oil channels such as are found in cress and carrot.[75] Nash[113] has reviewed the effect of environmental factors on plant uptake of chemicals from the soil.

Uptake and translocation of chemicals which reach the plant surface from the atmosphere depend upon the efficiency of cuticle penetration, adsorption into the symplast, and subsequent transport. Uptake from vapour in the atmosphere is related to the volatility of chemicals. For DDT and some cyclodiene insecticides, laboratory studies suggest that surface uptake is more important than root uptake.[113] However, this may not be true under field conditions.

Rates of uptake of chemicals by higher plants measured by laboratory methods[100,114] are not easily related to those under normal environmental conditions since uptake rates from small pots are always greater than those from large boxes or fields. Field tests[103,115] give information relevant to environmental behaviour but which is difficult to extrapolate to conditions other than those applying in the test system.

Chemicals volatilized from plants may make a substantial contribution to atmospheric contamination under certain conditions, for example, when it is hot and there is little air movement. Laboratory tests to measure volatilization from plants are very similar to those for assessment of volatilization from soil. The methods may be indirect, with the application of the chemical to leaf surfaces and subsequent determination of residues after a defined time,[116] or direct, air being passed over treated plant surfaces and then through a trap from which the volatilized chemical can be quantitatively assayed.[117] Laboratory ecosystems can also be used[66] but, as with all these laboratory systems, extrapolation to the natural environment depends on very careful consideration of meteorological and other conditions in the test.

Field tests for volatilization of chemicals from plants are exactly the same as for volatilization from soils, the chemicals having initially been sprayed on the plants.[118,119] These tests usually involve indirect determination of volatilization by measurement of the amount of chemical left on the plant as this is more important agriculturally than atmospheric contamination.

4. Biodegradation

Degradation of chemicals by living organisms, especially micro-organisms, plays a major part in removing the chemicals from the natural environment.[120] Many factors can affect the nature and kinetics of biodegradation. Water and soil must contain in an assimilable form adequate concentrations of all the nutrients needed for relevant micro-organisms to flourish. Such nutrients include nitrogen, phosphorus, sulphur, calcium, potassium, other elements in trace quantities, growth promoters, and, for aerobic metabolism, oxygen. In general, the concentration of assimilable carbon, nitrogen, and phosphorus should be in a C:N:P ratio of 100:5:1 for optimum microbial action. Optimum activity also usually requires a pH between 6 and 8 although there are micro-organisms that flourish at pH's as low as 3 and as high as 11. Similarly, optimum activity usually occurs over relatively narrow temperature ranges for different groups of micro-organisms, between 4°C and 35°C for aerobes and between 10°C and 65°C for anaerobes. Salinity and water activity are also important. Turbulence in aqueous environments ensures availability of oxygen. Light is, of course, essential for phototrophic organisms such as algae or cyanobacteria; it has an indirect effect on other micro-organisms through the fixation of carbon dioxide and release of oxygen by the phototrophs.

Not all organic molecules are equally susceptible to biodegradation. The following generalizations have been proposed.[120] Aliphatic chains are broken down more rapidly than aromatic or polycyclic compounds. Straight chains are more rapidly degraded than branched chains. The presence of heteroatoms in a ring structure and of halogenated or nitrated substituents greatly reduces biodegradability. The presence of double bonds favours biodegradation. Polymers break down slowly or not at all, but the coresponding monomers can usually be degraded more readily. Such generalizations must not be given too much weight because the metabolic capacity of micro-organisms is extensive, and the presence of a molecule in quantity in the environment will tend to favour the growth of strains of micro-organism that can degrade it.

Assessment of the biochemical oxygen demand (BOD) of chemicals in suitable test systems gives a measure of the capacity of micro-organisms to metabolize these chemicals in the presence of oxygen and also of the potential of the chemical to cause oxygen depletion in water to which it is discharged.[6] A measure of chemical oxygen demand (COD) gives a figure for the amount of oxygen required to oxidize the chemical completely or nearly so.[6] The ratio of the value for COD to that for BOD gives an indication of the ease with which a chemical may be biodegraded. This approach is usually supplemented by a ready aerobic biodegradability test. There are a number of such tests (five are described in the OECD Guidelines for Testing of Chemicals);[4] in all of these, positive results indicate rapid biodegradability but negative results may be due to poor acclimatization of the bacteria in the system. Other tests that may be carried out are those of inherent biodegradability (to find whether there is any potential at all for biodegradation of the chemical), sewage simulation tests, and tests of biodegradability in soil.[4]

5. Bioaccumulation, Bioconcentration and Biomagnification

Bioaccumulation is defined as the uptake of xenobiotics and their retention by living organisms. Bioconcentration is the term applied to the uptake and retention of xenobiotics by organisms from their immediate environment. Biomagnification is the term applied to uptake and retention of xenobiotics from food. Hence, bioaccumulation is a combination of bioconcentration and biomagnification. Most tests proposed as measures of bioaccumulation are for aquatic species. The OECD guidelines[4] for testing of chemicals list five tests for bioaccumulation; all five use fish. The same guidelines suggest that the n-octanol/water partition coefficient gives a good indication of bioaccumulation by terrestrial plant species. This measure of lipid solubility reflects an aspect of xenobiotic uptake which is very important, but metabolism of xenobiotics can also have a profound effect on bioaccumulation and the partition coefficient is little help in assessing this. One possible way of predicting likely bioaccumulation of a chemical in the terrestrial environment is to set up a model ecosystem.[121] Such a model might consist of a relatively small container with soil, water, and a selected range of species; the xenobiotic is added and, after a suitable time, its distribution and that of its metabolites are determined. However, interpretation of the results obtained in such a system is difficult.[122-125]

For aquatic organisms,[126] it appears that bioconcentration factors can be established with some precision if conditions are carefully standardized with regard to the following: choice of test organisms (species and sex), source of organisms, maintenance of organisms (temperature, feeding, water quality and water/biomass ratio, frequency of water changes), and mode of application of the test substance. Predictions made from bioconcentration factors must be related to particular ecosystems at risk. Many factors can alter the bioavailability of a substance. In rivers, lakes, and estuaries, adsorption to particulates or sediments and microbial transformations may be particularly important. In the sea, these factors may be less significant and predictions of bioaccumulations appear to be more reliable. For substances which are not readily excreted by aquatic organisms, biomagnification makes a significant contribution to bioaccumulation. For bottom-living organisms, sediments may be as important as the surrounding water as a source of xenobiotics.

Bioaccumulation by terrestrial plants is involved in the partition of xenobiotics between soil and air, and has already been discussed in this context. However, much more needs to be known about this as an essential part of our understanding of the environmental pathways which contribute to the exposure of terrestrial animals and human beings. For terrestrial animals[127] food and water are the main sources of xenobiotics. Therefore, the first area to be examined in predicting likely bioaccumulation by an animal is its dietary pattern. This is difficult because feeding varies with age, size, season, and location. The trophic level of the animal is also important. It is generally believed that successive predators will contain successively higher concentrations of persistent xenobiotics. However, this can only be true for a given substance if all the predators in a chain or web have a fairly high percentage assimilation of that substance and a fairly poor rate of excretion together with a reasonably constant exposure.

6. Conclusions

Interpretation of ecotoxicity is still more of an art than a science. Perhaps the greatest advances in this area will come from increasing understanding of structure-activity correlations for chemicals. However, at the present time, further effort is necessary to integrate the wide range of different types of assessment currently employed, which range over almost every scientific discipline. In the strictest sense, the laboratory methods described can permit only comparative evaluation of chemicals under the test conditions. Extrapolation from laboratory results to predict what will happen when chemicals enter the natural environment must be improved. This means that obtaining detailed knowledge of the environmental pathways followed by different types of chemicals must be a matter of immediate priority.

No matter how good our evaluation techniques may become, we must recognize that, because of the natural variation of test populations, high precision can never be obtained in the laboratory or field tests used in the interpretation of ecotoxicity. Thus, ecotoxicity assessments should be expressed in terms of the probability, at a stated confidence level, at which an expected adverse efect may occur. This is very rarely done, and may be statistically difficult, but it is essential if objective interpretation of ecotoxicity is to replace the subjective approach which currently predominates.

7. References

1. 'Methods for Estimating Risk of Chemical Injury: Human and Non-human Biota and Ecosystems', ed. V. B. Vouk, G. C. Butler, D. G. Hoel, and D. B. Peakall, SCOPE 26, SGOMSEC 2, John Wiley, Chichester, 1985.

2. OECD, 'Preliminary Report of the Step System Group, 28-30 November 1979', Organization for Economic Cooperation and Development, Paris, 1979.

3. OECD, 'Report on the Assessment of Potential of Environmental Effects of Chemicals; The Effects on Organisms Other than Man and on Ecosystems', published for the Organization for Economic Cooperation and Development by TNO Department of Biology, Study and Information Center for Environmental Research, Delft, 1980, Vols. 2 and 3.

4. OECD, 'Guidelines for Testing of Chemicals', Organization for Economic Co-operation and Development, Paris, 1981, with addenda 1983 and 1984.

5. CEC, 'Council Directive, 79/831/EEC', September 18, 1979, Commission of the European Communities, Brussels, 1979.

6. Health and Safety Executive, 'Approved Code of Practice, Methods for the Determination of Ecotoxicity, Notification of New Substances Regulations 1982', COP 8', HMSO, London, 1982.

7. D. B. Peakall, in 'Environmental Dynamics of Pesticides', ed. R. Haque and V. H. Freed, Plenum Press, New York, 1975, p.343.

8. T. Kozlowski, in 'Methods for Estimating Risk of Chemical Injury: Human and Non-human Biota and Ecosystems', ed. V. B. Vouk, G. C. Butler, D. G. Hoel, and D. B. Peakall, SCOPE 26, SGOMSEC 2, John Wiley, Chichester, p.573.

9. N. S. Fisher, E. J. Carpenter, C. C. Remsen, and C. F. Wurster, *Microb. Ecol.*, 1974, 1, 39.

10. P. Van Voris, R. V. O'Neill, W. R. Emanual, and H. H. Shugart, Jr., *Ecology*, 1980, 61, 1352.

11. 'Ecotoxicological Test Systems. Proceedings of a Series of Workshops'. ed. A.S. Hammons, EPA-560/6-81-004. US Environmental Protection Agency, Washington DC, 1981.

12. 'Terrestrial Microcosms and Environmental Chemistry, Proceedings, Corvallis, Oregon, June 13-14, 1977', ed. J. M. Witt, J. N. Gillett, and J. Wyatt, Report NSF/RA 790026, National Science Foundation, Washington DC, 1979.

13. 'Microcosms in Ecological Research', ed. J. P. Giesy, Jr., DOE Symposium Series 52, NTIS CONF-781101, US Department of Energy, Springfield, Virginia, 1980.

14. 'Microcosms as Potential Screening Tools for Evaluating Transport and Effects of Toxic Substances', ed. W. F. Harris, EPA-600/3-80-042, US Environmental Protection Agency, Washington DC, 1980.

15. 'Methods of Ecological Toxicology, A Critical Review of Laboratory Multispecies Tests', ed. A. S. Hammons, J. M. Giddings, G. M. Suter, II, and L. W. Barnthouse, EPA-560/11-80-026, US Environmental Protection Agency, Washington DC, 1980.

16. N. S. Fisher, E. J. Carpenter, C. C. Remsen, and C. F. Wurster, *Microb. Ecol.*, 1974, 1, 39.

17. A. H. Fielding and G. Russell, *J. Ecol.*, 1976, 64, 871.

18. J. P. Bennett and V. P. Runeckles, *J. Appl. Ecol.*, 1977, 14, 877

19. M. Kochhar, U. Blum, and R. A. Reinert, *Can. J. Bot.*, 1980, 58, 241.

20. J. Gerritsen and J. R. Strickler, *J. Fish Res. Bd. Can.*, 1977, 34, 73.

21. R. W. Drenner, J. R. Strickler, and W. J. O'Brien, *J. Fish Res. Bd. Can.*, 1978, 35, 1370.

22. C. B. Huffaker, K. P. Shea, and S. G. Herman, *Hilgardia*, 1963, 34, 305.

23. G. W. Suter, in 'Strategies for Ecological Effects Assessment at DOE Energy Activity Sites', ed. F.S. Sanders, ORNL/TM-6783, Oak Ridge National Laboratory, Oak Ridge, Tennessee, 1980, p.107.

24. J. M. Giddings and G. K. Eddlemon, *Water Air Soil Pollut.*,

1978, 9, 207.

25. J. Harte, D. Levy, J. Rees, and E. Sargebarth, 'Microcosms in Ecological Research', ed. J. P. Geisy, Jr., NTIS CONF-781101, US Department of Energy, Springfield, Virginia, 1980, p.105.

26. J. H. Rodgers, Jr., J. R. Clark, K. L. Dickson, and J. Cairns, Jr., 'Microscosms in Ecological Research', ed. J. P. Geisy, Jr., NTIS CONF-781101, US Department of Energy, Springfield, Virginia, 1980, p.625.

27. C. A. Oviatt, K. T. Perez, and S. W. Nixon, *Helgolander. Wiss. Meeresunters.*, 1977, 30, 30.

28. D. R. Jackson and J. M. Hall, *Pedobiologica*, 1978, 18, 272.

29. D. B. Porcella, V. O. Adams, and P. A. Cowan, 'Biostimulation and Nutrient Assessment', ed. E. Middlebrooks, D. H. Falkenborg, and T. E. Maloney, Ann Arbor Science Publications, Ann Arbor, Michigan, 1976, p.293.

30. 'Terrestrial Microcosms', ed. J. W. Gillett and J. M. Witt, National Science Foundation, Washington DC, 1979.

31. NRC, 'Testing for Effects of Chemicals on Ecosystems', A Report by the Committee to Review Methods for Ecotoxicology, Commission on National Resources, National Research Council, National Academy Press, Washington DC, 1981.

32. 'New Perspectives in Ecotoxicology', ed. S. A. Levin, S. Kimball, K. Kimball, and W. McDowell, Ecosystems Research Center, Cornell University, Ithaca, New York, 1984.

33. D. G. Coles and L. G. Ramspott, *Science*, 1982, 215, 1235.

34. D. W. Schindler, H. Kling, R. V. Schmidt, J. Prokopowich, V. E. Frost, R. A. Reid, and M. Caper, *J. Fish Res. Bd. Can.*, 1973, 30, 1415.

35. D. W. Schindler, R. Wagemann, R. B. Cook, T. Ruczynski, and J. Prokopowich, *Can. J. Fish Aquat. Sci.*, 1980, 37, 342.

36. R. J. Hall and G. E. Likens, 'Ecological Impact of Acid Precipitation', ed. D. Drablos and A. Tollan, Proceedings of an International Conference, SNSF-prosjektet, Oslo, Norway, 1980, p.375.

37. R. J. Hall, G. E. Likens, S. B. Fiance, and G. R. Hendrey, *Ecology*, 1980, 61, 976.

38. F. H. Bormann and G. E. Likens, 'Pattern and Process in a Forested Ecosystem: Disturbance, Development and the Steady-State Based on the Hubbard Brook Ecosystem Study', Springer Verlag, New York, 1979.

39. J. R. Geckler, W. B. Horning, T. M. Neiheisel, Q. H. Pickering, and E. L. Robinson, 'Validity of Laboratory Tests for Predicting Copper Toxicity in Streams', EPA-600/3-76-116, US Environmental Protection Agency, Washington DC, 1976.

40. R. W. Winner, J. S. Van Dyke, N. Caris, and M. P. Farrell, *Verh. Int. Verein. Limnol.*, 1975, 19, 2121.

41. R. W. Winner, M. N. Boesel, and M. P. Farrell, *Can. J. Fish. Aquat. Sci.*, 1980, 37, 647.

42. G. W. Barrett, *Ecology*, 1968, 49, 1019.

43. J. D. Moulding, *Auk*, 1976, 93, 692.

44. P. M. Irving and J.E. Miller, in 'Radiological and Environmental Division Annual Report', Part III, ANL-77-65, Argonne National Laboratory, Argonne, Illinois, 1977, p.24.

45. P. Van Voris and D. Tolle, in 'First Annual Report', RP 1224-5, Batelle Columbus Laboratories, Columbus, Ohio, 1979, p.4.

46. 'Distribution and Transport of Pollutants in Flowing Water Ecosystems', Ottawa River Project, Final Report, ed. D.R. Miller, Vols. 1 and 2, National Research Council of Canada, Ottawa, 1977.

47. C. W. Chen, in Proceedings, 'Ecological Effects of Acid Precipitation', EA-22-73, Electric Power Research Institute, Palo Alto, California, 1982, p.3/6.

48. W. B. Neely, in 'Dynamics, Exposure and Hazard Assessment of Toxic Chemicals', ed. R. Haque, Ann Arbor Science, Ann Arbor, Michigan, 1980, p.287.

49. D. Mackay, *Environ. Sci. Technol.*, 1979, 13, 1218.

50. D. Mackay and S. Paterson, *Environ. Sci. Technol.*, 1981, 15, 1006.

51. F. Schmidt-Bleek, W. Haberland, A. W. Klein, and S. Caroli, *Chemosphere*, 1982, 11, 383.

52. OECD Chemicals Testing Programme, 'OECD Guideline for Testing of Chemicals', 106: Adsorption/desorption', Umweltbundsamt Berlin, 1981, p.1.

53. D. Vockel, 'Colloquium 1980', GSF-Report 0-599, ed. Institut fur Okologische Chemie, Gesellschaft fur Strahlen- und Umweltforschung mbH Munich, Neuherberg, p.157.

54. G. W. Bailey and J. L. White, *Residue Rev.*, 1970, 32, 29.

55. G. G. Briggs, 'Proceedings of the 7th British Insecticide and Fungicide Conference', Nottingham, p.83.

56. A. Felsot and P. A. Dahm, *J. Agric. Food Chem.*, 1979, 27, 557.

57. C. T. Chiou, L. J. Peters, and V. H. Freed, *Science*, 1979, 206, 831.

58. G. G. Briggs, *Aust. J. Soil Res.*, 1981, 19, 61.

59. G. G. Briggs, *J. Agric. Food Chem.*, 1981, 29, 1050.

60. A. Sabljic and M. Protic, *Bull. Environ. Contam. Toxicol.*, 1982, **28**, 162.

61. M. Leistra, J. H. Smelt, and T. M. Lexmond, *Pestic. Sci.*, 1976, **7**, 471.

62. P. J. McCall, R. L. Swann, D. A. Laskowski, S. M. Unger, S.A. Vrona, and H. J. Dishburger, *Bull. Environ. Contam. Toxicol.*, 1980, **24**, 190.

63. C. J. Harris, *J. Agric. Food Chem.*, 1969, **17**, 80.

64. T. T. Liang and E. P. Lichtenstein, *J. Econ. Entomol.*, 1980, **73**, 204.

65. M. T. Virtanen, A. Roos, A. V. Arstila, and M. L. Hattula, *Arch. Environ. Contam. Toxicol.*, 1980, **9**, 491.

66. K. Figge and J. Klahn, *GIT-Fachz. Lab.*, 1982, **26**, 680.

67. P. Moza, I. Weisgerber, and W. Klein, *Chemosphere*, 1972, **1**, 191.

68. J. Kohli, I. Weisgerber, W. Klein, and F. Korte, *Chemosphere*, 1973, **2**, 153.

69. R. Haque, I. Weisgerber, D. Kotzias, and W. Klein, *Pestic. Biochem. Physiol.*, 1977, **7**, 321.

70. F. P. W. Winteringham, 'Appraisal of Tests to Predict the Environmental Behaviour of Chemicals', ed. P. Sheehan, F. Korte, W. Klein and Ph. Bourdeau, SCOPE 25, Wiley, Chichester, 1985, 169.

71. K. S. La Fleur, B. A. Wojeck, and W. R. McCaskill, *J.Environ. Qual.*, 1973, **2**, 515.

72. G. H. Willis and W. A. Hamilton, *J. Environ. Qual.*, 1973, **2**, 463.

73. J. H. Caro, H. P. Freeman, and B. C. Turner, *J. Agric. Food Chem.*, 1974, **22**, 860.

74. W. K. Lewis and W. G. Whitman, *Ind. Eng. Chem.*, 1924, **16**, 1215.

75. I. Scheunert and W. Klein, 'Appraisal of Tests to Predict the Environmental Behaviour of Chemicals', ed. P. Sheehan, F. Korte, W. Klein and Ph. Bordeau, SCOPE 25, Wiley, Chichester, 1985, 285.

76. P. S. Liss and P. G. Slater, *Nature (London)*, 1974, **247**, 181.

77. D. Mackay and P. J. Leionen, *Environ. Sci. Technol.*, 1975, **9**, 1178.

78. D. Mackay, 'The Handbook of Environmental Chemistry', Vol.2, part A, ed. O. Hutzinger, 1980, p.31.

79. R. E. Rathburn and D. Y. Tai, *Water Res.*, 1981, **15**, 243.

80. R. E. Rathburn, D. W. Stephens, D. J. Schultz, and D. Y. Tai, *Chemosphere*, 1982, 11, 1097.

81. C. J. M. Wolff and H. B. van der Heijde, *Chemosphere*, 1982, 11, 103.

82. D. Mackay, W. Y. Shiu, and R. P. Sutherland, *Environ. Sci. Technol.*, 1979, 13, 333.

83. C. D. Metcalfe, D. W. McLeese, and V. Zitko, *Chemosphere*, 1980, 9, 151.

84. J. H. Smith, D. C. Bomberger, and D. L. Haynes, *Environ. Sci. Technol.*, 1980, 14, 1332.

85. J. H. Smith, D. C. Bomberger, and D. L. Haynes, *Chemosphere*, 1981, 10, 281.

86. Y. Cohen, W. Cocchio, and D. Mackay, *Environ. Sci. Technol.*, 1978, 12, 553.

87. R. M. Slater and D. J. Spedding, *Arch. Environ. Contam. Toxicol.*, 1981, 10, 25.

88. W. Klopffer, G. Kaufmann, G. Rippen, and H.-J. Poremski, *Ecotox. Environ. Safety*, 1982, 6, 545.

89. L. Kilzer, I. Scheunert, H. Geyer, W. Klein, and F. Korte, *Chemosphere*, 1979, 10, 751.

90. G. S. Hartley, *Adv. Chem. Ser.*, 1969, 86, 115.

91. W. F. Spencer, W. J. Farmer, and M. M. Cliath, *Residue Rev.*, 1973, 49, 1.

92. J. W. Hamaker, 'Organic Chemicals in the Soil Environment', ed. C. A. I. Goring and J. W. Hamaker, Marcel Dekker, New York, 1972, Vol. 2, p.341.

93. W. F. Spencer and M. M. Cliath, *J. Agric. Food Chem.*, 1972, 20, 645.

94. W. F. Spencer and M. M. Cliath, *J. Agric. Food Chem.*, 1974, 22, 987.

95. M. M. Cliath and W. F. Spencer, *Environ. Sci. Technol.*, 1972, 6, 910.

96. R. Haque, D. W. Schmedding, and V. H. Freed, *Environ. Sci. Technol.*, 1974, 8, 139.

97. Y. H. Atallah, D. M. Whitacre, and B. L. Hoo, *Bull. Environ. Contam. Toxicol.*, 1979, 22, 570.

98. N. Burhard and J. A. Guth, *Pestic. Sci.*, 1981, 12, 37.

99. W. Ebing and I. Schuphan, *Ecotox. Environ. Safety*, 1979, 3, 133.

100. R. Kloskowski, I. Scheunert, W. Klein, and F. Korte, *Chemosphere*, 1981, 10, 1089.

101.M. J. Beall, R. G. Nash, and P. C. Kearney, 'Proceedings of the Conference on Environmental Modeling and Simulation', Cincinnati, Ohio, 1976, p.790.

102.G. W. Ware, B. Estesen, W. C. Kronland, and W. P. Cahill, *Bull. Environ. Contam. Toxicol.*, 1977, **17**, 317.

103.N. S. Talekar, L.-T. Sun, E.-M. Lee, and J. S. Chen, *J. Agric. Food Chem.*, 1977, **25**, 348.

104.K. E. Elgar, 'Pesticides – Environmental Quality and Safety, Suppl. Vol. III', ed. F. Coulston and F. Korte, G. Thieme, Stuttgart, 1975, P.250.

105.I. Scheunert, 'Colloquium 1980', GSF-Report D-599, ed. Institut fur Strahlen- und Unweltforschung mbH, Munich, 1981, p.85.

106.P. Moza, I. Weisgerber, and W. Klein, *J. Agric. Food Chem.*, 1976, **24**, 881.

107.P. Moza, I. Scheunert, and F. Korte, *Arch. Environ. Contam. Toxicol.*, 1979, **8**, 183.

108.P. Moza, I. Scheunert, W. Klein, and F. Korte, *J. Agric. Food Chem.*, 1979, **27**, 1120.

109.G. G. Briggs, R. H. Bromilow, and A. A. Evans, *Pestic. Sci.*, 1982, **13**, 495.

110.P. Leroux and M. Gredt, *Neth. J. Plant Pathol.*, 1977, **83** (Suppl.1), 51.

111.M. G. T. Shone and A. V. Wood, *J. Exp. Bot.*, 1974, **25**, 390.

112.G. G. Briggs, R. H. Bromilow, R. Edmondson, and M. Johnston, in 'Herbicides and Fungicides', ed. N. McFarlane, Chemical Society Special Publication No.29, 1977, p.129.

113.R. G. Nash, 'Pesticides in Soil and Water', ed. W. D. Guenzi, Soil Science Society of America, Madison, Wisconsin, 1974, 257.

114.H. H. Cheng, F. Fuhr, and W. Mittelstaedt, 'Pesticides – Environmental Quality and Safety', Suppl. Vol. III, ed. F. Coulston and F. Korte, G. Thieme, Stuttgart, 1975, p.271.

115.I. Scheunert, J. Kohli, R. Kaul, and W. Klein, *Ecotox. Environ. Safety*, 1977, **1**, 365.

116.I. Weisgerber, W. Klein, and F. Korte, *Tetrahedron*, 1970, **26**, 779.

117.G.W. Burt, *J. Environ. Qual.*, 1974, **3**, 114.

118.K. Sandrock, I. Weisgerber, W. Klein, and F. Korte, *Food Chem. Microbiol. Technol.*, 1972, **1**, 206.

119.M. M. Cliath, W. F. Spencer, W. J. Farmer, T. D. Shoup, and R. Grover, *J. Agric. Food Chem.*, 1980, **28**, 610.

120. R. Cabridenc, 'Appraisal of Tests to Predict the Environmental Behaviour of Chemicals', SCOPE 25, ed. P. Sheehan, F. Korte, W. Klein and Ph. Bourdeau, Wiley, Chichester, England, 1985, p.213.

121. R. L. Metcalf, G. K. Sangha, and I. P. Kapoor, *Environ. Sci. Technol.*, 1971, **5**, 709.

122. F. Moriarty, 'Ecological Effects of Pesticides', ed. F. H. Perring and K. Mellanby, Linnean Society Symposium Series No.5, Academic Press, London, 1977, p.165.

123. R. L. Metcalf, *Ann. Rev. Entomol.*, 1977, **22**, 241.

124. 'Terrestrial Microcosms', ed. J. W. Gillett and J. M. Witt, National Science Foundation, Washington DC, 1979.

125. P. H. Pritchard, 'Environmental Risk Analysis for Chemicals', ed. R. A. Conway, van Nostrand Reinhold, New York, 1982. p.257.

126. W. Ernst, 'Appraisal of Tests to Predict the Environmental Behaviour of Chemicals', ed. P. Sheehan, F. Korte, W. Klein, and Ph. Bourdeau, SCOPE 25, Wiley, Chichester, 1985, p.243.

127. F. Moriarty, 'Appraisal of Tests to Predict the Environmental Behaviour of Chemicals', ed. P. Sheehan, F. Korte, W. Klein, and Ph. Bordeau, SCOPE 25, Wiley, Chichester, 1985, p.257.

11

The European Community Chemicals Notification Scheme and Environmental Hazard Assessment

By J.L. Vosser

DEPARTMENT OF THE ENVIRONMENT, 43 MARSHAM STREET, LONDON SW1P 3PY, UK

1. Introduction

The assessment of hazard is a feature of the 'Sixth Amendment' of the 1967 European Communities Directive on the Classification, Packaging and Labelling of Dangerous Substances.[1] This Amendment introduces the concept of pre-marketing notification, which requires the submission to a relevant national authority of a package of scientific and commercial information which must be adequate for the purposes of hazard assessment. This chapter discusses how the commercial information and the results of ecotoxicity and physico-chemical tests are used in making such an assessment.

2. Competent Authorities

The European Community scheme is implemented in the United Kingdom by regulations made under the Health & Safety at Work *etc.* Act 1974 for the safety of workers and human health aspects, and under the European Communities Act 1972 for the environment part.[2] Notifications are made in the United Kingdom to the Health & Safety Executive who, with the Department of the Environment, make up the United Kingdom Competent Authority. All the Member States of the European Community have a Competent Authority, and to eliminate multiple submissions, a valid notification made in one state is considered to cover the requirements of the Directive throughout the community. Arrangements exist for the exchange and circulation of certain information between the various Competent Authorities themselves and with the European Commission. Competent Authorities must be satisfied that the information submitted to them is in accordance with the requirements of the Directive and its implementing regulations and is adequate for hazard assessment purposes. Although a substance may have been notified correctly in accordance with the national regulations, it never really ceases to be of interest to the Competent Authorities. If the notifier discovers changes in the intended use of the substance, new or changed properties, or proposes to market an increased quantity, the Competent Authority to which the original notification was made must be informed so that a further assessment of the hazard potential of the substance can be made. Further tests may be asked for if the Competent Authority considers these are necessary to improve the assessment.

3. Action Arising from Assessment of Notified Data

In the event of a substance having properties dangerous to human health, then under the Health & Safety at Work *etc.* Act, steps can be taken directly by the Health & Safety Executive to control manufacture and use. If, however, the substance does not pose a hazard to human health but does to the environment, then the Department of the Environment can advise the Water & Waste Disposal Authorities who can control discharges using the powers available to them under the Control of Pollution Act 1974.

4. Definition of a New Substance

A substance is considered to be new if it was not marketed for commercial purposes on the European Community market at any time in the 10 years prior to 18th September 1981. The European Commission will be publishing in all community languages a list of all substances which were on the market for this period (European Inventory of Existing Chemical Substances, EINECS). At present only draft lists are available for consultation by intending notifiers. It has been the responsibility of manufacturers to ensure that their range of chemicals is included in this list, because if they are not then the chemicals will attract the requirements of the notification scheme.

5. Types of Notification

5.1. Limited Announcements. These are generally made for the marketing of substances in quantities of less than one tonne for a number of purposes, and make known to the Competent Authorities certain basic information. This includes identification, quantity to be placed on the market, recommendation for safe handling and use, and labelling based on its acute mammalian toxicity. There is no requirement for submission of ecotoxicological data.

A company may make a limited announcement for a substance which is to undergo full-scale field trials before it can be marketed even though the quantity is greater than one tonne. In this case the substance is excluded from the full testing and notification requirements for one year only. But the company must provide some limited information and give an undertaking that the substance will be handled only by the recipient company's personnel and not be available to the general public. After 12 months the derogation ceases and any further supply must be preceded by full notification.

5.2. Full Notification. This is required when the intention is to place on the European Market one tonne year^{-1} or more of a new substance. This notification must be made 45 days before marketing and be accompanied by a Technical Dossier of information on the substance. Certain classes of chemical are exempt from the notification requirements. If any part of this package is omitted, then full justification for doing so must be given or the notification may not be accepted. A notification cannot be rejected on the grounds of toxicity to man and/or the environment, since the Directive is concerned with the provision of information, not with the 'management' of a chemical. The information required in a technical dossier is discussed next.

5.3. Data Submitted for a Full Notification. The information that
has to be submitted for a full notification falls under five
headings:

Identity of substance	(Appendix 1)
Information on the substance	(Appendix 2)
Physico-chemical properties	(Appendix 3)
Toxicological studies	(Appendices 4, 8, and 9)
Ecotoxicological studies	(Appendices 5, 8, and 9)
Possibility of rendering the substance harmless	(Appendix 6)

Appendices 1 to 6 form Annex VII, and Appendices 8 and 9, Annex
VIII, respectively to the Directive. In addition the notifier
must, where the toxicological data indicate, classify the
substance as either 'very toxic', 'toxic', or 'harmful' according
to specified criteria (Appendix 7, Directive Annex VI). At
present there are no agreed criteria for the classification of the
substance as far as its potential to cause harm to the environment
is concerned.

Tests must be conducted according to methods published by the
European Commission which form Annex V to the Directive. If this
is not possible because a test for a particular study is not
included in Annex V then a notifier may use a national or
international standard method. In all tests the principles of
current good laboratory practice must be applied.

The amount of information to be submitted increases when
certain levels of quantity placed on the European market are
reached (Appendices 8 and 9). Renotification is required at these
'trigger points' and further data may, or will, be required to
comply with the regulations. The 'trigger points' are shown in
Table 1.

Table 1

'Trigger Point'		*Quantity* (t year^{-1})	*Information* *required as per* *Appendix No.*	*Directive* *Annex*
'Base set'		1 - 10	1 to 7	VII
Level I	(a)	10 - 100 or 50 tonnes total	8	VIII
	(b)	100 - 1000 or 500 tonnes total	8	VIII
Level II		1000 or 5000 tonnes total	9	VIII

It should be noted that whatever quantity is marketed the
information listed in Appendices 1 to 7 must always be submitted.
At Level I (a) the Competent Authorities may ask for one or more
of the additional tests listed in Appendix 8 to be carried out if
the 'Base set' data do not permit an adequate assessment to be
made. At Level I (b) the Competent Authority will ask for the
Appendix 8 tests to be carried out. In either case, however, the

choice of the additional test(s) will be governed by the circumstances surrounding, for example, the proposed use of the substance, and its pathways to man and the environment. At Level II the programme of test work will be agreed between the notifier and the Competent Authority and will include one or more of the studies listed in Appendix 9.

6. Hazard Assessment using Notified Data

The assessment of hazard to man is dealt with in Chapter 12 and will not be considered further here. In making an assessment of the hazard to the environment, the data/information given in Appendices 1 and 2, together with some of the physico-chemical data listed in Appendix 3, are utilized to estimate the 'exposure' and, depending on the quantity of the substance being notified, the appropriate data from Appendices, possibly 6, 8, and 9. The 'effects' data in these Appendices are then compared with the 'exposure' or predicted environmental concentration (PEC).

7. Estimation of Exposure

In addition to the information and data listed in Appendices 2, 3, and 5, and where appropriate 8 and 9, information on routes into and through the environment and a knowledge of data on sewage treatment and dilution factors are also necessary for the estimation of 'exposure'. The estimation of the distribution of the chemical between the environmental compartments is at present still a matter of professional judgement. Various approaches to the calculation of distribution are currently under discussion. It is suggested that, because these approaches assume an equilibrium being established, they have little application to the situation most likely to arise, which is that of short-term local effects. Such effects are most likely to occur if the new substance finds its way into water and it is for this reason that the remainder of this section and chapter is concerned with the aquatic environment.

It should be remembered that the notifier may, or will, be asked to carry out further tests to demonstrate any potential hazards to species in other environmental compartments particularly soil. Factors which would indicate the need for further tests include higher tonnages and/or pathways through the environment and/or the final disposal point taken together with data on physico-chemical properties, degradability, and bioaccumulation potential as indicated by the n-octanol/water partition coefficient.

8. Use of Notified Data in Exposure Assessment

8.1. Structure (Appendix 1). This may be compared with those of existing chemicals whose behaviour in the environment may already be known.

8.2. Proposed Uses and Production (Appendix 2). The quantity of wastage arising during production and use may be estimated from these data but, more often than not, further enquiries will have to be made to elicit the necessary information.

8.3. Physico-chemical Properties (Appendix 3). The most useful data are those on hydrolysis, water solubility, boiling point, and vapour pressure. These will enable a prediction to be made of the environment compartment in which the chemical is most likely to be found. In the aquatic compartment the n-octanal/water partition coefficient will give an indication of the bioaccumulation potential of the chemical if it does not ionize in solution.

8.4. Degradation (Appendix 5). This is the most important process available either to reduce the concentration of the chemical in the environment or eliminate it. The regulations call for aerobic biotic and abiotic degradation, the latter only when degradation by biotic means is poor. It is unfortunate that the Biochemical Oxygen Demand (BOD) and its ratio to the Chemical Oxygen Demand (COD) are asked for as minimum evidence of biodegradation. The BOD is a measure of the microbial depletion of oxygen from water and the COD a measure of the amount of oxygen required to convert the compound, chemically to carbon dioxide, water, and other oxidation states of other elements such as sulphur and nitrogen that may be present in the molecule. These tests were developed for measuring, amongst other things, the strength and efficiency of treatment of sewage. If the ratio of COD to BOD is more than 2:1 then the sewage becomes progressively more difficult to treat as the ratio increases. If high ratios are found for new chemicals this may suggest that they will not biodegrade. This is not necessarily so because it may be that in the test system there are either no micro-organisms present which can metabolize the chemical or that the chemical may be toxic to them under the test conditions. Biodegradation is better determined using tests in which the ratio of organic carbon to micro-organisms is lower and more time is given for acclimatization to develop. Although anaerobic degradation tests have yet to be incorporated into the regulations, they are more appropriate than aerobic tests for certain types of chemicals, particularly those which adsorb strongly to sewage and other solids. The sewage sludges containing these solids are often subjected to the anaerobic digestion process in many sewage treatment plants. Apart from determining the degree of degradability, some of the tests described above will give an indication of the effect of the chemical on the efficiency of the process represented by the test.

Abiotic degradation test procedures have not been agreed internationally except for hydrolysis as a function of pH. Although this is considered to be a 'Base set' test (Appendix 5) it is called for only if biodegradation is poor. Photodegradation tests as well as those for degradation in soil are not yet internationally agreed and therefore data are not available for new substances.

9. Additional Information on Flow in Rivers, Sewage Treatment *etc.*

The notified data described in the previous paragraphs are taken into account with information on river flow and sewage treatment to predict the environmental concentration under realistic 'worst case' conditions. There are two routes by which the waste containing the chemical can reach the aquatic environment. These are described briefly below.

9.1. Direct Discharge to Rivers. In this case the waste substance is diluted in the process water which is in turn diluted on discharge into the river. The relevant concentration can be calculated if the outfall volume and river flow rate are known, but usually these figures must be estimated from figures on

average outfalls for each industry and river flow rates found in
the River Pollution Survey. The estimation of realistic values is
very difficult, as the volumes of effluent from different
factories vary considerably as, of course, would the amount of
substance present in the outfall at any time. The choice of a
realistic river flow rate is also quite difficult, because the
assessor has to decide what is the likely flow rate for a river
into which waste will discharge. In the UK most discharges are
into larger rivers, with flow rates of 5 to 20 m^3 s^{-1} (cumecs),
but the majority of river lengths (just over 55%) have flow rates
of between 0 and 0.62 cumecs. A reasonable 'worst case' which may
be used for initial calculations could use 0.62 cumecs as a basis
bearing in mind that approximately half the river flows are
greater than this.

9.2.   Disposal *via* Sewage Works. All the factors relating to
river flow mentioned above also apply in this case. In addition
to the variation associated with river flow, trade effluent is
present in flows to large and small sewage works and it is
difficult to carry out calculations for each type of works.
Therefore, to simplify matters, assessment is based on dilution
and treatment of waste by a typical medium-sized sewage works
serving 10,000 people. Typically, these works would have a small
amount of industrial waste and larger volumes of domestic waste.
Table 2 shows the sewage flow and sludge production parameters for
a works serving 10,000 people.   When calculating the exposure
concentration of a substance it is assumed that sludge and sewage
dilute the trade waste.

Table 2

Typical sewage flow and sludge production parameters
for a sewage works serving 10,000 people

Dry weather flow of waste water per head	240 l/head/day
Domestic input	180 l/head/day
Industrial input	60 l/head/day
Population	10000 people
Total sewage volume per day (outflow = inflow)	2400 m^3/day
Solids produced	90 g dry solid /head/day
Total solids	900 kg/day

     In UK sewage treatment there is usually a sedimentation stage
followed by aerobic treatment of the settled effluent by activated
sludge or biological filters.   Sludge produced in the
sedimentation tanks and as a result of the aerobic treatment is
either disposed of on land or to sea or in about half of the cases
subjected to an anaerobic digestion treatment before disposal.
When calculating the dilution of the waste substance the different
retention times in the different systems are ignored.   However,
the age of the aerobic sludge may have to be considered if the
rate of biodegradation of the substance is low. Also, it must be
remembered that a chemical not readily adsorbed by sludge may pass
through the sewage works unchanged if it does not aerobically
degrade and also that the data available for an assessment may not

include information on the anaerobic biodegradability of the chemical. In the absence of appropriate degradation data the first 'worst case' analysis assumes that the substance does not biodegrade in a sewage works.

9.3. Disposal *via* sewage sludge to land. Although the UK disposes of sewage sludge by incineration, dumping at sea, and disposal to land, this section considers only sludge used in agriculture, ploughed into grazing or arable land. Sludge is incorporated in the soil until specified toxic metal or nitrogen levels are reached. Between 2 and 10 tonnes of dry solid per hectare are added at a time, up to five times a year for grazing land. A soil depth of 20 cm for arable and 3 cm for grazing land is assumed.

10. Application of the Procedure

To illustrate the application of the stages in the estimation of exposure (predicted environmental concentration, PEC), the cases of a xenobiotic and a speciality chemical are considered next.

10.1. Xenobiotic Chemical. A typical example of this type of chemical is the linear alkyl benzenesulphonates (LAS) used in detergents. In 1978 the annual quantity placed on the UK domestic detergent market was about 70,000 tonne and it is assumed that the population as a whole (55 million) used them and discharged the whole quantity to waste after use. The *per capita* daily consumption of LAS is thus 3.5 g, which, taken with an average sewage works flow of 240 l per head per day, results in a concentration of LAS in crude sewage of about 14 mg l^{-1}. The LAS molecules degrade very readily in aerobic sewage treatment – usually at least 95% – and thus the concentration of LAS in treated sewage is generally less than 1 mg l^{-1}. On dilution of this effluent with river water, further reduction in the concentration of LAS is obtained, depending on the degree of dilution available. Data presented in the various reports of the Standing Technical Committee on Synthetic Detergents[3] show that in a well run sewage treatment plant it is possible to achieve upwards of 97% biodegradation of the LAS, giving concentrations in works effluents of about 0.3 mg l^{-1} at these levels of treatment efficiency. The concentration of LAS in crude sewages was reported to be generally in the range 12 – 20 mg l^{-1}, the variation being due to a number of factors, principally industrial activities and the flow of sewage.

10.2. Low Tonnage Speciality Chemical. For xenobiotic chemicals the prediction of the environmental concentration is relatively simple to carry out and probably leads to a value which is likely to be found in surface waters. For speciality chemicals such as dyestuffs, the estimation is much more difficult because a number of scenarios can be developed not all of which are likely to occur in practice: the difficulty from the Competent Authority's point of view lies in choosing the most appropriate. For a dyestuff, although the notifier may be able to give information on the wastage during production and the number of customers he intends to supply the product to, he may well have no information on how often his customers will use it or the efficiency of their operations. Further enquiries on these matters by the Competent Authority may be fruitful as will a knowledge of the actual dye bath efficiencies for the various types of dye. As an example, consider the supply of a dyestuff of 3 tonne year^{-1} to six customers. The other information supplied with the notification

shows that 90% of the dye is taken up by the fabric, it is
strongly adsorbed by sewage sludge (say 70%), is not degradable by
aerobic processes, and does not affect the normal sewage works
treatment processes. All this information is valuable in itself,
but it cannot be used unless the Competent Authority knows the
frequency of dyeing. Assuming that this is occasional and takes
place over ten days in the year, 2 batches a day with the effluent
being discharged to a small sewage works treating the sewage from
a population of 10,000 where the flow is 240 l/head/day, the
predicted environmental concentration is calculated as follows:

1.	Quantity per customer	0.5	tonne year^{-1}
2.	Quantity per working day (quantity per batch 0.025 tonne)	0.05	tonne
3.	Efficiency in dye bath: Wastage to sewer	90% 5.0	kg day^{-1}
4.	Concentration in crude sewage	2.1	mg l^{-1}
5.	Adsorption onto solids: Concentration of non-adsorbed dye	70% 0.63	mg l^{-1}

Liquid Phase

6.	Degradability: Concentration of dye in effluent	effectively nil 0.2	mg l^{-1}

Sludge Phase

7.	Adsorption: Concentration in sludge	90% 3888	mg kg^{-1}
8.	Sludge spread on arable land Concentration of dye in soil	19.4	mg kg^{-1}

The concentration of dye in the sludge may well be reduced if
the anaerobic sludge digestion process degrades the dye. The
concentration of the dye in the sewage works efluent (step 6) will
be reduced still further when it is diluted by the receiving
water. The final concentration may well be below that at which
the dye colours the water. Most pollution control authorities
restrict the concentration so that coloration does not occur.

If the Competent Authority considers that the concentration of
the dyestuff in sewage effluent is likely to lead to coloration of
surface water and to the concentration in sludge being too high,
then it is within its remit to suggest that wastewater containing
the dye is pretreated before it is discharged to sewer. The above
is but one of several scenarios that could be devised. They will
only be realistic if the information used is of the same quality
as above.

11. Predicting Environmental Concentrations - Conclusions

The two examples emphasize the need for the Competent Authority to
have available relevant information which must be accurate enough

to allow the calculation of a realistic estimate of the environmental concentration. It is the small-volume speciality chemicals which are likely to cause most difficulty, partly because the notifier very often has no knowledge of how his customers will use the chemical, whether wastage arising from their processes will be treated before discharge, *etc*. These difficulties may also arise with the larger-volume chemicals and in general the Competent Authority will try to envisage a reasonable 'worst case' approach. In carrying this out in practice, it is perhaps fortunate that there are a finite and, in fact, fairly small number of routes for a chemical to reach the environment.

12. Environmental Effects Data

The significance of the ecotoxicological data and the use to which they can be put is discussed in some detail in Chapter 10. This chapter is concerned with the comparisons made by the Competent Authority between the predicted environmental concentration (PEC) and the 'effects' data to determine whether or not the chemical gives cause for concern.

The 'Base set' data listed in Appendix 5 will indicate the short-term acute effects on a fish and a member of the aquatic food chain and the potential of a chemical to degrade by biological and/or non-biological processes. This limited amount of information clearly can only give a preliminary indication of hazard potential.

12.1. Effects *versus* Exposure.

The results of the fish and *Daphnia* tests will normally be accompanied by a graphical presentation of the results from which it may be possible in some cases to gain an indication of possible longer-term delayed effects (see also Chapter 10 for a more detailed discussion). From the degradation tests and the n-octanol/water partition coefficient (Appendix 5) it should be possible to have an indication of whether or not the chemical is persistent (poor degradability) and/or likely to bioaccumulate (n-octanol/water partition coefficient >1000). If these additional pieces of information do not suggest particular caution then the lower of the values for fish LC_{50} or *Daphnia* EC_{50} is compared with the PEC taking into account a safety factor. This factor lies somewhere between 10 and 1 million and available data[4-6] suggest that a value of 100 could be supported as adequate for the majority of cases. If the data suggest that the chemical could bioaccumulate and/or persist, then further tests would be called for to demonstrate whether this is likely to be the case. This action is allowed for in this Directive.

The safety factor is applied to the PEC and LC_{50}, or EC_{50} values so that if the PEC is 1/100 or less of the lower of the two effects values, then, having taken into account the data on persistence and bioaccumulation potential, the chemical should not give cause for concern. This description over-simplifies the whole process of determining whether there is cause for concern. In fact the assessor of the hazard potential will be taking into account as well such matters as possible routes to the environment and mammalian toxicology if there is a chance that mammals will come into contact with the chemical.

12.2. Further Testing.

Although the UK regulations require further

testing as the quantity of the chemical increases (Appendices 8 and 9), there is nothing to prevent the Competent Authority asking for these higher tests or any other information at the 'Base set' level if, on the evidence of the 'Base set' data or predicted environmental concentration, there is a need to confirm the implications of the preliminary assessment.

At the higher tonnage levels of Levels I and II there exists the possibility of choosing the most appropriate species from those listed in Appendices 8 and 9, the choice reflecting the pathways through and the final disposal point in, the environment.

13. Actions Arising from Hazard Assessment

If, after the Competent Authority has considered and assessed all of the information available it is still concerned that there could be a hazard to the environment arising from the use of the chemical, it will consider notifying the appropriate Pollution Control Authority (PCA). It will do so only after discussion with the notifier. The PCA has the responsibility of taking up the matter with the notifier so that it can decide what monitoring of discharges is needed and, if necessary, what conditions should apply to discharges.

14. Summary

The assessment of hazard to the environment posed by chemicals based on data notified to a Competent Authority under statutory schemes is not a precise science. The environment is an extremely complex system with many synergistic or antagonistic factors at play, and in the future it may be possible to incorporate these kinds of factors in mathematical models. Until these are developed and validated, hazard assessment will continue to involve judgements made by professionals on the basis of rather less information than is ideally desirable and applying safety factors which it is hoped err on the side of caution. Monitoring of the environment for the presence or otherwise of new chemicals and their effects is the only way to prove whether the original assessment was right; this may take many years.

15. Acknowledgement

This chapter is reproduced by permission of the Controller of Her Majesty's Stationery Office.

16. References

1. Council Directive 79/831/EEC Amending for the Sixth Time Directive 67/548/EEC on the Classification, Packaging and Labelling of Dangerous Substances OJ L259/10 15 October, 1979.

2. Statutory Instrument 1982 No. 1496 Health & Safety. The Notification of New Substances Regulations, HMSO, London, 1982, ISBN 0 11 027496 2.

3. 16th – 20th (Final) Annual Reports of the Standing Technical Committee on Synthetic Detergents, HMSO, London, 1976-1980.

4. J. B. Sprague, 'Measurement of pollutant toxicity to fish III: sub-lethal effects and "safe" concentrations', *Water Res.*, 1971, 5, 245 – 266.

5. C. M. Lee, 'Application and value of toxicity tests for predicting the acceptability of chemicals in the aquatic environment'. Presented at IAES/SECOTOX Symposium 'Testing of chemical substances for ecotoxicological revaluation', Munich, 1979.

6. P. A. Gilbert, and C. M. Lee, 'Tests appropriate to the aquatic and terrestrial environments'. Presented at IAES/SECOTOX Symposium 'The scientific basis for the assessment of hazards from chemicals associated with the environment', London, 1981.

Appendices 1-6: Base set information.

1.   IDENTITY OF THE SUBSTANCE

1.1.   **Name**

1.1.1. Names in the IUPAC nomenclature

1.1.2. Other names (usual name, trade name, abbreviation)

1.1.3. CAS number (if available)

1.2.   **Empirical and structural formula**

1.3.   **Composition of the substance**

1.3.1. Degree of purity (%)

1.3.2. Nature of impurities, including isomers and by-products

1.3.3. Percentage of (significant) main impurities

1.3.4. If the substance contains a stabilizing agent or an inhibitor or other additives, specify: nature, order of magnitude: ... ppm; ... %

1.3.5. Spectral data (UV, IR, NMR)

1.4.   **Methods of detection and determination**

     A full description of the methods used or the appropriate bibliographical references

2.   INFORMATION ON THE SUBSTANCE

2.1.   **Proposed uses**

2.1.1. Types of use

     Describe: the function of the substance ....
              the desired effects ....

2.1.2. Fields of application with approximate breakdown

     (a) closed system
         – industries ....
         – farmers and skilled trades ....
         – use by the public at large ....

     (b) open system
         – industries ....
         – farmers and skilled trades ....
         – use by the public at large ....

2.2.   **Estimated production and/or imports for each of the anticipated uses or fields of application**

2.2.1. Overall production and/or imports in order of tonnes per year 1; 10; 50; 100; 500; 1,000 and 5,000

     – first 12 months ....tonnes/year
     – thereafter ....tonnes/year

2.2.2. Production and/or imports, broken down in accordance with 2.1.1 and 2.1.2, expressed as a percentage
     – first 12 months ....
     – thereafter ....

2.3.   Recommended methods and precautions concerning:

2.3.1. handling ....

2.3.2. storage ....

2.3.3. transport ....

2.3.4. fire (nature of combustion gases or pyrolysis, where proposed uses justify this)

2.3.5   other dangers, particularly chemical reaction with water

2.4.   Emergency measures in the case of accidental spillage

2.5.   Emergency measures in the case of injury to persons (e.g. poisoning)

3.   PHYSICO-CHEMICAL PROPERTIES OF THE SUBSTANCE

3.1.   Melting point
         .... °C

3.2.   Boiling point
         .... °C .... Pa

3.3.   Relative density
         .... (D_4^{20})

3.4.   Vapour pressure
         .... Pa at .... °C
         .... Pa at .... °C

3.5.   Surface tension
         .... N/m (............... °C)

3.6.   Water solubility
         .... mg/litre (............... °C)

3.7.   Fat solubility
         Solvent—oil (to be specified)
         .... mg/100g solvent (............... °C)

3.8.   Partition coefficient
         n-octanol/water

3.9.   Flash point
         .... °C ☐ open cup ☐ closed cup

3.10. Flammability (within the meaning of the definition given in Part I of Schedule 2, paragraph 2(c), (*d*) and (*e*))

3.11. Explosive properties (within the meaning of the definition given in Part I of Schedule 2, paragraph 2(*a*))

3.12. Auto-flammability
         .... °C

3.13. Oxidizing properties (within the meaning of the definition given in Part I of Schedule 2, paragraph 2(b))

4.   TOXICOLOGICAL STUDIES

4.1.   Acute toxicity

4.1.1. Administered orally
         LD_{50} .... mg/kg
         Effects observed, including in the organs ....

4.1.2. Administered by inhalation
         LC_{50} .... (ppm) Duration of exposure .... hours
         Effects observed, including in the organs ....

4.1.3. Administered cutaneously (percutaneous absorption)
         LD_{50} .... mg/kg
         Effects observed, including in the organs ....

4.1.4. Substances other than gases shall be administered via two routes at least, one of which should be the oral route. The other route will depend on the intended use and on the physical properties of the substance.
         Gases and volatile liquids should be administered by inhalation (a minimum period of administration of four hours).
         In all cases, observation of the animals should be carried out for at least 14 days.

Unless there are contra-indications, the rat is the preferred species for oral and inhalation experiments.
The experiments in 4.1.1, 4.1.2 and 4.1.3 shall be carried out on both male and female subjects.

4.1.5. Skin irritation
The substance should be applied to the shaved skin of an animal, preferably an albino rabbit.
Duration of exposure .... hours

4.1.6. Eye irritation
The rabbit is the preferred animal.
Duration of exposure .... hours

4.1.7. Skin sensitization
To be determined by a recognized method using a guinea-pig.

4.2. **Sub-acute toxicity**

4.2.1. Sub-acute toxicity (28 days)
Effects observed on the animal and organs according to the concentrations used, including clinical and laboratory investigations ....
Dose for which no toxic effect is observed ....

4.2.2. A period of daily administration (five to seven days per week) for at least four weeks should be chosen. The route of administration should be the most appropriate having regard to the intended use, the acute toxicity and the physical and chemical properties of the substance.
Unless there are contra-indications, the rat is the preferred species for oral and inhalation experiments.

4.3. **Other effects**

4.3.1. Mutagenicity (including carcinogenic pre-screening test)

4.3.2. The substance should be examined during a series of two tests, one of which should be bacteriological, with and without metabolic activation, and one non-bacteriological.

5. ECOTOXICOLOGICAL STUDIES

5.1. **Effects on organisms**

5.1.1. Acute toxicity for fish
LC_{50} .... (ppm)
Duration of exposure ....
Species selected (one or more) ....

5.1.2. Acute toxicity for daphnia
LC_{50} .... (ppm)
Duration of exposure ....

5.2. **Degradation**
– biotic
– abiotic
The BOD and the BOD/COD ratio should be determined as a minimum

6. POSSIBILITY OF RENDERING THE SUBSTANCE HARMLESS

6.1. **For industry/skilled trades**

6.1.1. Possibility of recovery ....

6.1.2. Possibility of neutralization ....

6.1.3. Possibility of destruction:
– controlled discharge ....
– incineration ....
– **water purification station** ....
– others ....

6.2. **For the public at large**

6.2.1. Possibility of recovery ....

6.2.2. Possibility of neutralization ....

6.2.3. Possibility of destruction:
– controlled discharge ....
– incineration ....
– **water purification station** ....
– others ....

Appendix 7:   Criteria for classification as very toxic, toxic, or harmful

Substances shall be classified as "very toxic", "toxic" or "harmful" in accordance with the following criteria:—

    (a) classification as very toxic, toxic or harmful shall be effected by determining the acute toxicity of the commercial substance in animals, expressed in LD_{50} or LC_{50} values with the following parameters being taken as reference values:

Category	LD_{50} absorbed orally in rat mg/kg	LD_{50} percutaneous absorption in rat or rabbit mg/kg	LC_{50} absorbed by inhalation in rat mg/litre four hours
Very toxic	≤25	≤50	≤0·5
Toxic	>25 to 200	>50 to 400	>0·5 to 2
Harmful	>200 to 2000	>400 to 2000	>2 to 20

    (b) if facts show that for the purposes of classification it is inadvisable to use the LD_{50} or LC_{50} values as a principal basis because the substances produce other effects, the substances shall be classified according to the magnitude of these effects.

Appendix 8:   Level 1 requirements

LEVEL 1

Taking into account:

    – current knowledge of the substance,

    – known and planned uses,

    – the results of the tests carried out in the context of the base set,

the competent authority may require the following additional studies where the quantity of a substance placed on the market by a notifier reaches a level of 10 tonnes per year or a total of 50 tonnes and if the conditions specified after each of the tests are fulfilled in the case of that substance.

Toxicological studies

    – Fertility study (one species, one generation, male and female, most appropriate route of administration)

    If there are equivocal findings in the first generation, study of a second generation is required.

    It is also possible in this study to obtain evidence on teratogenicity.

    If there are indications of teratogenicity, full evaluation of teratogenic potential may require a study in a second species.

    – Teratology study (one species, most appropriate route of administration)

    This study is required if teratogenicity has not been examined or evaluated in the preceding fertility study.

    – Sub-chronic and/or chronic toxicity study, including special studies (one species, male and female, most appropriate route of administration).

    If the results of the sub-acute study in Schedule 1 or other relevant information demonstrate the need for further investigation, this may take the form of a more detailed examination of certain effects, or more prolonged exposure, e.g. 90 days or longer (even up to two years).

    The effects which would indicate the need for such a study could include for example:

    (a) serious or irreversible lesions;

    (b) a very low or absence of a 'no effect' level;

    (c) a clear relationship in chemical structure between the substance being studied and other substances which have been proved dangerous.

– Additional mutagenesis studies (including screening for carcinogenesis)

A. If results of the mutagenesis tests are negative, a test to verify mutagenesis and a test to verify carcinogenesis screening are obligatory.

    If the results of the mutagenesis verification test are also negative, further mutagenesis tests are not necessary at this level; if the results are positive, further mutagenesis tests are to be carried out (see B).

    If the results of the carcinogenesis screening verification test are also negative, further carcinogenesis screening verification tests are not necessary at this level; if the results are positive, further carcinogenesis screening verification tests are to be carried out (see B).

B. If the results of the mutagenesis tests are positive (a single positive test means positive), at least two verification tests are necessary at this level. Both mutagenesis tests and carcinogenesis screening tests should be considered here. A positive result of a carcinogenesis screening test should lead to a carcinogenesis study at this level.

Ecotoxicology studies

– An algal test; one species, growth inhibition test.

– Prolonged toxicity study with Daphnia magna (21 days, this study should also include determination of the 'no-effect level' for reproduction and the 'no-effect level' for lethality).

    The conditions under which this test is carried out shall be determined in accordance with the procedure described in Article 21 of the Directive.

– Test on a higher plant.

– Test on an earthworm.

– Prolonged toxicity study with fish (e.g. Oryzias, Jordanella, etc; at least a period of 14 days; this study should also include determination of the 'threshold level').

    The conditions under which this test is carried out shall be determined in accordance with the procedure described in Article 21 of the Directive.

– Tests for species accumulation; one species, preferably fish (e.g. Poecilla reticulata).

– Prolonged biodegradation study, if sufficient (bio)degradation has not been proved by the studies laid down in Schedule 1, another test (dynamic) shall be chosen with lower concentrations and with a different inoculum (e.g. flow-through system).

In any case, the notifier shall inform the competent authority if the quantity of a substance placed on the market reaches a level of 100 tonnes per year or a total of 500 tonnes.

On receipt of such notification and if the requisite conditions are fulfilled, the competent authority, within a time limit it will determine, shall require the above tests to be carried out unless in any particular case an alternative scientific study would be preferable.

Appendix 9: Level 2 requirements

LEVEL 2

If the quantity of a substance placed on the market by a notifier reaches 1000 tonnes per year or a total of 5000 tonnes, the notifier shall inform the competent authority. The latter shall then draw up a programme of tests to be carried out by the notifier in order to enable the competent authority to evaluate the risks of the substance for man and the environment.

The test programme shall cover the following aspects unless there are strong reasons to the contrary, supported by evidence, that it should not be followed:

- chronic toxicity study,
- carcinogenicity study,
- fertility study (e.g. three-generation study); only if an effect on fertility has been established at level 1,
- teratology study (non-rodent species) study to verify teratology study at level 1 and experiment additional to the level 1 study, if effects on embryos/foetuses have been established,
- acute and sub-acute toxicity study on second species: only if results of level 1 studies indicate a need for this. Also results of biotransformation studies and studies on pharmacokinetics may lead to such studies,
- additional toxicokinetic studies.

Ecotoxicology

- Additional tests for accumulation, degradation and mobility.

  The purpose of this study should be to determine any accumulation in the food chain.

  For further bioaccumulation studies special attention should be paid to the solubility of the substance in water and to its n-octanol/water partition coefficient.
  The results of the level 1 accumulation study and the physiochemical properties may lead to a large-scale flow-through test.

- Prolonged toxicity study with fish (including reproduction).

- Additional toxicity study (acute and sub-acute) with birds (e.g. quails): if accumulation factor is greater than 100.

- Additional toxicity study with other organisms (if this proves necessary).

- Absorption–desorption study where the substance is not particularly degradable.

12

The European Community Chemicals Notification Scheme: Human Health Aspects

By H.P.A. Illing

HEALTH AND SAFETY EXECUTIVE, MAGDALEN HOUSE, STANLEY PRECINCT, BOOTLE L20 3QZ, UK

1. Introduction

Under the European Community (EEC) '6th Amendment' Directive,[1] (in the United Kingdom the 'Notification of New Substances Regulations'[2]), data on the physical chemistry, toxicity, and ecotoxicity as well as the chemical identity and proposed use(s) of a new substance must be submitted to a 'Competent Authority' (the relevant Governmental Organization in each state). The data are then assessed by that 'Competent Authority' on behalf of all states in the Community. The process is outlined in Figure 1.

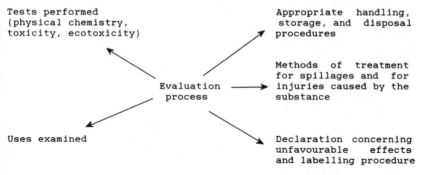

Figure 1 *Outline of assessment procedures for new substances*

Essentially the scheme is concerned with ensuring that Member States are alerted to the presence of a new substance* in the Community and that there is sufficient information on the hazardous properties of that substance to permit appropriate precautions to be taken for handling it, and to enable each Member State to take suitable control measures. In particular, the data are used to provide classification and labelling in respect of any dangerous properties. The classification and labelling are then accepted throughout the Community. This is part of the process of harmonization for labelling of all 'Dangerous Substances'.[3,4]

*A full examination of what constitutes a new substance is given in Chapter 11.

2. What Testing and When?

The amount of testing required for a new substance depends on the amount of the substance being placed on the market, its uses, and those toxicological properties already established.

If only small amounts of a new substance are to be marketed (<1 tonne year^{-1}) or a substance is being developed commercially, a separate procedure (involving a 'Limited Announcement') is employed. This 'Limited Announcement' procedure has not been harmonized. In the United Kingdom, the amount of toxicity data required varies according to the nature and use of the substance.

For substances subjected to a 'Full Notification' there are three major levels of use (*i.e.* amounts placed on the market). These are shown in Table 1. They are used to help determine the amount of toxicity testing required. The tests required are noted in Annexes VII and VIII of Directive 79/831/EEC[1] (see also Chapter 11). The 'Base Set' includes data on chemical identity, physical chemistry, use, toxicity, and ecotoxicity, and is designed to provide sufficient information on a substance for an initial assessment of its hazardous properties. It may also help in the selection of appropriate further tests. The toxicity tests are given in Table 2.

'Level 1' testing is considered in two parts (see Chapter 11). At the lower level a Competent Authority may require further testing and at the higher level they must (unless there are good scientific or other reasons for not doing the additional tests) ask for them. The toxicological studies which may be needed include a one-species, one-generation study on male and female fertility, a teratology study in one species, a sub-chronic (90 day) toxicity study, and additional genotoxicity studies. If there is a special need, other tests can be asked for. Current knowledge on existing and planned use, as well as the results of the 'Base Set' (and any other) tests will be taken into account when planning these studies.

Unless there are strong reasons for the contrary, the toxicity testing programme at Level 2 is substantial. It may include chronic toxicity and carcinogenicity studies, further studies on fertility and teratogenicity, and acute or sub-acute studies in additional species. At this stage, toxicokinetic studies are considered although they could be relevant at a much earlier stage. The need for studies not performed earlier may be reviewed by the Competent Authority at this stage.

Table 1

Levels set for examining data on new substances

Amount placed on the EC market (tonnes)		Level of testing required
p.a.	*Total*	
< 1	-	Limited announcement
1-< 10	< 50	Base Set
10-< 100	< 500	Level 1 (may)
100-<1000	<5000	Level 1 (must)
>1000	>5000	Level 2

Table 2

'Base Set' tests for toxicity

Test for	Comment
Acute toxicity	Oral and one other relevant route (for gases, inhalation). Normally conducted in rat; rabbit is an alternative for dermal studies.
Skin and eye irritancy	Normally conducted in rabbit.
Skin sensitization	Guinea pig maximization test preferred, but other adjuvant and non-adjuvant tests, described in the Directive, are acceptable.
Mutagenicity	At least one bacterial test (for point mutations) and one non-bacterial test using a different end point (*e.g.* micronucleus test, *in vitro* or *in vivo* cytogenetics).
Sub-acute toxicity	A 28 day study using an appropriate route after considering possible routes of human exposure. Normally conducted in rat; rabbit is an acceptable alternative for dermal studies.

Other testing, notably in the field of ecotoxicity, is required at all of these levels. The ecotoxicological testing requirements are given in Chapter 11, together with a more detailed explanation of the trigger levels for testing.

Once the 'Base Set' dossier has been prepared, there is considerable scope for discussion on what tests are actually required and when. On occasion it may be more useful (and less expensive in the long run) to start with a more sophisticated test than immediately necessary, especially if the substance is likely to require further testing soon afterwards. For example, it may be worthwhile to conduct a 90 day sub-acute toxicity study in place of the 28 day study normally performed as part of the Base Set.

Comparisons of the toxicity (or, indeed, of any property) of substances and, in consequence, their hazards, are much easier if standard test procedures are used. For this reason, amongst others, approved test methods have been published.[5-9] These test guidelines are in close agreement with those of the OECD.[10] They should be followed where possible, as Competent Authorities can refuse to accept a notification containing tests performed using non-standard protocols. If, however, another procedure is more relevant to a particular substance and gives more and/or better information (*i.e.* its use is scientifically justified), then the Competent Authority will consider the justification and the results obtained. Normally, it would be useful to 'sound out' a Competent Authority and discuss the merits of using an unusual test before carrying it out.

The EC Directive requires that all tests should be carried out in compliance with the principles of 'Good Laboratory Practice'.[1],[11] The purpose behind these principles is that Regulatory Authorities can be confident that the data are genuine, that they were generated using qualified, trained personnel and satisfactory techniques, and that the report accurately represents the data. This should be true irrespective of where (either within or outside the Community) the tests were performed.

All toxicity data must be obtained as humanely as possible. Nobody is absolved from the need to consider the welfare of animals. International guidelines should reduce the necessity for repeating studies using slightly different protocols, but these can only cover the generalities. When considering whether a specific test is required on a particular new substance, it is essential to consider also whether the test will cause unnecessary suffering. Any test involving animals should be performed only if, after taking into account developments in animal welfare and the need to minimize the number of animals used in toxicity testing, it is justified. This is particularly important in areas of toxicity where single 'limit dose' studies are alternatives to the more usual requirement for several dose levels.

3. Assessing the Result: What Next?

Generally, the results of toxicity tests are intended to lead to an assessment of the hazard to man; they are also relevant to hazard in the environment. The argument that the toxic effects of a substance which is acutely toxic to rat, and must be presumed so for man, is then presumed non-toxic to horses, pigs, voles, *etc.* is untenable. Tests must be considered as a whole as well as individually. Physico-chemical factors, such as structure, lipid solubility, and n-octanol/water partition are relevant to toxicity assessment. All the information available on a substance is relevant for the next stage, the assessment of possible hazards posed by the substance and how they should be handled.

In addition to decisions on classification and labelling (considered in the next section), results should lead to a consideration of methods for storage, handling, and disposal of the chemical and the appropriate treatment of accidental spillages and of injuries caused by the substance. All of these clearly depend on the nature of the hazard posed by the substance and its likely uses. Hence, there is a need for a 'Declaration concerning unfavourable effects of the substance in terms of the various uses envisaged', and the proposals for classification and labelling and for handling the substance should be consistent with this declaration.

4. What Should Appear on the Label?*

If a substance is shown in toxicity tests to be 'dangerous' (see Table 3), as defined in article 2(2) of the Directive[1] or schedule 2 of the regulations,[2] then,

(i)   the packaging for a substance should be labelled with the symbol appropriate to the nature of the hazard.

(ii)   the appropriate risk (R-) phrase(s) must appear on the label.

*See also Chapter 23.

(iii)   the appropriate safety (S-) phrase(s) must also appear on the label.

A list of all current agreed risk and safety phrases is given in Appendix 1. The criteria for classification and labelling are in a Directive which has been reproduced by the Health and Safety Executive as an Approved Code of Practice.[12,13]

Some of the risk and safety phrases are derived solely from the physical or chemical properties of the substance, and criteria for their use will not be examined here. Criteria relevant to the assessment of acute toxicity data are relatively straightforward; more difficulty is encountered in interpreting data from repeated-dose studies, genotoxicology studies, and studies on reproductive effects.

Only a few Level 1 studies have been submitted to Competent Authorities so far; thus there is little shared experience at interpreting tests at Level 1 or Level 2.

5. Interpreting Individual Tests

5.1. Acute Toxicity.
A most important part of testing for acute toxicity is the acquisition of data on sub-lethal symptoms, and these are relevant to the 'Declaration concerning unfavourable effects'. However, classification of a substance is based principally on lethality. There are three levels of classification as dangerous: very toxic, toxic, and harmful, and there is very little scope for interpretation (Table 3). The risk and safety phrases then indicate the probable route of exposure.

Table 3

Criteria for classification of a substance as 'Dangerous' as a result of acute lethality

Classification	LD_{50}(mg kg^{-1}) Oral (rat)	LD_{50}(mg kg^{-1}) dermal (rabbit)	LC_{50}(mg l^{-1};4 h) inhalation (rat)
Very toxic*	25	50	0.5
Toxic	200	400	2
Harmful	2000	2000	20

*These are qualified as follows: 'if swallowed' for tests by the oral route; 'in contact with skin' for tests by the dermal route; or 'by inhalation'.

Other phrases which may follow from acute toxicity studies are 'Danger of very serious irreversible effects' and 'Possible risk of irreversible effects'. For these phrases to be used, there should be 'Strong evidence that irreversible damage, other than that due to carcinogenic, mutagenic, or teratogenic effects, is likely to be caused by a single exposure by an appropriate route', generally in the appropriate dose range (that for toxic and above and that for harmful, respectively). They are therefore likely mainly when damage to an organ is seen at sub-lethal doses in animals at autopsy after recovery from the acute symptoms. Examples might include irreversible hepatotoxicity or nephrotoxicity following a single dose of the substance. These phrases are unlikely to be frequently used.

5.2.   Corrosion and Irritancy.   Specific tests are used to
investigate skin and eye irritancy.   Skin corrosion can also be
assessed in these tests.

    Corrosion is defined as 'full thickness destruction of skin
tissue'.   The phrases 'causes severe burns' and 'causes burns' are
used if corrosion occurs in 3 minutes or 4 hours respectively, or
if such a result can be predicted.   When the corrosive properties
of a substance can be predicted, to do the test may cause
unnecessary suffering to animals.   In such cases, the experiments
are likely to be ethically unacceptable, and the phrases may be
applied without actually performing the test.

    Skin irritants are classified on the basis of scoring systems
for erythema/eschar formation and oedema (Table 4).   The
definition of an irritant is 'if, when applied to animal skin for
up to 4 hours, significant inflammation is caused which is present
for 24 hours or more after the end of the exposure period'.
'Significant inflammation' occurs when an average score of 2 or
higher for all scores at each reading at 24, 48, and 72 h are
recorded for all animals when four or more animals are examined,
or for two of three animals when only three animals are tested.
Thus the average values for each animal are calculated first and
either the worst two average values (of three) or the average of
all values (for four or more values) is then used.

Table 4

Criteria for grading skin and eye irritants

Test for	Response examined	Grading scores*
Skin	Erythema and eschar formation	0-4
	Oedema	0-4
Eye	Cornea	0-4
	Iris	0-2
	Conjunctiva - redness	0-3
	- chemosis	0-4

*Detailed descriptions for these gradings can be found in the
literature.[5,6]

    The factors assessed in eye irritation studies are given in
Table 4.   A similar system of score averaging is used; averages
for four or more animals or the worst two of three animals
determine the classification.   There are, however, two levels of
classification.   The more severe classification uses the phrase
'Risk of serious damage to eyes' and is for substances for which
'if, when applied to the eye of the animal severe ocular lesions
are caused and which are present 24 hours or more after
instillation of the test material'.   Corneal lesion scores of 3 or
more and iris lesion scores of 1.5 or more are considered severe.
The phrase 'irritating to eyes' is appropriate when 'significant
ocular lesions occur within 72 hours after exposure and persist
for at least 24 hours'.   Score values for this classification are:
cornea 2 or more, iris 1 or more, conjunctiva redness 2.5 or more,
conjunctiva chemosis 2 or more.   When eye irritancy, particularly
severe irritancy, can be predicted then it may not be necessary to
carry out the test.   If there is doubt, a test using a single
animal could be used to give preliminary information.

Scoring systems are designed in order to reduce subjectivity in assessments which depend on skilled judgement, so it is essential that trained and experienced staff perform them.

There is no specific test for the phrase 'irritant by inhalation'. Use of this phrase is most likely to be the result of practical experience in man.

5.3. Sensitization. Skin sensitization is examined in the Base Set tests. Methods available include those in which an adjuvant is employed and those without adjuvant. For assessment purposes, a response rate of 30% of animals tested is required for a positive result when adjuvant is used; 15% is sufficient in non-adjuvant tests. A positive result leads to the phrase 'may cause sensitization by skin contact'.

The risk phrases 'may cause sensitization by skin contact' and 'may cause sensitization by inhalation' can be applied following practical experience in man. There is no suitable agreed animal test for the latter phrase; thus use of this phrase will be as a result of data obtained in man.

5.4. Repeated Dose Toxicity. As a 28 day repeated dose toxicity study is normally expected in a Base Set notification, there may be evidence relevant to the use of the risk phrase 'Danger of serious damage to health on prolonged exposure'. 'Serious damage' is clear functional disturbance or morphological change which has toxicological significance. If the phrase is to be used the levels at which these effects are seen are based on 90 day studies, but they also serve as a guide when other studies are being considered. These levels are: oral, 50 mg kg^{-1} body weight (b.w.) day^{-1}; dermal, 100 mg kg^{-1} b.w. day^{-1}; inhalation, 0.6 mg l^{-1}, 6 h day^{-1}. This definition allows considerable room for interpretation and it is discussed further in the next section.

The phrase 'Danger of cumulative effects' may also be worth considering when assessing sub-acute studies. The phrase is for substances which are likely to accumulate in the body and which cause some concern as a result of that accumulation.

5.5. Mutagenicity, Carcinogenicity, and Teratogenicity. Although the Directive refers to mutagenicity (including carcinogenicity pre-screening tests),[1] the tests included under this heading address mutagenicity ('point' or 'gene location' mutations), clastogenicity (occurrence of chromosomal (or chromatidal) breaks resulting in gain, loss, or rearrangement of pieces of the chromosome) and potentially aneuploidy (gain or loss of one or more chromosome).[14] Thus the term genotoxicity might be more appropriate.[14]

The definitions used for carcinogenicity, mutagenicity, and teratogenicity are given in Table 5. Generally they will be inappropriate for Base Set data. Any positive mutagenicity results will need to be considered carefully and this may result in a decision to ask for further testing. Possible considerations when assessing mutagenesis are examined in the next section.

Table 5

Criteria for classification as carcinogenic, mutagenic, or teratogenic

Effect and category	*Definition*	*Classification and risk phrases*
Carcinogenicity I	Substances known to be carcinogenic to man. There is sufficient evidence to establish a causal association between human exposure to a substance and the development of cancer.	Toxic May cause cancer
Carcinogenicity II	Substances which should be regarded as if they are carcinogenic to man. There is sufficient evidence to provide a strong presumption that human exposure to a substance may result in the development of cancer, generally on the basis of: appropriate long-term animal studies other relevant information.	Toxic May cause cancer
Carcinogenicity III	Substances which cause concern for man owing to possible carcinogenic effects but in respect of which the available information is not adequate for making a satisfactory assessment. There is some evidence from appropriate animal studies, but this is insufficient to place the substance in Category II.	Harmful Danger of irreversible effects
Mutagenicity I	Substances known to be mutagenic to man. There is sufficient evidence to establish a causal association between human exposure to a substance and heritable genetic damage.	Toxic May cause heritable genetic damage

Mutagenicity II

Substances which should be regarded as if they are mutagenic to man. There is sufficient evidence to provide a strong presumption that human exposure to the substance may result in the development of heritable genetic damage, generally on the basis of:

    appropriate animal studies
    other relevant information

Harmful
May cause heritable genetic damage

Mutagenicity III

Substances which cause concern for man owing to possible mutagenic effects but in respect of which the available information does not satisfactorily demonstrate heritable genetic damage. There is evidence from appropriate mutagenicity studies, but this is insufficient to place the substance in Category II.

Harmful
Possible risk of irreversible effects

Teratogenicity I

Substances which cause concern for man owing to possible mutagenic effects but in respect of which the available information does not satisfactorily demonstrate heritable genetic damage. There is evidence from appropriate mutagenicity studies, but this is insufficient to place the substance in Category II.

Toxic
May cause birth defects

Teratogenicity II

Substances which should be regarded as if they are teratogenic to man. There is sufficient evidence to provide a strong presumption that human exposure to the substance may result in non-heritable birth defects in offspring, generally on the basis of:

    appropriate animal studies
    other relevant information.

Harmful
May cause birth defects

6. Problem Areas

There are several important questions which need to be kept in mind when assessing data on toxicity. These include 'was this the right study?', 'was the study properly conducted?',* and 'were the results adequately interpreted?'. Where there is room for several opinions, different possible interpretations can emerge. Some of the areas where this may occur are discussed in this section. The opinions given are the author's personal views.

The potential problems are divided into two groups, those associated with toxicity testing and those associated with genotoxicity testing.

6.1.   Toxicity Testing. Dosing. In both acute and sub-acute studies at Base Set level there is scope for choice in respect of the route by which the substance will be administered. The advice on choice of route in the Directive uses the phrases 'will depend on the intended use and the physical properties of the substance' for acute studies and 'the most appropriate route having regard to the intended use, the acute toxicity and the physical and chemical properties of the substance' in sub-acute studies.[1,2] In the sub-acute studies there is also the choice of 5 day/week or 7 day/week dosing.

With gases and volatile liquids, the inhalation route is clearly useful, as it is likely to be a major potential route of exposure in man. Choosing between the oral route and the dermal route of exposure for solids and involatile liquids is less easy. Generally, exposure by either route is possible under conditions of poor industrial hygiene. However, dosing by the oral route is somewhat easier to perform, the dose received is better controlled, and absorption is maximized. When local irritancy occurs in dermal studies without systemic toxicity, oral studies may yield more information, either by confirming the lack of systemic effects or by giving an indication of what systemic effects may occur.

In short-term sub-acute studies, gavage dosing (or its equivalent in large animals) is preferable. The test methods indicate that constant-volume dosing is preferred, and thus a vehicle will be needed. The concentrations used at the high dose (*e.g.* 25% for the 5 g kg^{-1} top dose in an acute study) will mean that solids will often have to be administered in suspension, either in aqueous vehicles or in vegetable oil, even though it is desirable to use a solution. Suspending agents, such as carboxymethylcellulose, polyethylene glycol, starch, or guar gum may therefore be needed for homogeneity.

Admixture in diet or drinking water may be easier for long-term studies. Palatability of the diet/water and stability of test substance in the medium need to be checked, otherwise the amount of the substance actually administered may be very different to that intended.

---

* In this sense, a properly conducted study would be in accordance with GLP. However, a poorly conducted study, perhaps using a flawed protocol, might also be thought to comply with GLP, which is essentially concerned with staffing, facilities, and documentation, rather than the scientific aspects of a study.

If analytical confirmation of the amount dosed is required, then some method development is worthwhile in order to ensure that the data generated are useful. Knowledge of the specificity, sensitivity, precision, and accuracy[15] of the analytical procedure is desirable. For example, in an imprecise analytical procedure where the coefficient of variance is 25% at the level of substance being measured, could an analytical value of 1500 mg kg^{-1} for the dose solution actually mean that the predicted 2000 mg kg^{-1} had been administered? In this case the method is too imprecise to be worthwhile but, more often, basic quality control data to help decide this question are absent. Much fuller discussion of how to obtain such data is given elsewhere.[15]

<u>Assessment and interpretation of data in sub-acute studies</u>. The 28 day study in the Base Set is a very limited study. As the number of animals in each dose group is small (five/sex/dose level), relying on statistics alone in interpreting results can be misleading. A total of over 20 haematological and biochemical measurements are obtained in the standard study; thus some positive statistical significances are to be expected when a *p* value of 0.05 is used. Statistical significances of no biological importance may occur because particular groups had low standard deviations even though the means for both groups lie within normal variation; *i.e.* the power of the test can, on occasion, be too high.

When there are only five animals in each group (or fewer if animals are lost owing to dosing errors, *etc.*) one or two extreme values can lead to the opposite problem, statistical non-significance because of a high standard deviation. As Confidence Intervals are wide, the statistical test is not powerful enough and false negatives occur. When this occurs, it is often worthwhile to look for clustering of data. If several parameters which may be affected by a particular toxicity are all altered simultaneously in the group of animals or in the individuals showing extreme values, then it is likely that a biologically important effect is occurring.

A 'no-effect level' is asked for in all sub-acute studies. Zbinden called the no-effect level an 'old bone of contention';[16] on one occasion three separate groups of assessors came to three different conclusions as to what was the no-effect level and only three dose levels were studied! This lack of consensus may have arisen partly because there were no common criteria for judging what is a no-effect level and partly as a result of a feeling that if an effect is present at a certain level, its presence automatically triggers the phrase 'serious damage to health on prolonged exposure.' Both Zbinden[16] and the Royal Society[17] have pointed out that the no-effect level depends on the parameters measured in a study, *i.e.* it is study- and parameter-specific. In a presentation at an international colloquium on this subject, it was suggested that 'the exposure level at which no adverse effects are detected in an animal experiment is often as little as one-tenth that of the nearest higher dose at which minimal adverse effects, or effects of doubtful toxicological significance, are found.[18] Serious damage (clear functional disturbance or morphological change which has toxicological significance) is not the same statement. Thus there can be distinct gaps between the claimed no-effect level, the level at which mild, possibly transient but biologically significant effects were seen, and the level at which serious damage was found (and labelling is needed).

An effect may be a toxicologically significant preliminary indication that serious damage will occur following higher or more prolonged dosing, and hence potentially useful in the biological monitoring of human exposure. Example 1 is a summary of such a study.

Example 1

Summary of a hypothetical 28 day study illustrating the problem of defining a no-effect level and the level at which severe damage was seen. The substance would not be considered dangerous

4.2.1. Sub-acute toxicity (28 days)

TEST ANIMAL: Rat, Sprague-Dawley

NUMBER/SEX: 5 males and 5 females at each dose level

DOSE LEVELS: 0, 50, 100, and 1000 mg kg^{-1} b.w.

NATURE OF DOSE MATERIAL, ROUTE, FREQUENCY:

VEHICLE: Water

DOSING REGIME: Aqueous solution administered orally, by
                        gavage, 5 days a week.

NOTE: One male, from high-dose group, was killed *in
                        extremis*, due to misdosing.

SIGNS OF TOXICITY: There were no toxicologically important effects
                        on body weight or food consumption. Black
                        faeces were noted for all animals in the high-
                        dose group from day 16.

                        Mean corpuscular volume and mean cor-
                        puscular haemoglobin were slightly, but statis-
                        tically, significantly (p <0.05), increased in
                        high-dose males; other haematological
                        parameters were within normal limits.

                        The principal effect on clinical chemical
                        examination concerned M^{2+} levels. In plasma;
                        these were reduced (p <0.05) in high-dose
                        animals (males 36%, females 69%); in urine,
                        they were elevated (up to 95-fold) in a
                        dose-dependent manner with significant (p <
                        0.05) increases at 50 mg kg^{-1} day^{-1} or above
                        (males) or 100 mg kg^{-1} day^{-1} or above
                        (females). Other effects were only seen in
                        high-dose animals; they included reduced
                        alkaline phosphatase activity (possibly due to
                        low M^{2+} levels) and elevated bilirubin levels
                        (possibly the result of increased iron
                        absorption) in males and slightly (males 1.7%,
                        females 3.2%) but significantly (p <0.05)
                        reduced chloride levels in serum.

EFFECTS IN ORGANS: Statistically significant (p <0.05) increases
                        (18%) in liver:body weight ratio occurred in
                        high-dose males and in liver weight (low 9%,

middle 7%, high 16%) and liver:body weight ratio (low 10%, middle 5%, high 19%), kidney weight (high-dose 8%) and kidney:body weight ratio (high-dose 9%) in females.

On histological examination, high haemosiderin levels occurred in macrophages or reticulo-endothelial cells of lung, liver, spleen, and large intestine of animals from the high-dose male and lung and large intestine of animals from the middle- and high-dose female groups.

Parakeratosis (oesophagus, tongue) in 2 males, enteritis (degenerative and inflammatory changes) in 1 male, atrophy of the thymus (1 male, 1 female) or spleen (1 male, 1 female), hyperplasia of the bone marrow (1 male) or lymph node (1 female), renal congestion (2 males) occurred only in high-dose animals. Most of these lesions were probably due to iron excess and/or M^{2+} deficiency.

DOSE SHOWING NO TOXIC EFFECT:
Less than 50 mg kg^{-1} day^{-1}

METHOD: Annex V

COMMENTS: Although effects were seen at all dose levels, 'clear functional disturbance or morphological change which has toxicological significance' only occurred at 1000 mg kg^{-1}/day^{-1}. Thus the substance should NOT be classified as 'Harmful' as a result of this test.

6.2. Genotoxicity Testing. There is general agreement amongst genetic toxicologists that a battery of mutagenicity assays is a useful tool for the preliminary screening of potential human mutagens and carcinogens.[19] False positives and false negatives will occur; thus an absolute minimum requirement is for two tests and this is reflected in the 'Base Set' testing. With such testing, separate test systems using different end points are essential. Thus the point mutation test in bacteria (*e.g.* the Ames test) should be followed up with a test in a non-bacterial system, preferably using an end point dependent on chromosomal damage.

Tests can be conducted *in vitro* or *in vivo*. *In vitro* tests allow an examination of potential genotoxicity of the parent substance. However, a metabolizing system, when present, may not provide the same spectrum of metabolites as *in vivo*, particularly if reductive metabolism occurs. Thus, possible mutagenic metabolites may not be detected. Positive results *in vitro* may also be of limited value. They may not be relevant if the substance (or its metabolites produced by microfloral metabolism) are not absorbed or do not reach an appropriate site of action.

There may also be disadvantages in using *in vivo* tests, some of which are very time- and animal-consuming. Usually effects are measured at only one site, for example the bone marrow. The site may not be relevant if the chemical or the active metabolite never reaches the critical concentration there, or is ineffective there because the appropriate interaction with endogenous material does not take place.

Problems of interpretation usually arise when there are mixed results. Negative results in satisfactorily conducted appropriate tests cause few labelling problems. Careful consideration is required if a study gives positive results. Further testing is then normally required, principally to establish whether the positive result is relevant to human health. A supplementary test using the same end point might indicate that the first test gave a false positive. An *in vivo* test may provide evidence that a positive result *in vitro* is not relevant, particularly if the substance is also shown to be poorly absorbed. Normally, a single positive result from an *in vitro* test is insufficient evidence on which to classify a substance as a category 3 mutagen or carcinogen.

7. Final Word

Any attempt to reduce the need to test substances several times with slightly different protocols for different authorities is welcome. The EEC '6th Amendment' scheme for assessing the hazards which arise from new substances represents one such effort at international commonality of assessment. Although there are bound to be initial difficulties, these will lessen as experience with the scheme increases.

8. Acknowledgements

The author thanks Mr. J. Vosser and Mr M. van den Heuvel for helpful discussions. However, the views expessed in this chapter are the author's own and are not necessarily those of any of the 'Competent Authorities'.

9. References

1. Directive 79/831/EEC; *Official Journal*, No. L259, 15.10.1979, p.10.

2. Statutory Instrument SI 1982/1496, HMSO, London, 1982.

3. Directive 67/548/EEC; *Official Journal*, No. L196, 16.8.1967, p.1.

4. Statutory Instrument SI 1984/1244, HMSO, London, 1984.

5. Directive 84/449/EEC; *Official Journal*, No. L251, 19.9.1984, p.1.

6. Annex V, EEC Directive 79/831, Part B. Toxicological Methods of Annex VIII (Draft July 1983).

7. Approved Code of Practice: COP 8, Methods for the determination of ecotoxicity properties, HMSO, London, 1982.

8. Approved Code of Practice: COP 9, Methods for the determination of physico-chemical properties, HMSO, London, 1982.

9. Approved Code of Practice: COP 10, Methods for the determination of toxicity, HMSO, London, 1982.

10. OECD Guidelines for the testing of chemicals, OECD, Paris, 1981.

11. Approved Code of Practice: COP 7, Principles of Good Laboratory Practice, HMSO, London, 1982.

12. Directive 83/467/EEC; *Official Journal*, No. L259, 16.9.1983.

13. Approved Code of Practice: Classification and Labelling of Substances dangerous for supply and/or conveyance by road, HMSO, London, 1982.

14. A. W. Hayes, Ed. 'Principles and Methods of Toxicology', Raven Press, New York, 1982.

15. J. Chamberlain, 'The Analysis of Drugs in Biological Fluids', CRC Press, Boca Raton, FL, 1985.

16. G. Zbinden, 'The No Effect Level – an Old Bone of Contention in Toxicology', *Arch. Toxicol.*, 1979, **43**, 3.

17. 'Risk Assessment: A study group report', Royal Society, London, 1983, p.62.

18. M. Sharratt, 'Uncertainties Associated with the Evaluation of Health Hazard of Environmental Chemicals from Toxicological Data', in 'The Evaluation of Toxicological Data for the Protection of Public Health', Proceedings of an International Colloquium, Luxembourg, 1976, p.113.

19. D. Anderson, 'Mutagenicity Testing', *Rev. Environ. Health*, 1981, 3, 369.

Appendix: List of Risk and Safety Phrases

Risk (R-) Phrases

1: Explosive when dry
2: Risk of explosion by shock, friction, fire, or other sources of ignition
3: Extreme risk of explosion by shock, friction, fire, or other sources of ignition
4: Forms very sensitive explosive metallic compounds
5: Heating may cause an explosion
6: Explosive with or without contact with air
7: May cause fire
8: Contact with combustible material may cause fire
9: Explosive when mixed with combustible material
10: Flammable
11: Highly flammable
12: Extremely flammable
13: Extremely flammable liquified gas
14: Reacts violently with water
15: Contact with water liberates highly flammable gases
16: Explosive when mixed with oxidizing substances
17: Spontaneously flammable in air
18: In use, may form flammable/exposure vapour-air mixture

19: May form explosive peroxides
20: Harmful by inhalation
21: Harmful in contact with skin
22: Harmful if swallowed
23: Toxic by inhalation
24: Toxic in contact with skin
25: Toxic if swallowed
26: Very toxic by inhalation
27: Very toxic in contact with skin
28: Very toxic if swallowed
29: Contact with water liberates toxic gas
30: Can become highly flammable in use
31: Contact with acids liberates toxic gas
32: Contact with acids liberates very toxic gas
33: Danger of cumulative efects
34: Causes burns
35: Causes severe burns
36: Irritating to eyes
37: Irritating to respiratory system
38: Irritating to skin
39: Danger of very serious irreversible effects
40: Possible risk of irreversible effects
41: Risk of serious damage to eyes
42: May cause sensitization by inhalation
43: May cause sensitization by skin contact
44: Risk of explosion if heated under confinement
45: May cause cancer
46: May cause heritable genetic damage
47: May cause birth defects
48: Danger of serious damage to health by prolonged exposure

Safety (S-) Phrases

1: Keep locked up
2: Keep out of reach of children
3: Keep in a cool place
4: Keep away from living quarters
5: Keep contents under ... (appropriate liquid to be specified by the manufacturer)
6: Keep under ... (inert gas to be specified by the manufacturer)
7: Keep container tightly closed
8: Keep container dry
9: Keep container in a well ventilated place
12: Do not keep the container sealed
13: Keep away from food, drink, and animal feeding stuffs
14: Keep away from ... (incompatible materials to be indicated by the manufacturer)
15: Keep away from heat
16: Keep away from sources of ignition - No Smoking
17: Keep away from combustible material
18: Handle and open container with care
20: When using do not eat or drink
21: When using do not smoke
22: Do not breathe dust
23: Do not breathe gas/fumes/vapour/spray (appropriate wording to be specified by the manufacturer)
24: Avoid contact with skin
25: Avoid contact with eyes
26: In case of contact with eyes, rinse immediately with plenty of water and seek medical advice
27: Take off immediately all contaminated clothing

28: After contact with skin, wash immediately with plenty of ...
   (to be specified by the manufacturer)
29: Do not empty into drains
30: Never add water to this product
33: Take precautionary measures against static discharges
34: Avoid shock and friction
35: This material and its container must be disposed of in a safe
   way
36: Wear suitable protective clothing
37: Wear suitable gloves
38: In case of insufficient ventilation, wear suitable respiratory
   equipment
39: Wear eye/face protection
40: To clean the floor and all objects contaminated by this
   material use ... (to be specified by the manufacturer)
41: In case of fire and/or explosion do not breathe fumes
42: During fumigation/spraying wear suitable respiratory equipment
   (appropriate wording to be specified)
43: In case of fire use ... (indicate in the space the precise
   type of fire-fighting equipment. If water increases the risk,
   add - Never use Water)
44: If you feel unwell, seek medical advice (show the label where
   possible)
45: In case of accident or if you feel unwell, seek medical advice
   immediately (show the label where possible)
46: If swallowed seek medical advice immediately and show this
   container or label
47: Keep at temperature not exceeding ...°C (to be specified by
   the manufacturer)
48: Keep wetted with ... (appropriate material to be specified by
   the manufacturer)
49: Keep only in the original container
50: Do not mix with ... (to be specified by the manufacturer)
51: Use only in well ventilated areas
52: Not recommended for interior use on large surface areas
53: Avoid exposure. Obtain special instructions before use

Combinations of some of these Risk- and Safety-Phrases have also
been agreed.

13

Identification of the Carcinogenic Risk of Chemicals to Humans

By H. Vainio and J. Wilbourn

WORLD HEALTH ORGANIZATION, INTERNATIONAL AGENCY FOR RESEARCH ON CANCER, 150 COURS ALBERT THOMAS, 69372-LYON CEDEX 08, FRANCE

1. Introduction

The identification of carcinogenic risk factors is an essential requisite for the primary prevention of cancer. The term 'risk' has been used differently in Europe and the USA. Throughout this paper, 'risk' refers only to qualitative risk or hazard* and not to quantitative risk*, which is an estimate of the probability of a hazard at a given exposure level and often includes statistical modelling for extrapolations from high to low doses.

During the 1970s, various scientists stated that 80-90% of all human cancers could be attributed to environmental factors.[1,2,3] These statements were often misinterpreted to mean that 80-90% of human cancers are caused by man-made synthetic chemicals. Recently, Doll and Peto[4] have provided estimates of the proportions of human cancers caused by environmental factors in one industrialized country (USA). The 'human environment', as defined by those authors, includes not only occupational, iatrogenic, and environmental exposures to chemicals, but also human behaviour patterns that are influenced by cultural habits (e.g. tobacco smoke, alcohol), diet, sexual habits, and other lifestyle factors (e.g. exposure to sunlight). Calculated proportions of cancer deaths attributed to tobacco and diet far exceed those due to occupation, drugs, and environmental pollution.

Some estimates of attributable risks are based upon the fact that there is a wide variation in the risk of cancer among different populations, and the degree of variation depends upon the cancer type. For example, there is a more than 100-fold difference in the incidence of some cancers, such as those of the lung and nasopharynx, between different populations[5] (see Table 1). For other cancers, e.g. nephroblastoma, there is only a two-fold difference between the highest and lowest rates recorded in 78 cancer registries throughout the world.

Strong additional evidence that environmental factors are responsible for differences in cancer morbidity and mortality comes from studies on migrants, which show that populations that move from their country of origin to another country develop the pattern of cancers common to that country within one or two generations. This has been shown clearly in Japanese who migrated to California: breast cancer mortality rates in American-born Japanese women are now similar to those of American whites, whereas breast cancer rates in Japan have increased only slightly in recent years.[6] The original Japanese migrants tended to

*See Glossary of Terms

Table 1

Highest and lowest standardized incidence rates (per 10^5 male population per year) for cancers at common sites, for all cancer registries in Cancer Incidence in Five Continents, Vol. IV (from Waterhouse et al.[5]

Site	Highest rate	Lowest rate
Lip	22.8	0.2
Tongue	13.7	0.5
Mouth	13.0	0.5
Nasopharynx	32.9	0.3
Oesophagus	24.7	1.1
Stomach	100.2	3.7
Colon	32.3	3.1
Rectum	22.6	1.5
Liver	34.4	0.6
Pancreas	18.3	1.0
Larynx	18.9	1.3
Bronchus, trachea	107.2	1.1
Melanoma of the skin	17.2	0.2
Prostate	100.2	0.8
Penis	5.7	0.2
Bladder	30.2	2.4
Brain, nervous system	10.1	1.0
Lymphosarcoma	7.5	1.2

maintain their previous lifestyles and habits, and thus changes in cancer rates did not occur immediately, but only in the first or second generation. If genetic determinants had been involved, very little change in cancer rates would have occurred. Stomach cancer rates, which remain high in Japan, have also decreased.

In 1968, the International Agency for Research on Cancer (IARC) received an urgent request from the scientific community and regulatory agencies to provide, for the primary prevention of cancer, lists of chemicals that are carcinogenic to humans and to experimental animals. This idea was quickly abandoned, however, since it was evident that the quality and quantity of data on individual chemicals varied considerably. It would have been relatively easy to list chemicals for which there were clear and unquestionable results, but it would have been highly misleading or impossible to include within the same list chemicals for which the quality of the available information was less reliable. Therefore, preparation of such lists was postponed and, instead, a programme was initiated to produce monographs on individual chemicals.

2. The IARC Monographs Programme

The programme, initiated in 1970, underwent considerable expansion from 1972, particularly with the financial support of the National Cancer Institute of the USA, and began to produce three volumes a year - a system maintained to the present. The objective of the programme has been to collect all available relevant epidemiological and experimental data on chemicals to which humans are known to be exposed and for which some evidence of carcinogenicity exists, to evaluate those data in terms of human risk with the help of international working groups of acknowledged experts in chemical carcinogenesis and related fields and in

epidemiology, to publish the conclusions of those groups as a
series of monographs, and to ensure their dissemination to health
authorities and to the scientific community. Several reviews of
the programme have been published previously by Agthe and
Tomatis,[7] Tomatis *et al.*,[8] and Tomatis.[9]

Each monograph comprises several sections: chemical and
physical characteristics; production, use, occurrence, and
analytical methods; biological data relevant to the evaluation of
the carcinogenic risk of the chemical, covering all available data
on carcinogenicity studies in animals, data on the metabolism of
the chemical in experimental animals and humans, the
carcinogenicity of the metabolites (if tested), essential data on
toxicity, data on mutagenicity; and observations in humans
covering epidemiological studies and case reports. The last
section of each monograph has summaries of the pertinent findings
on human exposures, carcinogenicity in experimental animals,
effects on reproduction, mutagenic and related effects, and case
reports and epidemiological studies in humans. Finally, the
degree of evidence of the carcinogenicity of the exposure to
humans and to experimental animals is evaluated.

More recently, the programme has been expanded to include
evaluations on exposures to complex mixtures, such as those that
occur in occupational settings (*e.g.* aluminium production, coke
production, leather tanning) or through cultural habits (*e.g.*
tobacco chewing, tobacco smoking). In these monographs, the
sections on chemistry, production, and use are modifed so that
historical perspectives, a description of the industry, or habit
and exposures can be fully summarized.

It is important to note that these criteria for evaluation
were not established at the beginning of the programme, but have
been built up by the various working groups whose deliberations
resulted in the first 35 volumes of the *IARC Monographs* together
with the reports of *ad hoc* groups convened to recommend criteria
for evaluation of the carcinogenic risks of chemicals.[10]

3. Working Procedures

Approximately one year in advance of a meeting of a working group,
a list of the substances or complex exposures to be considered is
prepared by IARC staff in consultation with other experts.
Subsequently, all relevant biological data are collected by IARC;
recognized sources of information on chemical carcinogenesis and
on-line systems such as CANCERLINE, MEDLINE, and TOXLINE (see also
Chapter 2) are used in conjunction with US Public Health Service
Publication No. 149. Representatives from industrial associations
may assist in the preparation of sections describing industrial
processes. Bibliographical sources for data on mutagenicity and
teratogenicity are the Environmental Mutagen Information Center
and the Environmental Teratology Information Center, both located
at the Oak Ridge National Laboratory, TN, USA.

The collection of data and the preparation of first drafts for
the sections on chemical and physical properties, on production
and use, on occurrence, and an analysis are currently carried out
by Tracor Jitco, Inc., and its subcontractor, Technical Resources,
Inc., both in Rockville, MD, USA, under a separate contract with
the US National Cancer Institute. Most of the data so obtained
refer to the USA and Japan; IARC attempts to supplement this

information with that from other sources in Europe.

Six months before the meeting, articles containing relevant biological data are sent to an expert(s), or are used by IARC staff, to prepare first drafts of the sections on biological effects. The complete drafts are then compiled by IARC staff and sent, before the meeting, to all participants of the Working Group for their comments.

The Working Group meets in Lyon for seven to eight days to discuss and finalize the texts of the monographs and to formulate the evaluations. After the meeting, the master copy of each monograph is verified by consulting the original literature, edited by a professional editor, and prepared for reproduction. The aim is to publish monographs within nine months of the Working Group meeting. Each volume of monographs is printed in 4000 copies for distribution to governments, regulatory agencies, and interested scientists.

With regard to biological data, only reports that have been published or accepted for publication are reviewed by the working groups, although a few exceptions have been made: in certain instances, reports from government agencies that have undergone peer review and are widely available are considered. The monographs do not cite all the literature on a particular chemical or complex exposure: only those data considered by the Working Group to be relevant to the evaluation of carcinogenic risk to humans are included.

From time to time there has been heavy pressure to accept for evaluation unpublished material made available *confidentially* to the IARC and its expert consultants. There are several reasons for not considering unpublished material in evaluations of carcinogenicity: (a) Only published material is universally available, so that any reader of the *Monographs* can check the conclusions of the Working Group against the original sources of information. (b) Published data are formally definitive, and their significance cannot be altered by subsequent publications (except, for example, when a subsequent publication showed that the carcinogenic behaviour of a compound was, in fact, due to an impurity); unpublished material often concerns on-going studies and may consist of progress reports giving results that are no longer valid once the study has been terminated. (c) Unpublished reports may be supplemented subsequently by additional experiments or experimental groups, thus altering both the design and/or the significance of the provisional results. Unpublished results may even accumulate until, and unless, they provide the desired information.

This does not imply that the publication of results makes them, or the study from which they are derived, automatically excellent. It is a commonly shared experience that many published results are obtained from experiments that are far from excellent and that some do not comply with basic requirements for accuracy. However, since all the published material on biological data that are used in the evaluation of carcinogenicity within the *Monographs* is analysed carefully and critically, studies that do not satisfy the basic requirements of accuracy can be, and are discarded.

One strong argument that has been raised in favour of using

unpublished material is that most experiments that result in negative findings remain unpublished. Scientific journals used not to be interested in the publication of such results; as a consequence, scientists were not stimulated to finalize results that seemed to be less exciting than those indicating a carcinogenic effect and were therefore unlikely to be suitable for publication. The situation today is radically different, however, and most journals now accept sound negative results for publication.

4. Criteria for the Evaluation of Data

4.1. Epidemiology. Evidence of carcinogenicity to humans can be derived from case reports, descriptive epidemiological studies, and analytical epidemiological studies.

An analytical study that shows a positive association between an exposure and a cancer may be interpreted as implying causality to a greater or lesser extent, on the basis of the following criteria: (a)   There is no identifiable positive bias. (By 'positive bias' is meant the operation of factors in study design or execution that lead erroneously to a more strongly positive association between an exposure and disease than in fact exists. Examples of positive bias include, in case-control studies, better documentation of the exposure for cases than for controls, and, in cohort studies, the use of better means of detecting cancer in exposed individuals than in individuals not exposed.)   (b) The possibility of positive confounding has been considered.   (An example of positive confounding is the association between coffee consumption and lung cancer, which results from their joint association with cigarette smoking.)   (c) The association is unlikely to be due to chance alone.   (d) The association is strong.   (e) There is a dose-response relationship.

In some instances, a single epidemiological study may be strongly indicative of a cause-effect relationship; however, the most convincing evidence of causality comes when several independent studies done under different circumstances result in 'positive' findings.

Analytical epidemiological studies that show no association between an exposure and a cancer ('negative' studies) should be interpreted according to criteria analogous to those listed above: (a) there is no identifiable negative bias; (b) the possibility of negative confounding has been considered; and (c) the possible effects of misclassification of exposure or outcome have been weighed.   In addition, it must be recognized that the probability that a given study can detect a certain effect is limited by its size.   This can be perceived from the confidence limits around the estimate of association or relative risk.   In a study regarded as 'negative', the upper confidence limit may indicate a relative risk substantially greater than unity; in that case, the study excludes only relative risks that are above the upper limit.   This usually means that a 'negative' study must be large to be convincing.   Confidence in a 'negative' result is increased when several independent studies carried out under different circumstances are in agreement.   Finally, a 'negative' study may be considered to be relevant only to dose levels within or below the range of those observed in the study and is pertinent only if sufficient time has elapsed since first human exposure to the agent.   Experience with human cancers of known etiology suggests that the period from first exposure to a chemical carcinogen to development of clinically observed cancer is usually measured in decades and may be in excess of 30 years.

The evidence for carcinogenicity from studies in humans is assessed by the Working Group and judged to fall into one of four groups, defined as follows:

1. <u>Sufficient evidence</u> of carcinogenicity indicates that there is a causal relationship between the exposure and human cancer.

2. <u>Limited evidence</u> of carcinogenicity indicates that a causal interpretation is credible, but that alternative explanations, such as chance, bias, or confounding, could not adequately be excluded.

3. <u>Inadequate evidence</u> of carcinogenicity, which applies to both positive and negative evidence, indicates that one of two conditions prevailed: (a) there are few pertinent data; or (b) the available studies, while showing evidence of association, do not exclude chance, bias, or confounding.

4. <u>No evidence</u> of carcinogenicity applies when several adequate studies are available which do not show evidence of carcinogenicity.

4.2. Animal Carcinogenicity Data.   In evaluating studies from experimental animals, both the interpretation and evaluation of a particular study as well as the overall assessment of the carcinogenic activity of a chemical (or complex mixture) involve several considerations of qualitative importance, including: (a) the experimental parameters under which the chemical was tested, including route of administration and exposure, species, strain, sex, age, *etc.*; (b) the consistency with which the chemical has been shown to be carcinogenic, *e.g.* in how many species and at which target organ(s); (c) the spectrum of neoplastic response, from benign neoplasms to multiple malignant tumours; (d) the stage of tumour formation in which a chemical may be involved: some chemicals act as complete carcinogens and have initiating and promoting activity, while others may have mainly promoting activity; and (e) the possible role of modifying factors.

Many chemicals induce both benign and malignant tumours. Among chemicals that have been studied extensively, there are few instances in which the neoplasms induced are only benign. Benign tumours may represent a stage in the evolution of a malignant neoplasm or they may be 'end-points' that do not readily undergo transition to malignancy.

Dose-response studies are important in the evaluation of carcinogenesis: the confidence with which a carcinogenic effect can be established is strengthened by the observation of an increasing incidence of neoplasms with increasing exposure.

The evidence of carcinogenicity in experimental animals is assessed by the Working Group and judged to fall into one of four groups, defined as follows:

1. <u>Sufficient evidence</u> of carcinogenicity is provided when there is an increased incidence of malignant tumours: (a) in multiple species or strains; or (b) in multiple experiments (preferably with different routes of administration or using different dose levels); or (c) to an unusual degree with regard to incidence, site, or type of tumour, or age at onset. Additional evidence may be provided by data on dose-response effects.

2. <u>Limited evidence</u> of carcinogenicity is available when the data suggest a carcinogenic effect but are limited because: (a) the studies involve a single species, strain, or experiment; or (b) the experiments are restricted by inadequate dosage levels, inadequate duration of exposure to the agent, inadequate period of follow-up, poor survival, too few animals, or inadequate reporting; or (c) the neoplasms produced often occur spontaneously and, in the past, have been difficult to classify as malignant by histological criteria alone (*e.g.* lung adenomas and adenocarcinomas and liver tumours in certain strains of mice).

3. <u>Inadequate evidence</u> of carcinogenicity is available when, because of major qualitative or quantitative limitations, the studies cannot be interpreted as showing either the presence or absence of a carcinogenic effect.

4. <u>No evidence</u> of carcinogenicity applies when several adequate studies are available which show that, within the limits of the tests used, the chemical or complex mixture is not carcinogenic.

It should be noted that the categories *sufficient evidence* and *limited evidence* refer only to the strength of the experimental evidence that these chemicals or complex mixtures are carcinogenic and not to the extent of their carcinogenic activity nor to the mechanism involved. The classification of any chemical may change as new information becomes available.

<u>4.3. Mutagenicity and Other Short-term Test Data</u>. Most established short-term tests for mutagenicity employ as end-points well defined genetic markers in prokaryotes and lower eukaryotes and in mammalian cell lines. The tests can be grouped according to the end-point detected:

1. Tests of *DNA damage*. These include tests for covalent binding to DNA, induction of DNA breakage or repair, induction of prophage in bacteria, and differential survival of DNA repair-proficient/-deficient strains of bacteria.

2. Tests of *mutation* (measurement of heritable alterations in phenotype and/or genotype). These include tests for detection of the loss of alteration of a gene product, and change of function through forward or reverse mutation, recombination and gene conversion; they may involve the nuclear genome, the mitochrondrial genome, and resident viral or plasmid genomes.

3. Tests of *chromosomal effects*. These include tests for detection of changes in chromosome number (aneuploidy), structural chromosomal aberrations, sister chromatid exchanges, micronuclei, and dominant-lethal events. This

classification does not imply that some chromosomal effects are not mutational events.

4. Tests for *cell transformation*, which monitor the production of preneoplastic or neoplastic cells in culture, are also of importance because they attempt to simulate essential steps in cellular carcinogenesis. These assays are not grouped with those listed above since the mechanisms by which chemicals induce cell transformation may not necessarily be the result of genetic change.

The data from short-term tests are summarized by the Working Group and the test results tabulated as positive, negative, or equivocal according to the end-points detected and the biological complexities of the test systems (see the example shown in Table 2).

Table 2

Overall assessment of data from short-term tests: ethylene oxide

| | Genetic activity | | | Cell |
	DNA damage	Mutation	Chromosomal effects	transformation
Prokaryotes	+	+		
Fungi/green plants		+	+	
Insects		+	+	
Mammalian cells (*in vitro*)	+	+	+	
Mammals (*In vivo*)	+		+	
Humans (*in vivo*)			+	

Degree of evidence in short-term tests
for genetic activity: *Sufficient*   No data

An overall assessment of the evidence for *genetic activity* is then made on the basis of the entries in the Table, and the evidence is judged to fall into one of four categories, defined as follows:-

1. Sufficient evidence is provided by at least three positive entries, one of which must involve mammalian cells *in vitro* or *in vivo* and which must include at least two of three end-points - DNA damage, mutation, and chromosomal effects.

2. Limited evidence is provided by at least two positive entries.

3. Inadequate evidence is available when there is only one positive entry or when there are too few data to permit an evaluation of an absence of genetic activity, or when there are unexplained, inconsistent findings in different test systems.

4. No evidence applies when there are only negative entries;
    these must include entries for at least two end-points and two
    levels of biological complexity, one of which must involve
    mammalian cells *in vitro* or *in vivo*.

It is emphasized that the above definitions are operational,
and that the assignment of a chemical or complex mixture into one
of these categories is thus arbitrary.

In general, emphasis is placed on positive results; however,
in view of the limitations of current knowledge about mechanisms
of carcinogenesis, certain cautions should be respected. (a) At
present, short-term tests should not be used by themselves to
conclude whether or not an agent is carcinogenic, nor can they
predict reliably the relative potencies of compounds as
carcinogens in intact animals. (b) Since the currently available
tests do not detect all classes of agents that are active in the
carcinogenic process (*e.g.* hormones), one must be cautious in
utilizing these tests as the sole criterion for setting priorities
in carcinogenesis research and in selecting compounds for animal
bioassays. (c) Negative results from short-term tests cannot be
considered as evidence to rule out carcinogenicity, nor does lack
of demonstrable genetic activity attribute an epigenetic or any
other property to a substance.

5. Supplements to the IARC Monographs

Ten years after the initiation of the programme an attempt was
made to respond to the initial request to produce lists of
carcinogens. This resulted in Supplement 1 to the *Monographs*.
The exercise was repeated in February 1982, when a Working Group
was convened to re-evaluate the evidence of carcinogenicity to
humans of 155 chemicals, groups of chemicals, or industrial
exposures that had ben considered in the first 29 volumes of *IARC
Monographs* for which some data on carcinogenicity in humans had
been reported in the published literature. Re-evaluations were
made on the basis of studies summarized in the monographs and of
all relevant data published subsequently. The results of this
meeting were published as Supplement 4 to the *IARC Monographs*.[11]

Descriptive evaluations of the available epidemiological,
animal carcinogenicity, and short-term test data were prepared and
the degrees of evidence of carcinogenicity to humans and to
animals and for activity in short-term tests were decided upon.
An overall evaluation of the carcinogenic risk to humans was then
made according to the criteria given in Table 3.

In addition, an Appendix was prepared listing all chemicals
evaluated in the first 29 volumes of *IARC Monographs* for which
there is considered to be *sufficient evidence* of carcinogenicity
in experimental animals.

Table 3

Evaluation of carcinogenicity to humans

Group	Evidence
Group 1, exposure carcinogenic	Evidence of carcinogenicity to humans from epidemiological studies is *sufficient*.
Group 2, exposure probably carcinogenic to humans	Evidence of carcinogenicity to humans from epidemiological studies ranges from limited to inadequate. To reflect this range, the category is subdivided into higher (Group 2A) and lower (Group 2B) degrees of evidence. Data from studies in experimental animals played an important role in assigning compounds to Category 2, particularly to Group 2B; thus, the combination of *sufficient evidence* in animals and *inadequate evidence* in humans usually resulted in a classification of 2B. In a very few cases, the known chemical properties of the compounds and the results from short-term tests allowed the classification of compounds to be upgraded (from 3 to 2B or from 2B to 2A).
Group 3, exposure cannot be classified as to its carcinogenicity to humans	Evidence of carcinogenicity to humans and to experimental animals is inadequate to make an evaluation.

For many of these chemicals (or complex mixtures) for which there is *sufficient evidence* of carcinogenicity in animals, data relating to carcinogenicity for humans are non-existent. In the absence of adequate data on humans, it is reasonable, for practical purposes, to regard chemicals for which there is *sufficient evidence* of carcinogenicity in animals as if they presented a carcinogenic risk to humans. The use of the expressions 'for practical purposes' and 'as if they presented a carcinogenic risk' indicates that, at the present time, a correlation between carcinogenicity in animals and possible human risk cannot be made on a purely scientific basis, but only pragmatically. Such a pragmatic correlation may be useful to regulatory agencies in making decisions related to the primary prevention of cancer.

6. Results to Date

By the end of 1985, 40 volumes of *IARC Monographs* had been published or were in press. Evaluations had been made of the carcinogenicity of over 700 chemicals, groups of chemicals, industrial processes, and exposures to complex mixtures. Since one of the criteria for inclusion of a chemical or an exposure in the *Monographs* is that there is some evidence of suspicion that the exposure is carcinogenic to humans and/or to experimental

animals, application of these criteria necessarily results in a
bias towards selecting chemicals or exposures that may be
carcinogenic and that have been studied in long-term animal
bioassays.   However, not all chemicals selected for evaluation
have been shown to be carcinogenic in animals or humans.

Table 4

*Chemicals, groups of chemicals, complex mixtures, and industrial
processes causally associated with cancer in humans (Group 1)*

4-Aminobiphenyl	Coal-tar pitch
Analgesic mixtures containing phenacetin	Conjugated oestrogens
Arsenic and arsenic compounds	Cyclophosphamide
Asbestos	Diethylstilboestrol
Azathioprine	Melphalan
Benzene	Methoxsalen with ultraviolet A therapy (PUVA)
	Mineral oils (certain)
Benzidine	Mustard gas
Betel quid with tobacco	2-Naphthylamine
N,N-Bis(2-chloroethyl)-2-naphthylamine (Chlornaphazine)	
Bis(chloromethyl) ether and technical-grade chloromethyl methyl ether	Shale-oils
Butane-1,4-diol dimethanesulphonate (Myleran)	Smokeless tobacco products
Certain combined therapy for lymphomas (including MOPP)	Soots
Chlorambucil	Tobacco smoke
Chromium and certain chromium compounds	Treosulphan
Coal-tar	Vinyl chloride
Auramine manufacture	Furniture manufacture (wood dusts)
Boot and shoe manufacture and repair (certain exposures)	Isopropyl alcohol manufacture (strong-acid process)
Coal gasification (older processes)	Nickel refining
Coke production (certain exposures)	Rubber industry (certain occupations)
	Underground haematite mining (with exposures to radon)

    Epidemiological data or case reports are available for less
than 200 of these exposures. For 30 chemicals or complex mixtures
and nine industrial processes, the available data provide causal
evidence of carcinogenicity to humans (see Table 4).   Causal
evidence can exist only when epidemiological studies of
carcinogenicity are available; experimental data are important
when the evidence provided by epidemiological studies is less than
sufficient.   A further 63 chemicals or mixtures of chemicals and
five industrial processes are probably carcinogenic to humans on
the basis of the available evidence.   This group includes
exposures for which the evidence provided by human studies ranges

from almost sufficient to inadequate; in most instances, data from long-term animal studies provided sufficient experimental evidence of carcinogenicity and data from short-term tests indicated genotoxic activity. Table 5 lists the chemicals and industrial processes in Group 2A (*i.e.* for which there is higher degree of probability that they are carcinogenic to humans).

Table 5

Group 2A: Exposures highly suspected of being human carcinogens

Compound	Use or exposure
Acrylonitrile	Plastics monomer
Aflatoxins	Food contaminant
Benzo[*a*]pyrene	Coal- and petroleum-derived products, tobacco smoke
Beryllium and beryllium compounds	Metal industry
Combined oral contraceptives	Drug
Diethyl sulphate	Chemical intermediate
Dimethyl sulphate	Chemical intermediate
Nickel and nickel compounds	Nickel refining, welding fumes
Nitrogen mustard	Cancer chemotherapy
Oxymethalone	Drug
Phenacetin	Drug
Procarbazine	Cancer chemotherapy
o-Toluidine	Dyestuffs manufacture

Process

Aluminium production (certain exposures)
Iron and steel founding (certain exposures)
Manufacture of magenta

Of the chemicals evaluated in the first 40 volumes of the *Monographs*, there is *sufficient evidence* for the carcinogenicity in animals of 214. For 127 of these, no data on humans are available, and, in this situation, these chemicals should be regarded for practical purposes as if they presented a carcinogenic risk to humans.

Of the 30 chemicals or groups of chemicals that are causally associated with cancer in humans, no experimental data from long-term carcinogenicity bioassays were available for two, namely, certain combined chemotherapy for lymphomas and treosulphan (Table 6). Of the remaining 28, 18 have been tested in experimental animals with results that provided *sufficient evidence* of carcinogenicity. Furthermore, results of recent experiments on benzene in mice and rats indicate a clear carcinogenic effect; these would provide *sufficient evidence* for benzene, also, bringing the total to 19. Of the remaining nine exposures, three have been inadequately studied (arsenic and arsenic compounds, conjugated oestrogens, smokeless-tobacco products), and for six (analgesic mixtures containing phenacetin, azathioprine, betel quid with tobacco, chlornaphazine, myleran and mustard gas) the evidence for carcinogenicity is *limited*. When the recent data on benzene are taken into consideration, either *sufficient* or *limited evidence* of carcinogenicity in experimental animals is thus available for 85% of the established human carcinogens.

Table 6

Chemicals, groups of chemicals, and complex mixtures for which there is sufficient evidence of carcinogenicity to humans (IARC Monographs Volumes 1 – 39)[1]

Chemical or exposure	Evidence for Carcinogenicity in humans	Evidence for Carcinogenicity in animals	Evidence for activity in short-term tests
4-Aminobiphenyl	Sufficient	Sufficient	Sufficient
Analgesic mixtures containing phenacetin	Sufficient	Limited	No data
Arsenic and arsenic compounds	Sufficient	Inadequate	Limited
Asbestos	Sufficient	Sufficient	Inadequate
Azathioprine	Sufficient	Limited	Sufficient
Benzene	Sufficient	Limited[2]	Limited
Benzidine	Sufficient	Sufficient	Sufficient
Betel quid with tobacco	Sufficient	Limited	Sufficient
N,N-Bis(2-chloroethyl)-2-naphthylamine (Chlornaphazine)	Sufficient	Limited	Limited
Bis(chloromethyl) ether and technical-grade chloromethyl methyl ether	Sufficient	Sufficient	Limited
Butane-1,3-diol dimethanesulphonate (Myleran)	Sufficient	Limited	Sufficient
Cetain combined chemotherapy for lymphomas [including MOPP (procarbazine, nitrogen mustard, vincristine and prednisone)]	Sufficient	Not tested	Inadequate
Chlorambucil	Sufficient	Sufficient	Sufficient
Chromium and certain chromium(VI) compounds	Sufficient	Sufficient	Sufficient
Coal-tars	Sufficient	Sufficient	Inadequate[3]
Coal-tar pitches	Sufficient	Sufficient	Sufficient
Conjugated oestrogens	Sufficient	Inadequate	Inadequate
Cyclophosphamide	Sufficient	Sufficient	Sufficient
Diethylstilboestrol	Sufficient	Sufficient	Inadequate
Melphalan	Sufficient	Sufficient	Sufficient
Methoxsalen and ultraviolet A therapy (PUVA)	Sufficient	Sufficient	Sufficient
Mineral oils (certain)	Sufficient	Sufficient	Inadequate
Mustard gas	Sufficient	Limited	Sufficient
2-Naphthylamine	Sufficient	Sufficient	Sufficient

Shale oils	Sufficient	Sufficient	Sufficient[4]
Smokeless-tobacco products	Sufficient	Inadequate	Sufficient[5]
Soots and soot extracts	Sufficient	Sufficient	Limited
Tobacco smoke	Sufficient	Sufficient	Sufficient[6]
Treosulphan	Sufficient	Not tested	Inadequate
Vinyl chloride	Sufficient	Sufficient	Sufficient

1. Based upon Supplement 4 to *IARC Monographs* and subsequent volumes.

2. New National Toxicology Program studies indicate *sufficient evidence*.

3. *Sufficient evidence* for all coal-tar-derived products.

4. Low-temperature shale-derived crude oils.

5. Ethanol extracts.

6. Cigarette-smoke condensate.

7. Conclusions

Although conventional approaches using cancer epidemiology have
provided a wealth of information on the causes of human cancer,
these have serious limitations in terms of identifying specific
causative factors, particularly for human cancers that are of
multifactorial origin. In addition, such studies are more often
retrospective than predictive and, often, not highly sensitive.

Long-term animal bioassays are of critical importance in
determining carcinogenicity. The first demonstration that
chemicals could induce cancer in animals was made by Yamagiwa and
Ichikawa in 1918 who produced skin cancer in rabbits after
repeated applications of coal-tar.[12] In 1922, Passay[13] reported
the carcinogenicity of soot extract in mice. These studies were
the first to confirm the clinical findings of Percival Pott much
earlier in history (1775), and they heralded an intensive period
of testing, particularly of polycyclic aromatic compounds, by skin
application. Another landmark in experimental carcinogenicity
testing was the demonstration by Hueper, Wiley, and Wolfe in 1938
that bladder tumours could be produced in dogs by administration
of 2-naphthylamine,[14] confirming observations of an increased
occurrence of these neoplasms in workers exposed to
2-naphthylamine and other aromatic amines. For several chemicals
(*e.g.*, aflatoxins, 4-aminobiphenyl, diethylstilboestrol,
melphalan, vinyl chloride) evidence of carcinogenicity in
experimental animals preceded evidence obtained from
epidemiological studies or case reports. Thus, long-term
carcinogenicity tests in animals can be used to identify direct
cause-effect relationships.

One of the major difficulties in the experimental approach,
however, is that a number of extrapolations are inherent in the
process, including that from laboratory animals to humans and that
from high to low doses. The epidemiological approach relates
directly to humans and thus obviates the necessity for
extrapolation from one species to another. Nevertheless, in
drawing conclusions from epidemiological studies, certain other
important and difficult extrapolations have to be made, *e.g.* from
high-level occupational exposure to low-level community exposure.

8. References

1. E. Boyland, *Prog. Exp. Tumor Res.*, 1969, **11**, 222.

2. E. Wynder and G. Gori, *J. Natl. Cancer Inst.*, 1977, **58**, 825.

3. J. Higginson and C. S. Muir, *J. Natl. Cancer Inst.*, 1979, **63**,
    1291.

4. R. Doll and R. Peto, *J. Natl. Cancer Inst.*, 1981, **63**, 1191.

5. J. Waterhouse, C. S. Muir, J. Powell, and K. Shanmugaratnam,
    'Cancer Incidence in Five Continents, Vol. IV', IARC
    Scientific Publications No. 42, International Agency for
    Research on Cancer, Lyon, 1982.

6. J. E. Dunn, *Natl. Cancer Inst. Monograph*, 1977, **47**, 157.

7. C. Agthe and L. Tomatis, in 'The Physiopathology of Cancer',

ed. F. Hornburger, Vol. 2, pp. 345-358, Basel, S. Karger AG, 1976.

8. L. Tomatis, C. Agthe, H. Bartzch, J. Huff, R. Montesano, R. Saracci, E. Walker, and J. Wilbourn, *Cancer Res.*, 1978, 38, 877.

9. L. Tomatis, *J. Cancer Res. Clin. Oncol.*, 1984, 108, 6.

10. IARC Internal Technical Report, 77/002.

11. IARC, 'Chemicals, Industrial Processes and Industries Associated with Cancer in Humans', IARC Monographs, Volumes 1 to 19, Lyon, 1982, Supplement 4, IARC Monographs on the Evaluation of the Carcinogenic Risk of Chemicals to Humans.

12. K. Yamagiwa and K. Ichikawa, *Cancer Res.*, 1918, 3, 1.

13. R. D. Passey, *Brit. Med. J.*, 1922, 1112.

14. W. C. Hueper, F. H. Wiley, and J. D. Wolfe, *J. Ind. Hyg.*, 1938, 20, 46.

14

Organic Colorants - Interpretation of Mammalian-, Geno-, and Eco-toxicity Data in Terms of Potential Risks

By R. Anliker

ECOLOGICAL AND TOXICOLOGICAL ASSOCIATION OF THE DYESTUFFS MANUFACTURING INDUSTRY, CLARASTRASSE 4/6, P O BOX 4058, CH-4005 BASLE 5, SWITZERLAND

1. Introduction

A realistic assessment of the potential risk arising from colorants depends primarily on an evaluation of the available toxicological and ecotoxicological data. However, a further prerequisite is a knowledge of the physico-chemical, coloristic and application properties of the colorant together with information on the exposure situation. This aspect will be discussed initially.

Synthetic organic colorants are not only encountered in all aspects of daily life as technical and functional necessities, but they also make a decisive contribution to the quality of life. The remarkably wide range of uses of colorants, the extraordinary variety of materials which need to be coloured, and the rigorous performance requirements demand a broad pallete of products: nowadays there are about 3000 individual chemical structures. Doubts are frequently expressed that such a large number of products is necessary, but this totally neglects the complexity of the application technology and the coloristic requirements.

The structures of organic colours belong to widely diverse chemical classes, consisting mainly of substituted aromatic and heteroaromatic groups. More than half of all colorants are azo-compounds, in which such groups are joined by one or several azo-bonds. Other significant classes are characterized by a common structural element, *e.g.* the anthraquinone, triphenylmethane, and phthalocyanine colorants. However, in view of the enormous heterogenecity of their detailed structures, and their physico-chemical and application properties, it is totally erroneous to regard these as uniform groups merely on the grounds of a single structural element.

This consideration also applies with regard to their biological activity, which can vary dramatically in spite of close structural relationships. There are single limitations to the conclusions which can be drawn on the basis of structure-activity relationships, but there are an increasing number of examples where seriously faulty judgements and over-hasty suspicions have resulted from applying otherwise useful methods either superficially or too broadly, in an attempt to deduce toxicological properties from chemical structure. It is not possible to generalize regarding the toxicological properties of whole colorant groups, regardless of whether this classification arises on the basis of structural or application similarities.

On the other hand, it is possible to identify narrow groups of colorants for which it is possible to attribute a certain toxic effect to a common structural characteristic. Untested substances with the same structural element (*e.g.* a nitro-group), or with similar metabolism (*e.g.* formation of benzidine from benzidine-based dyes), can provide a much sounder basis for forecasting their probable toxicological behaviour. Experimental work is needed to obtain reliable answers and to permit the evaluation of individual substances.

For the purpose of risk estimation it must be remembered that this is not only a function of the potential harmful effects due to unfavourable toxicological properties of a product and the extent of its bioavailability in the organism, but it is also a function of exposure and the probability of its occurrence. Risk is generally expressed in terms of statistical probability, *e.g.* a risk of 10^{-6} means that a particular harmful effect will affect one person in a million exposed persons under a specific set of exposure conditions. It is therefore evident that the risk of a harmful effect can be reduced by lowering exposure and this approach affords a means of improving safety which is of fundamental importance. This conclusion in no way detracts from importance of toxicological testing of products. But testing alone does not make products safer. The test data provide the knowledge regarding potential product hazards and help to define the requirements for safe handling and use, whereas by following good working practice, *i.e.* by reducing exposure as far as is reasonably possible, it is possible to increase the margin of safety.

2. Classification and Application of Colorants

2.1. Definition. Dyes and organic pigments (referred to collectively as organic colorants) are substances which, applied to substrates, selectively reflect or transmit incident daylight. Substances which evoke the sensation of black or white are also regarded as colorants. The characteristic property of pigments is their extremely low solubility in water and in the substrate. They generally exhibit very low solubility in organic solvents and for this reason they remain almost totally in the solid form during the coloring process and when applied to the substrate. Colorants which do not meet these criteria are termed dyes. Most commercial products are mixtures of one or several colorants with a combined colorant content lying in the range of 10-95%. The other components help to confer the necessary technical application properties.

2.2. Classification and Identification. The Colour Index,[1] the most widely used catalogue of commercial colorants, has gained worldwide recognition. It classifies colorants under two different systems, one of which is based on application characteristics (Table 1). Each colorant is given a Colour Index Generic Name, which indicates to which application class each colorant belongs, for example Direct Yellow 106. A second classification is based on grouping together colorants with certain common structural elements *e.g.* azo, anthraquinone, triphenylmethane, and phthalocyanine types. Both classifications show many unavoidable overlaps. Thus, azo-compounds can also contain structural elements characteristic of other structural classes, or a vat dye can find applications as a pigment, disperse, or solvent dye. A further somewhat coarse classification can be made on the basis of solubility in water and in organic solvents, or of ionic activity.

Table 1   *Colour Index Generic Names*

Acid dyes	Mordant dyes
Azoic components and compositions	Natural dyes
Basic dyes	Oxidation bases
Developers	Pigments
Direct dyes	Reactive dyes
Disperse dyes	Reducing agents
Fluorescent brighteners	Solvent dyes
Food dyes	Sulphur dyes
Ingrain dyes	Condense sulphur dyes
	Vat dyes

If the chemical structure is known, a five-digit number, the Constitution No., is also allocated to the Colour Index. This number provides some indication of the structural class since different classes are allocated specific number ranges (Table 2). The Colour Index Generic Name, *i.e.* Direct Yellow 106) and the Constitution No. (C.I. 40300) provide a simple and unambiguous means of identifying the colorant. The advantage of the system lies in this simple and unambiguous description of extremely complicated compounds which under the IUPAC nomenclature system could require several lines of text (Figure 1). Such complex names may well be comprehensible to certain nomenclature experts but they mean little to medical practitioners and consumers. Apart from being extraordinarily cumbersome in text, such complicated names can lead not only to errors and serious confusions, but they also hinder rapid communication of information under emergency conditions.

For these reasons such nomenclature is an unsuitable means of identification of hazardous substances on warning labels and has been postulated in certain EEC regulations without regard to practical reality.[2]

Table 2   *Colour Index classes based on chemical constitutions with Colour Index numbers*

	C.I.Numbers		*C.I.Numbers*
Nitroso	10000-10299	Indamine	49400-49699
Nitro	10300-10999	Indophenol	49700-49999
Monoazo	11000-19999	Azine	50000-50999
Disazo	20000-29999	Oxazine	51000-51999
Trisazo	30000-34999	Thiazine	52000-52999
Polyazo	35000-36999	Sulfur	53000-54999
Azoic	37000-39999	Lactone	55000-55999
Stilbene	40000-40799	Aminoketone	56000-56999
Carotenoid	40800-40999	Hydroxyketone	57000-57999
Diphenylmethane	41000-41999	Anthraquinone	58000-72999
Triarylmethane	42000-44999	Indigoid	73000-73999
Xanthene	45000-45999	Phthalocyanine	74000-74999
Acridine	46000-46999	Natural	75000-75999
Quinoline	47000-47999	Oxidation bases	76000-76999
Methine	48000-48999	Inorganic pigments	77000-77999
Thiazole	49000-49399		

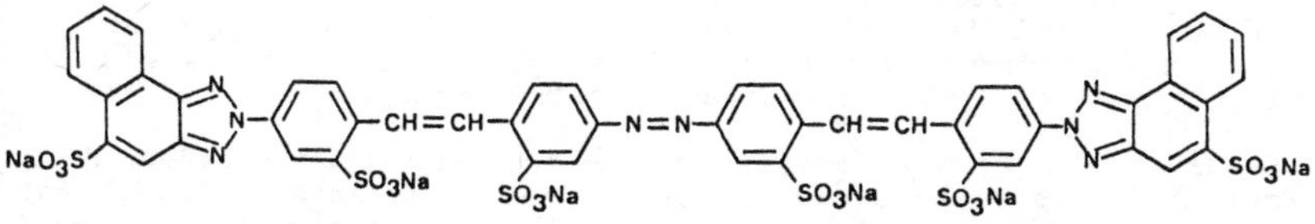

Figure 1

Direct Yellow 106 C.I. 40300
4,4'-bis[4-(5-sulphonaphtho[1,2-d]-triazol-2-yl)-2-sulphophenyl
vinyl]-azobenzene-3,3'-disulphonic acid hexa-sodium salt

2.3. Quality of the Tested Colorant.

A further important requirement for the evaluation of toxicological data is a knowledge of the purity of the tested substance.

Colorants are tailored to achieve a certain coloristic effect on a particular substrate, and to fulfil also the technical application requirements. It is one of the peculiarities of dye technology that colorants must generally be offered not in highly purified form but as formulations (preparations). There are cases where the dye, by design or accident, is a mixture of different structural isomers and in order to achieve the desired coloristic effect it often also contains small quantities of shading dyes. For control of the colour strength, and thus the application properties, the dye concentrates obtained in the production plant are converted into the desired commercial quality by formulation with various diluents and other additives. It is these formulated products which are handled and used by the processor. Consequently many toxicological data are determined for the commercial grade material and recorded in the corresponding safety data sheets.[3]

The commercial form is tested rather than the highly purified active ingredient because the commercial products of various suppliers, although containing the same basic structure, can differ significantly in their content of by-products owing to differences in manufacturing and purification processes. This situation can lead to significant differences in safety-related data. It is therefore not correct to assume that two commercial products are identical with respect to their health and environmental properties, merely because they are based on the same Colour Index structure. This useful classification system is not sufficient for the full characterization of an individual commercial product: it merely indicates the structure of the colorant involved. It is only possible to interpret data sensibly or to compare them when the compositions of the tested products have been adequately defined analytically. There are no generally valid rules to determine which form of a product should be tested, and this must be decided on a case-by-case basis. For the study of chronic effects it is preferable to use a highly concentrated or technically pure form in order to determine the effect of a specific chemical structure and to exclude influences due to possible toxic impurities.

3. Mammalian and Genotoxicity Database

It must be recognized that the majority of chemical substances, be they of natural or synthetic origin, may exhibit biological activity and hence may under certain circumstances be harmful. Whether or not this occurs depends not only on the potency of their toxic effects but also on the exposure conditions and the extent to which they are available to the organism.

Given the wide diversity of chemical structures referred to earlier, it is to be expected that organic colorants, although not designed to be biologically active, will exhibit a fairly broad spectrum of toxicological properties and this is in line with the available data. The extensive literature dealing with such data has been comprehensively reviewed elsewhere.[4-8]

3.1. Acute and Sub-acute Toxicity.
Dyes, and even more markedly organic pigments, generally exhibit low acute oral toxicity. In a survey[9] of about 4500 commercial products the LD_{50} (rat, oral) was greater than 5000 mg kg^{-1} for 82% of the products; in the range 2000-5000 mg kg^{-1} for 10%, and less than 250 mg kg^{-1} for fewer than 1% of the products. Of this more toxic 1%, none of the values was less than 100 mg kg^{-1}, and only 15 individual chemical structures were involved. In the case of organic pigments, only one result was below 2000 mg kg^{-1}. No certain conclusions can be drawn from the data regarding structure-activity relationships. A relatively high proportion of the dyes in the <250 mg kg^{-1} and 250-2000 mg kg^{-1} ranges were Basic dyes but this is not surprising as many quaternary ammonium compounds of various chemical classes are biologically active. Meanwhile the number of tested commercial colorants has exceeded 8000 and a similar pattern was observed. Testing with repeated doses (10-90 days) has generally been limited to those dyes which have been used as food, cosmetic, or drug additives or which have been used as biocides. The majority of the products tested belong to the Azo, Acid, Direct, Solvent, and Triphenylmethane dyes. In general, they have been found to be of medium or low sub-acute toxicity. For only a few products tested is the 'no observable effect level' less than 25 mg kg^{-1} body weight.[4]

Often of more interest in the case of products exhibiting low acute toxicity is the question of cumulative or irreversible toxic effects of repeated dosing, and such a sub-chronic test now forms part of the Base Set of tests for notification of new substances under the EEC 6th Amendment. In a 28-day study in the rat of eight selected colorants with LD_{50} > 2000 mg kg^{-1} no adverse effects were noted even at high dose levels of 1000 mg kg^{-1} b.w.[10]

3.2. Allergic Effects.
Allergic effects, involving cases of both skin and pulmonary sensitization, have been reported for some colorants belonging to various structural classes.[11-14] Some Reactive dyes have been associated with occupational asthma (pulmonary sensitization).[15,16] Allergic reaction has also been provoked after ingestion of certain food and drug dyes.[17] Considering the very large numbers of persons exposed, such effects occur rarely, but the consequences for the individuals involved can be distressing and early recognition of such problems is sought through monitoring of workplace exposure and of consumers.

3.3. Chronic Toxicity. In the debate on the toxic effects of chemicals, there is no doubt that carcinogenic and more recently mutagenic effects are perceived by the public as the most threatening phenomena. The chemicals sector remains a focus for this concern in spite of the weight of evidence that individual chemicals contribute to only a small extent to the incidence of cancer. Today the generally accepted estimate of cancer causation based on mortality indicates that 4% are related to occupational exposures and an additional 1% to other exposures to industrial products.[18]

In the case of organic colorants, there is no conclusive evidence of any carcinogenic effect in man. However, some dyes have shown such effects in long-term animal studies. Examination of the available test data using criteria by, for example, the International Agency for Research on Cancer (IARC)[19] and Longstaff[20] indicates that so far sufficient evidence of carcinogenicity in animals is only available in the case of about a dozen dyes. For a number of colorants the available evidence justifies a strong suspicion of such an effect, *e.g.* for the benzidine-based dyes for which the metabolism in man and animals has been shown to involve reductive cleavage of the azo-bonds. The use of such dyes in high-exposure applications, *e.g.* hair dyeing, children's and finger paints, do-it-yourself dyeing, or under poor working conditions in processing plants, poses an elevated risk. The benefits of good working practice have been demonstrated in a remarkably conclusive study[21] of workers involved in the manufacture and processing of benzidine-based dyes. Under good working conditions no benzidine or its primary metabolites, *N*-monoacetyl- and *N*-diacetyl-benzidine, were detectable in the urine of the exposed workers.[21,22] In contrast such metabolites were detected in the urine of workers in plants where good industrial hygiene was not practised. Such potential risks were an important factor in the decision by the major dye producers voluntarily to cease the manufacture of benzidine and benzidine-based dyes in the early 1970s.[23]

In general it can be concluded that azo-compounds which would form a carcinogenic amine on reductive cleavage of the azo-bonds are more likely to be carcinogenic. However, owing to the extraordinary complexity of the dye molecules it is to be expected that such reduction will in many cases only be partial, and that there will also be various other competing deactivating and activating processes which add to the uncertainty of any prediction.

Appropriate safety measures should be taken in those cases where reductive bleaching or stripping of dyes leads to the formation of carcinogenic amines.

In a situation where the public is calling for a drastic reduction of animal testing, the available laboratory capacity is limited, and only a minority of organic colorants for technical use can economically support the high cost of long-term animal bioassay; the efforts of the National Cancer Institute (NCI), in collaboration with the regulatory agencies and industry, to select representatives of the various structural classes for testing is a logical approach. There are, however, several other problems.

The dye sector, like others, is increasingly confronted with the controversial interpretation of carcinogenic effects observed

at extremely high dose levels. As a result of the relatively low
general toxicity of colorants, it is possible for test animals to
tolerate such very high doses. Effects, usually only observed at
these high dose levels, must then be evaluated in terms of risk at
the trace exposure levels which pertain under practical conditions
(these are often many orders of magnitude lower). Uncertainty
about the relevance of such test results also exists because it is
known that in many cases the use of excessively high doses can
provoke a carcinogenic effect by overwhelming the deactivating
metabolic process and/or as a result of extensive damage to vital
organs or tissues (*e.g.* hepatotoxicity, or tissue damage at the
site of repeated subcutaneous injections).

The enactment of such major chemical control laws as the Toxic
Substances Control Act in the USA and the so called 6th Amendment
in the EEC has prompted similar activities worldwide at both
national and provincial level. This has resulted in lists of
carcinogenic substances shooting out of the ground like mushrooms,
as a basis for mandatory safety measures, product labelling, or
the limitations of use of products. However, in the absence of
recognized criteria for the designation of products as
carcinogens, or any attempt to take into account the potency of
carcinogenic action, this approach inevitably leads very often to
questionable risk assessments. Such lists make no distinction
between, for example, the highly carcinogenic Aflatoxin B_1 and a
weakly carcinogenic textile dye with a potency several hundred
thousand times less. In this context it is worthwhile to mention
the view expressed by the American Conference of Governmental
Industrial Hygienists (ACGIH)[24] 'No substance is to be considered
an occupational carcinogen of any practical significance which
reacts by the gastrointestinal route at or above 500 mg kg^{-1} b.w.
day^{-1} for a life time, equivalent to about 100 g total dose for
the rat and 10 g total dose for the mouse'.

In view of the large number of substances and the limited
resources for comprehensive toxicity testing, attempts are being
made   to   predict   toxic   effects   from   structure-activity
relationships (SARs) using various modern computerized correlation
processes. The Quantitative Structure Activity Relationship
(QSAR) is undoubtedly a useful tool in the hands of experts, but
it cannot serve as a reliable decision basis because of the
inherent uncertainty of the conclusions. In a recent paper[25] some
examples from the field of dyes illustrate how varied and also
unforeseeable the type and potency of toxic effects of very
similar products can be.

Thus, incorrect conclusions drawn from studies involving
overdosage, use of dose regimens not relevant to practical
exposure, or SAR considerations may cause unjustified and
premature suspicion of whole classes of substances. This can lead
to demands of excessive testing and divert already scarce
resources from real priorities. A cause of concern worth
mentioning in the field of colorants is the question of whether
sulphonated aromatic amines, important intermediates for colorants
and known to be primary metabolites of azo-colorants, pose an
equally potent carcinogenic risk as some of their unsulphonated
analogues, especially 2-aminonaphthalene. This suspicion was
fostered by allegedly positive results in two out of seven
1-amino- and 2-amino-naphthalenesulphonic acids in the lung
adenoma test in the mouse using intraperitoneal application.[26]
The statistics and the conclusions of this study scarcely stand up

to critical examination. Furthermore, the doses were applied to an excess which is only fully appreciated when expressed in terms of humans. Within 8 weeks the equivalent of 4.2 kg product was injected into the abdominal cavity. Based on the recently available toxicological and physico-chemical data there is no good scientific reason to suspect the sulphonated aromatic amines referred to of having a carcinogenic potential of any practical significance. For example, a recent bioassay[27] in the mouse on 2-aminonaphthalenesulphonic acid (Tobias acid), the closest analogue to the human carcinogen 2-aminonaphthalene, produced no carcinogenic effect even at the high dose level of 1000 mg kg^{-1} b.w. There is also indirect, but nevertheless convincing, evidence of their low chronic toxicity and low carcinogenic potential. It is well established that the most important metabolic route for azo dyes in mammalian species involves the formation, often in high yields, of aromatic amines by reductive cleavage of the azo-group.[4] Therefore, feeding tests with such azo-compounds, especially at high dose levels, also tend to reflect toxicological properties of these aromatic amine metabolites. Most azo colorants yielding only sulphonated aromatic amines have not shown any carcinogenic effect of significance, if at all. For this reason these studies can be considered as indirect evidence of an apparent lack of carcinogenic potential of any significance in the sulphonated aromatic amines involved. The two examples shown in Figure 2 are illustrative.

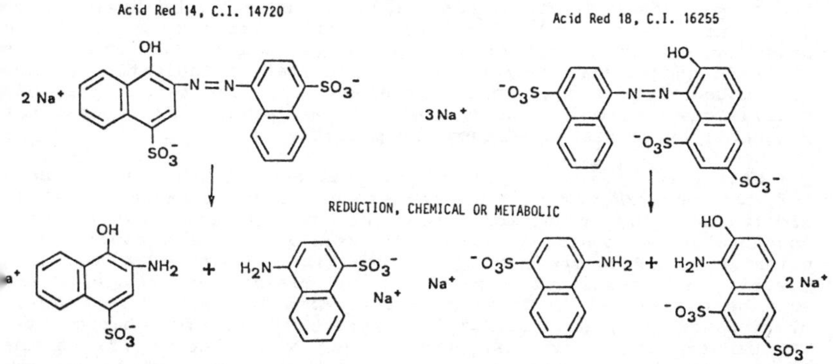

Figure 2 Formation of sulphonated aromatic amines from Acid dyes by reductive cleavage of the azo-group

Acid Red 14, C.I. 14720, and Acid Red 18, C.I. 16255, both not carcinogenic in bioassays,[28,29] are metabolized into 4-amino-naphthalene-1-sulphonic acid and 7-hydroxy-8-aminonaphthalene-1,3-disulphonic acid or 3-amino-4-hydroxynaphthalene-1-sulphonic acid respectively. It can be concluded that these sulphonated aromatic amines have a similarly low toxicity profile.

In addition the four 2-aminonaphthalene-monosulphonic acids with the sulphonic acid group in position 2, 5, 6, or 8 were not mutagenic in the *Salmonella*/mammalian-microsome test.[30] Negative results were reported in the same test also for 2-amino-5-chloro-4-methyl-, 3-amino-5-chloro-4-hydroxy-, and 3-amino-4-hydroxy-benzenesulphonic acids.

In cases where sulphonated aromatic amines are used as intermediates for the synthesis of azo colorants, they may be present in small amounts as impurities in the end-product. Considering the low exposure occurring in the technical use of colorants; the potential risk arising from such impurities or from the amounts of such products possibly formed by metabolic reduction is certainly negligible.

3.4. The Problems of Short-term Mutagenicity Testing. The known limitations of the expensive and time-consuming long-term testing encouraged an intensive search for a simpler test to establish the carcinogenic effect of substances. In recent years a large number of such tests have been developed and described and it is beyond the scope of this chapter to review the wide variety of tests now available.

Most short-term tests used today for the identification of potential carcinogens are based on the determination of a mutagenic effect or the ability of the chemical to react with genetic material. Although prediction of a carcinogenic effect on the basis of such a mutagenic action still involves uncertainties, the dye-making industry, especially in view of the great number of colorants (*ca*.3000) on the market, is interested in perfecting test systems which could enable the rapid predictive screening of a large number of substances. At the present stage it is important to recognize their inherent limitations in terms of implications for regulatory control. The first shortcoming is that the majority of methods have not been fully assessed to determine their predictive value for detecting carcinogenic potential. The second is that these tests must be regarded as giving a qualitative rather than a quantitative indication of the carcinogenic potential and therefore, at least so far, no conclusions concerning the potency may be drawn. This means that no quantitative risk assessment is possible.

Short-term tests are, however, included within the EEC Base Set for notification of new substances, and because of their limitations any regulatory response to adverse results will undoubtedly be controversial (with regard to their carcinogenic potential). Products giving two positive responses in two test systems with unrelated end-points are classed as mutagenic and must be classified in Category 3 according to Annex VI (part II) of the amended Council Directive 67/548/EEC (see also Chapter 11). The danger symbol and Risk-phrases (see also Chapter 12) on the label (see also Chapters 12, 22, and 23) of such products are the same as for Category 3 carcinogens, *i.e.* suspected carcinogens, and suitable safety measures to minimize exposure must be indicated on warning labels by appropriate precautionary phrases. The need for certain restrictions in the use of such products should be considered. For example, such a product should not be used in situations involving a higher exposure potential, *e.g.* in home-dyeing products, in finger paints, or for colouring materials intended to come into contact with food.

3.5. Diversity of Use and Exposure Levels. The almost universal use of colour in most articles of daily life creates a diversity of exposure situations which is unique. The use of colorants in food, drugs, and cosmetics is strictly regulated in most industrial countries and this is appropriate because for those products exposure is deliberate and widespread.

The rapid development of knowledge in the field of toxicology has caused a significant tightening of safety requirements, especially in the case of food colorants. For a product to be permitted nowadays it must show no signs of a carcinogenic potential even at extremely high dose levels, and this has led to a noticeable reduction in acceptable candidates. That a range of products can still be offered is due to the fact that the application requirements (*e.g.* stability, shade range) are not extensive. Recently recommended acceptable daily intake (ADI) values lie in the range 0.15–15 mg kg^{-1} b.w., *i.e.* a maximum permissible daily uptake of 10.5–1050 mg per person (70 kg body weight).[29] These values were calculated using safety factors of between 100 and 1000. The no observable effect level (NOEL) in long-term animal studies of some of these colorants were as high as 2000 mg kg^{-1} b.w. The permitted amounts are sufficient for the desired coloristic effect in foodstuffs.

In the case of coloured cosmetics the possible exposure, *i.e.* the uptake of colorants, is several orders of magnitude lower than in the case of food colorants, whereas the number of exposed persons is still very large. Colorants which have shown a weak carcinogenic potential under extreme dosing conditions have caused much controversy and uncertainty about their possible risk and acceptability. The uncertainty arises mainly from two sources.[31] Firstly, the various mathematical methods which are used to extrapolate from high- to low-dose response, after adjustment for difference in body size, may lead to risk estimates which differ by as much as six orders of magnitude. Secondly, the assumption is usually made that the susceptibility of test animal tissue, after allowance for dose level and animal size, is the same as that of human tissue. The fact that interspecies variation in response to many carcinogens can amount to several orders of magnitude illustrates the tenuous nature of this assumption.

In the case of colorants for technical applications, *i.e.* for the coloration of textiles, leather, paper, varnishes, and plastics, human exposure in terms of colorant uptake by the body is not inherent in the application and is involuntary. Such exposure is largely dependent on the handling conditions during manufacture, application, and use of the colorants. In theory, the exposure could be zero but this is not achievable in practice. Nonetheless, the objective must be to reduce all such exposures to the minimum practicable level, and at least to below the threshold of unacceptable risk.

As already mentioned the stringent technical requirements, the enormous variety of application processes, and the large number of substrates to be coloured, calls for a remarkably large number of products (*ca.* 3000). As a basis for the determination of risk it must be assumed that the colorants are properly handled and applied, and that appropriate plant and equipment are available. It is not appropriate to estimate the risk primarily on the basis of exposure values obtained under improper working conditions. If the calculated risk under proper conditions is too high, then it must be determined whether and how necessary exposure reductions can be achieved. Should this prove not to be possible then the use of such products must be limited or indeed relinquished.

The available epidemiological studies clearly indicate that the risk under proper conditions of handling and application cannot be very high. However, because of their somewhat limited sensitivity, such studies do not permit any reliable estimates of low or moderate risks. For further clarification of whether such products pose an acceptable risk it is necessary to estimate possible risks on the basis of toxicological data from animal studies, and practical exposure levels. The greatest exposure potential is in the manufacturing process but involves only a limited number of persons who are, in any case, trained in handling chemicals. Furthermore, in the manufacturing plant the necessary protective measures to reduce exposure are relatively easily implemented.

Most critical is the exposure in the processing of colorants, especially in weighing and mixing of dyes. Of particular concern are exposure situations involving inadequately designed plant facilities, poorly trained personnel, and lack of suitable protective equipment. The use of dusty products poses the most difficult problem. In spite of substantial efforts it has still not been possible to obtain all colorants in a liquid form. For this reason priority has also been given to the development of low-dusting forms. Nowadays low-dusting powders, and even free-flowing non-dusting granular forms, are available, and this trend will continue. In the case of dust exposure it can be assumed that dust particles greater than 7 μm will be trapped in the nose and trachea and from there reach the digestive tract, whereas particles less than 7 μm (alveolar dust) will reach the lungs and be absorbed. Exposure limits to dust vary from country to country but are generally 5-6 mg m^{-3} for fine dust (respirable dust) and 10-15 mg m^{-3} for total dust.

Dust measurements in manufacturing and dyeing plants[21,32-34] indicate that under good conditions of working practice atmospheric dust levels are typically less than 1mg m^{-3} total dust. With poor working practice and the use of dusty products levels far above the usual permissible limits for inert dust (*i.e.* greater than 15 mg m^{-3}) were observed.

In the study[21] mentioned earlier it was possible to show that in the case of benzidine dye powders the exposed workers excreted benzidine and related metabolites. Calculations indicate that inhalation of 2 mg m^{-3} of a benzidine-derived dye over an 8 h working day could result in the excretion of 14-24 p.p.b. benzidine in the urine over a full day. The author concluded that taking into account the limited historical data, it would appear that human urinary excretion of greater than 100 p.p.b. of benzidine and its derivatives contributes to a noticeably increased risk of cancer in man. In dye manufacturing and textile finishing plants observing good working hygiene these metabolites could not be detected in the urine of workers handling benzidine dyes using very sensitive analytical methods (detection limit 1 p.p.b.). It can be assumed that with excretion below this limit the possible risk should be insignificant. Analogous results were obtained in a similar study on occupational exposure with workers weighing benzidine-derived dyes.[22] The monitoring of amine excretion in urine of persons exposed to aromatic amines or compounds which are expected to be metabolized in the body to such amines can therefore generally offer the possibility to check the adequacy of the safety precautions taken.

Sensitization of the respiratory tract by various chemicals is a major cause of morbidity in industry. Concerning colorants several cases of pulmonary sensitization have been reported[14,16,35] involving also some Reactive dyes. As there are at present no accepted animal models of pulmonary sensitization which can aid the prediction of sensitizing potential the Reactive dyes should be handled with caution and inhalation of dust and aerosols must be prevented.

The wearing of dyed textiles probably represents the most widespread exposure of consumers to technical dyestuffs. With properly dyed textiles the exposure, *i.e.* the migration of the dyes onto the skin, seems to be very small. Based on tests[36] with simulated perspiration the potential exposure was estimated to be much less than 1 mg per person per day. Cutaneous penetration, especially in the case of ionic dyes, is very limited, usually less than 1% of the amount applied to the skin. Therefore the average uptake per person per day would amount to a few micrograms or even less. It can be concluded that the organotoxic risk is judged to be very small. Experience shows, however, that the wearing of garments coloured by certain dyes may cause allergic contact dermatitis. The dyes involved can be found in practically all dyestuff classes but a higher incidence has been recorded with the non-ionic Disperse dyes.[12] Although the reason for this higher incidence is not fully understood, it may be assumed that these non-ionic, lipophilic dyes penetrate the outer layer of the skin more easily, triggering off the allergic reaction.

The use of colorants in consumer goods such as coloured toys or food packaging materials is already regulated in many countries. The regulations require that no visible amounts of colorants migrate from the coloured material when measured by standard methods. In the case of food packaging materials the limit of non-visible migration is usually less than 0.05 p.p.m. (*i.e.* less than 50 μg of migrated colorant per kg of food) but this also depends on the colour strength of the particular colorant. Most suitable for this field of application are insoluble organic pigments which are of a very low order of toxicity.

The exposure in the case of finger paints, however, is considerably higher. The proposal to use certified food colorants is not practicable as these dyes stain clothing and decorative textiles. The only practical materials suitable for this purpose are insoluble pigments which do not possess any undue toxicological properties. To deter children from regular oral uptake the addition of bitter substances such as sucrose octa-acetate is recommended.

A further field of application representing a potentially higher, uncontrollable exposure is the home-dying of textiles. For this purpose only specially selected and sufficiently tested dyes should be used.

4. Environmental Data

Extensive reviews of the environmental properties of colorants have been published.[4-6,38] A great deal of additional information has been developed by manufacturing companies and some are included in their safety data sheets.

4.1. Fish Toxicity. A survey of fish toxicity data[39] on over 3000 commercial products by ETAD indicates that the majority of dyes are not very toxic to fish. Only 2% were found to be toxic at a level of less than 1 mg l^{-1}. These included 27 different structures (four Acid dyes, sixteen Basic dyes, and seven Metal-complex dyes). The LC_{50} of 59% was over 100 mg l^{-1}. 75% of dyestuffs tested by a major producer are not toxic at the maximum dose level of 100 mg l^{-1} with 3.5% toxic between 0.1 and 1 mg l^{-1}. It should be noted that visible pollution of a river would generally be observable for synthetic, water-soluble dyes at about 1 mg l^{-1}. In natural waters this would be unacceptable on aesthetic grounds. Thus, provided that there is no 'visual pollution', dyes are unlikely to have adverse acute effects on aquatic life. Although no systematic work on possible chronic effects of dyestuffs is known, fish have been exposed over an 8 week period to sub-acute levels of dyestuffs in bioaccumulation studies. These sub-acute levels were well above the visual threshold and no effects on the fish were observed.

4.2. Inhibition of Wastewater Bacteria. This parameter is measured using the respiration inhibition test developed by ETAD which is now an OECD Guideline[40] and is used to assess possible inhibitory effects of dyestuffs on aerobic wastewater bacteria. Of 200 dyes of all application types tested[41] only 18 showed IC_{50} less than 100 mg l^{-1}. These are mainly Basic dyes, three in the range 1-10 mg l^{-1} and 15 between 10 and 100 mg l^{-1}. These results indicate that the majority of dye types are unlikely to be significantly toxic to the aerobic sewage treatment process as their concentration on entry to effluent treatment plants rarely exceeds 100 mg l^{-1}. For the few dyestuffs showing a strong inhibitory effect and likely to reach the effluent treatment work in significant amounts a more detailed assessment of possible adverse effects is advisable.

4.3. Toxicity to Algae. A knowledge of the effect of dyes on algal activity is important to the evaluation of the possible impact of discharges to oxidation ponds or receiving waters. Algae are important components of aquatic ecosystems and algal photosynthesis is a critical source of oxygen. Growth inhibition tests of 56 dyestuffs[42] showed close parallels with fish toxicity data, with the exception of some Acid dyes highly toxic to fish which did not affect algae (*Selenastrum capricornutum*).

4.4. Toxicity to Higher Plants. A mechanism by which dyes are partly removed from aqueous effluents is adsorption onto the sludge during the biological treatment. If such sludges are dispersed on agricultural land the adsorbed dyes may be transferred to soil and may affect plant growth. Calculations based on the principles indicated by the OECD[43] indicate that in a reasonable 'worst case' the levels of dyes which might reach agricultural land would be *ca.* 1 mg kg^{-1} of soil (see also Chapter 11). In a study organized by ETAD[39] the effects of three dyestuffs (Acid, Basic, and Disperse dyes) on the germination and growth of three plant species, *i.e.* Sorghum, Sunflower, and Soya, were examined at levels of 1, 10, 100, and 1000 mg kg^{-1} of soil. No effects were observed at any of the three lower test levels. At the 1000 mg kg^{-1} test level there was no effect on germination, but a variable effect on growth, depending on the dye and the particular plant species. After 21 days growth period of the study, the plant foliage was analysed. Only at the dosage level of 1000 mg kg^{-1} were traces of dye found at the limit of detection, *i.e.* 0.1, 0.2, and 2 p.p.m. respectively.

4.5. Biodegradability and Elimination. In view of the necessarily high stability of dyes against a multitude of physical and chemical influences, it is not surprising that they show a similar stability against breakdown by many micro-organisms. Experimental work[44] has confirmed that in general dyestuffs are *not biodegraded* in the short retention time of aerobic treatment processes carried out during the conventional biological treatment of industrial or municipal wastewaters. For a few specific dyes breakdown under aerobic conditions has been demonstrated[45] with adapted micro-organisms and long retention times, but it is unlikely that such processes could be of practical use for dyes in general. In contrast to this behaviour, many dyes, especially the important azo dyes, are degraded under anaerobic conditions.[46] Such conditions are met in the anaerobic digestion process at sewage treatment works, and in anaerobic sediments and soils. Thus, although the rate of degradation varies with the structure of the dye and the biological conditions, dyes are not persistent. The relationship between rate and structure is not yet well understood.

In the case of azo dyes, which represent over 60% of all dyes, the principal initial breakdown step under anaerobic conditions involves a reductive cleavage of the azo-group to give either lipophilic or hydrophilic (mostly sulphonated) aromatic amines. It has been shown[47] that many of the lipophilic amines, some of which exhibit toxic effects (*e.g.* carcinogenicity) are generally biodegradable under aerobic conditions and thus unlikely to persist in the environment if released (*e.g.* from sediments) as a result of the anaerobic cleavage of azo dyes. Hydrophilic aromatic amines on the other hand do not appear to have any significant mammalian or aquatic toxicity.[48] Some such amines have been shown to be aerobically biodegradable, but it is probable that, for most, biodegradation will be slow. However, it seems virtually certain that no toxicologically significant concentration of these materials will be found, particularly bearing in mind that their high water solubility means that they have practically no bioaccumulation potential.

Many dyestuffs, because of their inherent high affinity to substrates, are adsorbed onto the sludges during sewage treatment and are thus eliminated from the final treated effluent. The degree of such adsorption is dependent on the precise nature and extent of the treatment process, the dye structure, and other components of the effluent, particularly as they influence pH.[49] Various studies by ETAD member companies[50] showed that, for instance, Acid dyes are more readily removed at lower pH, and for dyes in general elimination by adsorption may be between 20 and 90%. Certain dyes may require a combination of biological and physico-chemical treatment (*e.g.* precipitation with ferric salts in alkaline solution) to achieve an effluent of acceptable colour (say <1 mg l^{-1}) for discharge to a small river or stream. Of the relatively few dyes which are toxic to aquatic life at low concentrations most are of the Basic dyestuff class, and these dyes have a very high exhaustion level onto the substrate being dyed, and are also particularly well eliminated by adsorption during biological treatment. Thus, in spite of their toxicity, the probability of any hazardous effects will be minimal as they are unlikely to reach receiving water in any significant amount.

4.6. Bioaccumulation. The partition coefficient (P_{OW}) in n-octanol/water has been shown to be a useful indicator for the estimation of the bioaccumulation tendency of water-soluble dyes. If the P_{OW} value is below 1000 it can be confidently predicted that the bioaccumulation factor (BF) in fish is less than 100: *i.e.* the dye does not bioaccumulate.[51] For over 75 dyes investigated, including various lipophilic Disperse dyes, the BF values are well under the critical value of 100. Even Disperse dyes with P_{OW} values over 1000 were found not to accumulate. This is probably explicable by their pronounced aggregation tendency and their molecular size making transport across membranes difficult.

Another indicator for assessing the bioaccumulation tendency is the water solubility. No bioaccumulation is to be expected of chemicals, including dyes, having a water solubility above 2000 mg l^{-1}.[52] In a recent review on the relation between water solubility, P_{OW} and BF of organic chemicals a limit of >3mmol l^{-1} was considered to be more appropriate.[53]

For organic pigments it has been shown[54] that the P_{OW} has no practical value as an indicator of bioaccumulation tendency. As the determination of the P_{OW} practically water-insoluble pigments is extremely difficult the P_{OW} was calculated. The resulting very high P_{OW} values would suggest strong bioaccumulation tendencies. But no accumulation in fish as compared with the amount of pigments dispersed in the test water has been observed. The reason for this apparent inconsistency is the very limited storage potential of pigments in lipids as indicated by their low solubilities in n-octanol. Under practical conditions not even this limit will be reached because of the molecular weight effect as also suggested for Disperse dyes, *i.e.* a substantial inhibition of permeation through membranes and other biological materials due to the large molecular size of these organic pigments. Supporting evidence for this explanation was given in a recent study[55] on the relationship between bioaccumulation in fish and steric factor of lipophilic chemicals. A loss of membrane permeation was suggested for lipophilic molecules greater than 9.5 A, a dimension which certainly is surpassed by large pigment molecules.

4.7. Environmental Concentration. As far as any possible detrimental environmental effects are concerned the colorants must be divided into two groups: the water-soluble and dispersable dyes on the one hand and the water-insoluble pigments on the other. Of the latter group only negligible amounts reach the environment owing to their extremely low water solubility and their application in mostly non-aqueous systems.

As shown in Figure 3, dyes enter the environment mainly in wastewater from production, from dyeing and recycling processes, and from domestic sources. It is estimated that in this way about 10% of the world production, *i.e.* about 40,000 tonnes active-material per year are lost. Today the major part of wastewater containing colorants is treated in wastewater purification plants in the industrialized countries. As already mentioned dyestuffs are practically not biologically degraded during their short stay in the plant. A considerable part, in the range 10-90% is eliminated by adsorption on the sludge depending on the type of dye and the quality of the wastewater (pH, composition). The remaining part can be eliminated by precipitation, adsorption, or chemical processes. The amount reaching the receiving water depends on the efficiency of the total of purification processes. Owing to their high affinity a part of these dyes is adsorbed on the suspended solids in the receiving water and deposited in the aquatic sediment.

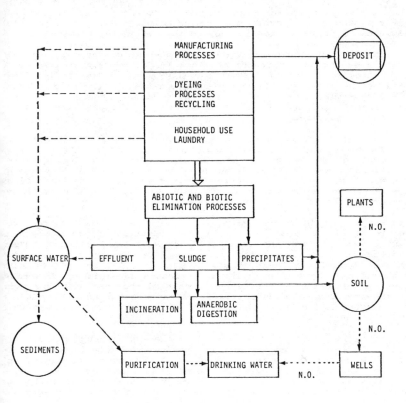

N.O. Not observed

Figure 3 *Distribution of dyes into the environmental compartments*

The fate of the dyes adsorbed on the sludge depends on its further use. More and more it is incinerated or digested anaerobically. Studies[46] proved that dyes can be biologically degraded under anaerobic condition. Untreated sludge is disposed of by storage in controlled deposits or by agricultural use. In view of the high heavy-metal content of the sludge this use is being increasingly reduced.

Concentrations in the water compartment. Calculations[39] of the total level of dyes in a small receiving river (4.5 cumecs) arising from a hypothetical medium sized textile dye-house using 30 tonnes of dyes per year and assuming a 50% removal of the dyes in the sewage plant, also receiving sewage from a population of 20,000, result in a maximum level of 0.01 p.p.m. A current study[56] should provide data on dye concentrations in a river system receiving effluents from major carpet dyeing industries using *ca*. 36,000 tonnes of commercial dyes per year. The data should also permit the assessment of possible risks in view of the river's use as a source for drinking water, and to initiate measures to reduce waste and improve the wastewater treatment.

The dye concentrations of the most-used Acid dyes found in all samples except in direct effluent samples of wastewater treatment facilities were generally less than 0.02 p.p.m. The average flow rate of the river was 192 cumecs with a range of 24 – 1840 cumecs. The triphenylmethane dye C.I. Acid Blue 1, widely used in toilet flush blocks has been found to be present in the Lee River system at concentrations up to 0.0017 p.p.m.[57]

The detection of such small amounts of dyes in wastewater is an extremely difficult analytical task all the more so as the effluent of a dye-house may contain traces of 20 or more individual dyes. A total level of 1 p.p.m. is likely to cause a visible coloration of the water. If the individual contaminating dye is known, a detection limit of 0.001 p.p.m. can usually be achieved under favourable circumstances.

Concentrations in river sediments. In view of the substantive nature of the dyes it is likely that substantial parts of them will be adsorbed on the suspended solids in the receiving water and then be found in aquatic sediments. In the course of a study in the Coosa River Basin,[58] dye levels of 0.1-3 mg kg^{-1} of dry sediment have been found. Investigations[59] on sediments from the lower Rhine (Dutch-German border) indicated that the major portion of the isolated coloured matter appeared to be of natural origin. The synthetic dye level was below 0.05 p.p.m. and no individual dye could be identified.

Concentration in soil. Dyes may enter the soil through the use of sludge from plants treating dye-containing wastewater. No data of possible levels are available so far. However, as shown in Section 4.4, calculations with 'worst case' assumption levels to be expected would be well below 1 mg kg^{-1} soil and unlikely to have adverse effects on the germination and growth of terrestrial plants.

Concentration in processed and well water. River systems serve as a major source for the preparation of potable water. The few studies available indicate that if dyes are present their concentrations seem to be in the sub-microgram region. In a finished potable water sample from the Lee River system a dye, C.I. Acid Blue 1, was found at the concentration of 0.4 μg l^{-1}.[57] The level of this dye in the unprocessed river water was 0.9-1.7 μg l^{-1}. Processed water from the Thames did not contain detectable amounts (detection limit 0.3 μg l^{-1}). In an extensive study[60] the concentrations of fluorescent whitening agents (FWA) in major European rivers and drinking water supplies of cities situated on these rivers were measured using a spectrofluorimetric method. FWAs are widely used in detergents and in textile finishing plants. They are classified as colorants and have similar dyeing properties as dyestuffs. Only in one situation where substantial quantities of a particular FWA were released into a small river were detectable levels (0.3-8 μg l^{-1} or 0.008

p.p.m.) found. None of the drinking water samples of all locations contained detectable amounts of FWA (detection limit 0.01 μg l^{-1} or 0.01 p.p.b.).

5. Conclusions

The available epidemiological studies and the enormous practical experience with colorants for dyeing textiles, leather, paper, and plastics confirm that they can be used without any unreasonable risk provided that adequate working conditions and safety precautions are observed. However, adverse health effects can result in situations where excessive exposure occured owing to inadequate working conditions or misuse of products. Apart from a few cases of allegedly enhanced cancer rate, the major adverse health effects are related to allergic reactions.

When colorants find uses, involving a higher exposure potential to consumers, such as for example in home-dying, children's paint, finger paints, and in the colouring of toys and food packaging materials, only colorants with an adequately defined toxicity profile and which meet more stringent quality requirements (*e.g.* non-migration, higher purity) should be used.

Hazard identification through product testing will undoubtedly remain an important part of the effort to improve the protection of man from adverse effects of the chemicals. However, the known limitations on the ability of animal tests to detect hazards for humans will persist, and, therefore, possible hazards may pass unrecognized. There can be no doubt that the emphasis must continue to be on the reduction of exposure, not merely to a control limit but as far below that limit as is reasonably practicable. This offers indisputable benefits in terms of reducing risks, already recognized or not, and avoids the numerous controversies which surround the assessment of many test results. In this way it would be possible to concentrate more of the scarce toxicological resources on investigative, rather than routine testing, with the objective of identifying real 'villains' which should be eliminated from the market place, or subjected to use restrictions, because they may pose unreasonable risks even under conditions of good working practice.

Once the hazardous properties of the products are determined and the possible sources of potential risks explored, it is most important that the personnel in the workplace handling the chemicals be aware of their hazardous properties and be advised on the precautions necessary for their safe handling. In this context attention must be drawn to the utmost importance of non-dusty and liquid forms. Workplace monitoring, *i.e.* medical surveillance of exposed persons and control of product emissions, must become an integral component of the safety measures.

With the progressive improvement of wastewater treatment and orderly waste disposal the burden on the environment of dyestuffs continues to decline. Dyestuffs are not biodegraded to any practical extent during their short passage through sewage treatment plants. However, they have been shown to biodegrade slowly under anaerobic conditions. The ultimate fate of dyestuffs discharged in effluents appears likely to be an anaerobic environment, *i.e.* partly in natural sediments, partly in soil.

The potential of any serious adverse environmental effects is judged to be low considering the relatively small amounts of dyestuffs reaching the receiving waters, their lack of bioaccumulation, and their biodegradability under anaerobic conditions. Nevertheless, investigations continue to extend the understanding of dyestuffs behaviour in the environment and to explore all sources of possible risks.

6. References

1. 'Colour Index', published by the Society of Dyers and Colourists, Bradford, England, and the American Society of Textile Chemists and Colorists, Research Triangle Park, NC, USA., 3rd Edn., 1971, revised 1975, 1982, and running Supplements Vol. 1-7.

2. A. J. Hinton, *J. Soc. Dyers Colour.*, 1986, **102**, 1.

3. R. Anliker, *Textilveredlung*, 1983, **18**, 130.

4. A. W. Burg and M. C. Charest, 'Azo Dyes: Evaluation of Data Relevant to Human Health and Environmental Safety', July 1980, Report by A. D. Little, Inc. to Dyes Environmental and Toxicology Organization, Inc., available through ETAD, P.O. Box, CH-4005 Basle 5, Switzerland.

5. C. C. Sigman, C. T. Helmes, P. A. Papa, D. L. Atkinson, M. K. Doeltz, and A. Winship-Ball, 'Anthraquinone Dyes and Related Chemicals: Review and Assessment of Potential Environmental and Health Aspects', January 1983, Report by SRI International, Menlo Park, CA, USA, to Dyes Environmental and Toxicology Organization, Inc. Scarsdale, New York, available through ETAD, P.O. Box, CH-4005 Basle 5, Switzerland.

6. E. A. Clarke and R. Anliker, 'Handbook of Environmental Chemistry', Springer, Berlin/Heidelberg, 1981, Vol. 3/Part A, Chapter 7, p.181.

7. R. Anliker, 'Atti del Convegno Materie Coloranti ed Ambiente di Lavoro', Istituto di Medicina del Lavoro, Milano, 1984, Chapter 5, p.73.

8. A. C. D. Cowley, *Polym. Paint Colour J.*, 1985, **175**, 580.

9. R. Anliker, *Textilveredlung*, 1981, **16**, 431.

10. H. K. Leist, *Ecotox. Environ. Safety*, 1982, **6**, 457.

11. P. L. Lywie, B. Herve-Bazin, J. Fousserau, C. Cavelier, and H. Coirier, 'Les Eczémas Allergiques Professionels dans l'Industrie Textile', Inst. Nat. Recherche Securité, Rapport No. 224/RI, Paris, 1977.

12. K. L. Hatch, *Textile Res. J.*, 1984, **54**, 664.

13. E. Cronin, 'Contact Dermatitis', Churchill Livingstone, London, 1980.

14. K. Alanko, H. Keskinen, F. Bjorsten, and S. Ojanen, *Clin. Allergy*, 1978, **8**, 25.

15. B. M. Hansen and K. H. Schulz, *Dtsch. Med. Wochenschrift*, 1984, 109, 1469.

16. Health & Safety Executive, News Release, HSE London, 1986, February 12.

17. E. G. Weirich, 'Farbemittel in der Pharmazie', Europ. Symp., *Acta Pharm. Techn. Suppl.*, 1979, 8, 7.

18. M. M. Heckler, Statement on Cancer Prevention Awareness Program. Washington, March 6, 1984.

19. IARC Monographs on the Evaluation of the Carcinogenic Risk of Chemicals to Humans, October 1982, IARC, Lyon, Suppl. 4, 11.

20. E. Longstaff, *Dyes Pigments*, 1983, 4, 243.

21. M. Boeniger, 'Carcinogenicity and Metabolism of Azo Dyes, Specially those Derived from Benzidine', DHHS (NIOSH), 1980, Publication No. 80-119.

22. P. F. Meal, J. Cocker, H. K. Wilson, and J. M. Gilmour, *Br. J. Ind. Med.*, 1981, 38, 191.

23. R. Anliker, *Ecotox. Environ. Safety*, 1977, 1, 211.

24. American Conference of Governmental Industrial Hygienists, 'TLVs Threshold Limit Values and Biological Exposure Indices for 1985-86', ACGIH, Cincinnati, 1985, p.44.

25. R. Anliker, *Textilveredlung*, 1983, 18, 92.

26. J. C. Theiss, M. B. Shimkin, and E. K. Weisburger, *J. Natl. Cancer Inst.*, 1981, 67, 1299.

27. G. Della Porta and T. A. Dragani, *Carcinogenesis*, 1982, 2, 647.

28. NTP. Carcinogenesis Bioassay of C.I. Acid Red 14. National Toxicology Program, Research Triangle Park, NC/Bethesda, MD, 1982, NTP-80-67, NIH Publication No. 82-1776.

29. Report of the Scientific Committee for Food, July 1983, III/1259/83-EN.

30. J. Le, R. Jung, and M. Kramer, *Fd. Chem. Toxic.*, 1985, 23, 695.

31. T. C. Campell, *Fed. Proc.*, 1980, 39, 2467.

32. A. M. Thiess and G. Wellenreuther, *Arbeitsmed. Sozialmed. Praventivmed.*, 1982, 17, 249.

33. R. Mona, 'Dust Concentration in the Atmosphere of Dyestuffs Standardizing Works', 1981, ETAD Report G 1001 Part I, Basle, Switzerland, unpublished.

34. R. Mona, 'Determination of the Dust Concentration in the Atmosphere of a Colour Store', 1981, ETAD Report G 1001 Part II, Basle, Switzerland, unpublished.

35. M. Ringenbach, *Arch. Mal. Prof. Med. Trav. Secur. Soc.*, 1985, 46, 219.

36. ETAD, 'Extractability of Dyestuffs from Textiles', March 16, 1983, Basle, Switzerland, unpublished.

37. ETAD, Analytical Method No. 221, 1980, available through ETAD, P.O. Box, CH-4005 Basle 5, Switzerland.

38. I. Holme, 'Review on Applied Chemistry', 1984, Vol. 7, p.111.

39. D. Brown, paper presented at the IAES(SECOTOX) International Symposium on 'Technical Organic Additives and Environment', May 24/25, 1984, Interlaken, Switzerland. To be published in *Ecotox. Environ. Safety.*

40. OECD, 'Activated Sludge Respiration Rate Text', Method 209, OECD, Paris, 1981.

41. D. Brown, H. R. Hitz, and L. Schafer, *Chemosphere*, 1981, 10, 245.

42. L. W. Little and M. A. Chillingworth, 'ADMI: Dyes and Environment', Vol. II, American Dye Manufacturer's Institute, Inc. New York, 1974, Chapter II.

43. OECD, Hazard Assessment Project: Working Party on Exposure Analysis. Final Report, Berlin, 1982, p.52.

44. U. Pagga and D. Brown, *Chemosphere*, 1986, 15, 479.

45. H. G. Kulla, 'Microbial Degradation of Xenobiotics and Recalcitrant Compounds', FEMS Symposium No.12, Academic Press Inc., London, 1981, p.387.

46. D. Brown and P. Laboureur, *Chemosphere*, 1983, 12, 397.

47. D. Brown and P. Laboureur, *Chemosphere*, 1983, 12, 405.

48. ETAD Project 'Environmental Behaviour of Sulfonated Aromatic Amines', ETAD, P.O. Box, CH-4005 Basle 5, Switzerland, unpublished data.

49. M. Nakuoka, S. Tamura, Y. Maeda, and T. Azumi, *Sen i Gakkaishi*, 1983, 39, 69.

50. H. R. Hitz, W. Huber, and R. H. Reed, *J. Soc. Dyers Colour.*, 1978, 94, 71.

51. R. Anliker, E. A. Clarke, and P. Moser, *Chemosphere*, 1981, 10, 263.

52. OECD, Expert Group on Degradation and Accumulation. Final Report, Vol. I, Part 1+2, Umweltbundesamt, FRG, Government of Japan, Berlin and Tokyo, Dec. 1979.

53. C.A.M. Van Gestel, K. Otermann, and J. H. Canton, *Regul. Toxicol. Pharmacol.*, 1985, 5, 422.

54. R. Anliker and P. Moser, Paper presented at the International Symposium on 'Bioavailability of Environmental Chemicals', Schmallenberg, FRG, 1984. To be published in *Ecotox. Environ. Safety*.

55. A. Opperhuizen, E. W. van der Velde, F. Gobas, D. A. K. Liem, and J. M. D. van den Steen, and O. Hutzinger, *Chemosphere*, 1985, **14**, 1871.

56. W. C. Tincher, 'Analysis for Acid Dyes in the Coosa River Basin', School of Textile Engineering, Georgia Institute of Technology, Atlanta, GA, Personal communication.

57. M. L. Richardson and A. Waggott, *Ecotox. Environ. Safety*, 1981, **5**, 424.

58. W. C. Tincher, 'Survey of the Coosa River Basin for Organic Contaminants from Carpet Processing'. Final Report. Contract No. E-27-630, Prot. Div., Dept. of Natural Resources, State of Georgia, USA, 1978.

59. ETAD, 'Sediment Analysis', Status Report Project E 3013, June 27, 1980, ETAD, Basle, Switzerland, unpublished.

60. G. Anders, 'Environmental Quality and Safety', G. Thieme, Stuttgart, Academic Press, New York, 1975, Suppl. Vol. IV, Chapter VI/2, p.143.

15
Interpretation of Epidemiological Data — Pitfalls and Abuses

By P.C. Elwood

MRC EPIDEMIOLOGY UNIT, 4 RICHMOND ROAD, CARDIFF CF2 3AS, UK

1. Introduction

The identification or the assessment of hazardous chemicals suggests a somewhat restricted use of epidemiology based on the examination of groups of workers or subjects affected by exposure to relatively high levels of a substance. This does represent a distinct and important problem and much of what follows focuses on this kind of situation. An example would be the examination of the association between the exposure of certain groups of textile workers to carbon disulphide, and the occurrence of ischaemic heart disease. The classical epidemiological strategies, however, are much more appropriate to the investigation of the chronic, low-level exposure of large groups of subjects to a possible hazard such as possible long-term effects of fluoride in a domestic water supply on the pattern of mortality or morbidity in a community, or the contribution of lead in household paint to the blood lead of young children.

These two kinds of situation have quite marked differences, though there is of course a considerable overlap both in the nature of the hazards and the possible effects, and also in the type of investigation which is appropriate. The various epidemiological strategies will therefore be described in some detail, but it may be helpful at this stage to state that, in general, the 'case-control' study and the 'retrospective follow-up' study are the most appropriate for the investigation of the more acute, or the more unusual hazard, while the 'cross sectional' and 'prospective' surveys are usually best for the study of the more widespread low-level exposure of a population to a possible hazard.

All epidemiological studies require very careful evaluation and interpretation and the sub-title of this chapter has been chosen in an attempt to distinguish between deficiencies in the design or conduct of a study (designated 'pitfalls') and the drawing of unwarranted conclusions or unreasonable extrapolations of conclusions ('abuses'). This distinction is however, somewhat arbitrary.

2. The Design of Epidemiological Studies

Basically, epidemiology is the making of measurements, of known reproducibility, in a bias-free manner, on representative samples of subjects drawn from defined communities. All elements in this definition are important and each will be amplified in what follows. First, however, the design of epidemiological studies must be considered in detail.

2.1. Case-control Studies.

These are of particular value in the study of a suspected hazard and so the design features, and the possible pitfalls and abuses, are most important in the context of this chapter. These have been discussed at great length elsewhere[1-4] and so only a brief review will be given here.

Strictly, case-control studies start with a disease or an effect, and work back towards possible causes. This is the 'mirror-image' of the prospective or 'cohort' study in which a group is defined and followed forward. Feinstein,[4] in a discussion of the scientific principles of case-control studies, has proposed the name 'trohoc' (*i.e.* cohort spelt backwards) for studies which start with the outcome and look backwards for causes!

Case-control studies are efficient and flexible, they are particularly well suited to the study of rare diseases or uncommon effects, and they are often the only study design ethically applicable to a problem. However, they are open to a wide range of biases and other methodological pitfalls which may lead to invalid conclusions. As one writer[4] has lamented 'case-control studies (are) planned with extraordinary latitude for the investigator, and with enormous scope in variability of methods and principles. The cases can be selected from whatever group of newly diagnosed patients is conveniently available: the controls can come from a diversity of clinical and community sources: the data about exposure can be acquired in whatever arbitrary way seems to suit the investigator, and the hypotheses about cause and effect may even be established after, rather than before, the data are analysed'.

Case-control studies are retrospective and are based on subjects who are, or who may be, affected. This identifies the first limitation in the strategy, namely the difficulty in identifying a representative sample of cases. In the nature of the design, studies usually have to be based on a selected group, such as those who have died, or those who have presented at hospital with advanced disease. While these may be suitable for the detection of a hazard, they constitute a misleading basis for the estimation of risk. It is possible of course, to examine a complete cohort of an exposed group, such as a factory population, or all residents in a defined area, and to detect all affected subjects and from these to make valid estimates of the risk of developing the disease or effect. However, while such an initial survey would give an ideal basis for the subsequent conduct of a case-control study, it constitutes in fact a cross-sectional survey (see later).

The case-control strategy has at best only a very limited place in the study of low-level exposures of communities such as are required in environmental research. The main limitation arises again from the fact that the study starts with cases, that is subjects, who exhibit some manifestation of clinical disease; as these are unrepresentative of all affected subjects, such studies yield results from which generalizations can seldom be made with confidence.

2.2. Cross-sectional Surveys.

These represent a basic epidemiological strategy. Their strength is their ability to describe the distribution of factors in the community, and associations between factors. They also allow a complete enumeration of cases of a disease or subjects showing a clinical or other effect and this gives a basis for reliable estimates of prevalence and estimates of the risk which can be generalized to other populations. One great advantage of the cross-sectional approach is the speed and relative ease with which a survey can be conducted. Its completion does not depend on the recovery of past details nor on future developments. It is therefore highly appropriate to the study of chronic effects on health such as a raised blood pressure in workers exposed to cadmium, or factors of possible relevance to health such as blood lead levels, or the association between disease and environmental factors such as coffee or tea drinking and evidence of ischaemic heart disease.

The results of a cross-sectional survey are critically dependent on representative population sampling. This necessitates a definition of the target population, the listing of all eligible subjects, and the selection at random of the number of subjects required for the survey. The main limitation of the cross-sectional survey, however, is that cause and effect cannot be distinguished. Because all the observations are made at a single point in time, the order in which the various factors manifest themselves is unknown. For example, a dilemma in the lead debate is whether or not there is an association between blood lead level and the intellectual performance of children. Even if conclusive cross-sectional evidence were to emerge demonstrating an association, cross-sectional evidence will shed no light on which factor is causal, that is whether high blood lead levels cause a lowering of IQ or whether children with lower IQs raise their own blood lead levels through *pica* and other habits relevant to lead absorption.

2.3. Prospective Studies. These are powerful, but very difficult to conduct. A group of subjects are identified and then observed to detect the development of a disease, or the incidence of a new disease. The group followed is usually known as a cohort and hence these are 'cohort studies'. In addition to testing hypotheses about diseases, changes in factors of possible relevance to disease, such as blood lead levels, or serum cholesterol, can be studied. The overall strategy is to relate events or changes to the levels of factors at the commencement of the follow-up period; for example, the occurrence of strokes or the development of heart disease can be related to previous blood pressure. Or changes in blood lead levels can be followed in a cohort of children and related to age and to changes in exposure to lead from various environmental sources.

Such an approach would, for example, be appropriate in sorting out the uncertainties in the association between cadmium and blood pressure.[5,6] In brief: on present evidence it cannot be stated whether raised levels of cadmium in some subjects with hypertension are causal and preceded the rise in blood pressure, or whether renal damage led to the hypertension and at the same time led to an accumulation of cadmium in the renal tissue. Indeed a third mechanism has been suggested in this situation which underlines the need for prospective evidence, namely the fact that some anti-hypertensive drugs can chelate metals and promote the elimination of cadmium from the body.[7,8]

Many aspects of a prospective study have to receive very careful consideration, above all the proportion of subjects which is lost to follow-up. Subjects who are not readily traced are likely to include, on the one hand, some who are fit and have moved to other jobs and, on the other hand, some who have moved because of illness or have died. While there is no single criterion for judging the acceptability of a prospective study, very high rates of follow-up can be achieved. An outstanding example was a thirty year follow-up by Cochrane and his colleagues of 8526 men in which only 11 (0.1%) were not traced.[9] In the UK the use of National Health Service records held centrally by the Office of Population Censuses and Surveys (OPCS) has made the task of follow-up relatively simple.

2.4. Retrospective Follow-up Studies.

These, as usually conducted, represent an attempt to obtain prospective data retrospectively. That is, a group of subjects, some of whom in the past were exposed to an environmental factor, are identified and death or survival is ascertained. Old factory rolls are a frequent basis for a retrospective follow-up study. One may of course attempt to relate past exposure to morbidity rather than mortality, but if this is done very great care is required because the retrospective ascertainment of illness is highly susceptible to subjective bias. The use of mortality as the measure of effect is therefore much to be preferred and it is reasonable to suppose that if any exposure is harmful to health then it will show in an increase in mortality and that this should be detected if the size of the exposed cohort is large enough. The size of follow-up studies is therefore of particular importance, because it is all too easy for a study to be based on whatever exposed subjects are readily available and then for conclusions to be stated in terms which imply that there is no effect.

Again, as in the prospective study, the most important single aspect of any follow-up study is the completeness of the identification of the members of the defined group. Subjects who are the most difficult to trace are likely to include those who are ill and have perhaps moved away to be cared for by relatives, or who have died, but will also include those whose work or interests have caused them to move, that is, subjects who are likely to be very fit. The balance between these various causes for moving away, and hence the reasons for failure to follow-up, will vary from one situation to another in a quite unpredictable way. Again, however, the use of records held at OPCS greatly facilitates follow-up of subjects within the UK.

One design aspect of the follow-up study deserves very careful consideration. This is the fact that the period for which estimates of exposure are made must not overlap with the period of follow-up. This is mandatory in studies in which death is the outcome measure, such as, for example, studies of heart disease mortality in workers exposed to carbon disulphide. Difficulties in interpretation can also arise if the clinical outcome of exposure is a serious illness which is likely to lead to the discharge or cessation of exposure of an affected subject. Very careful consideration should be given to situations where the effect of exposure may be relevant to the development of disease, such as would be the case in a follow-up study of changes in blood pressure in workers handling cadmium.

At first sight the ideal follow-up study would seem to be based on a defined cohort of subjects which is followed for a defined period of time, during which both exposure and deaths are measured. Such a design, however, would make it inevitable that those subjects who survive, or those who remain exposed and whose exposure is not terminated by illness, will have, on average, greater exposures to the hazard than those who succumb either to death or disease. It is necessary therefore, however inappropriate it may seem, to study the survival of a cohort *after* a defined period of exposure. For example, Tiller, Schilling, and Morris[10] defined groups of textile workers in terms of their exposure to carbon disulphide prior to 1949 and then examined mortality during the period 1950 to 1964.

2.5. Intervention Studies. These are the most powerful source of evidence in epidemiology and they are widely used in medical research, usually as the randomized controlled clinical trial. There is little place for such strategies in the study of possible hazards, because intervention would clearly have to be in terms of a reduction of an exposure rather than, as is more usual in medical research, the exposure of patients to an agent such as a drug.

Intervention studies may occasionally have a place in environmental research, but always in terms of the effect of reducing exposures to a hazard. Such an approach has ben used to evaluate the effect of the removal of lead pipes[11] on blood lead levels of residents in certain dwellings and the effect on blood lead of hardening a plumbo-solvent water supply.[12] The most powerful design for an intervention study, however, is the randomized controlled trial based on individual subjects. This study design is recognized in medicine as the most powerful investigative strategy available and the rules for its conduct have been well worked out.[13,14]

3. Pitfalls: Errors in Design

Epidemiology has become a relatively precise science and its value in medicine is widely appreciated. So too are its limitations: the difficulties in achieving a high response rate, in identifying and controlling confounding factors in the examination of an association, and the ultimate dificulties in distinguishing causation from association. While the value of community-based studies seems to be recognized outside medicine, the need for the strict application of epidemiological procedures, and the limitations imposed on conclusions drawn from studies in which these procedures have been compromised, does not seem to be widely understood. In fact, any compromise of the classical design features will limit the value of a study and must be taken into account in the interpretation of the results.

3.1. The Need for a Prior Hypothesis. Good epidemiological research always involves the testing of a prior hypothesis. The hypothesis must be carefully defined, the measures of outcome or effect defined, and the measurements of the hazard or environmental factor of possible relevance carefully planned. It is important that hypotheses be limited to one, or at most a very few, and if there are several these should be ranked in some order of clinical importance before analyses commence. Further, the assessment of statistical significance should allow for multiple testing if there are several related hypotheses. It is often not recognized that if a number of possible outcomes are defined then strictly each of these constitutes a separate hypothesis. Clearly, if there are 20 possible symptoms or effects of an exposure, and if these are independent of each other, then on chance grounds alone (at *p* < 0.05), one will show a significant association with the exposure. While this danger can readily be appreciated when outlined in this way, it is all too easy, if the prior hypothesis is stated in vague terms, such as that exposure to a hazard causes 'ill health', to examine one symptom or index of health after another until something turns up! Unfortunately, it is difficult to define 'data dredging' but great care must be exercised in interpreting the results of multiple analyses. However, Armitage[15] has commented that 'informed judgement is likely to be a safer guide (to the interpretation of multiple contacts of a rather disconnected nature) than any body of theory known to us at present'.

It is of course, permissible to examine a body of data after the prior hypothesis has been tested in order to see whether or not there are associations other than those which had originally been defined. However, this procedure must be recognized as an attempt to *generate* hypotheses. It is quite a distinct process from the *testing* of a hypothesis. Further, if new hypotheses are generated there is no alternative to the setting up of new studies to test hypotheses so generated.

An example of this occurred in a study of congenital malformation in infants and trace elements in the water supply. The concentrations of 12 trace elements were measured and related to the incidence of neural tube malformations in 48 defined areas. There were no prior hypotheses and in the event two trace elements showed consistent and significant associations with incidence.[16] This was regarded by the authors to have been a hypothesis-generating exercise and so the same workers went on to conduct a retrospective case-control study to test the hypotheses generated earlier. The hypotheses were not supported.[17]

Another danger with respect to the hypothesis under test is that detailed analysis leads to its being changed during the conduct of the analysis. This clearly happened in a study of chromosomal damage and occupational exposure to hair dyes.[18] The planned test of the hypothesis led to its rejection because 'no significant differences in chromosomal damage were found in peripheral-blood lymphocytes from 60 professional hair colourists compared with 36 control subjects closely matched for age and sex'. However, the authors proceeded to construct new sets of cases and controls on the basis of whether or not the subject's own hair was dyed or not and they present and interpret the results of this *'post hoc'* act of hypothesis-generation as though they had emerged from a valid act of hypothesis testing. This particular study has been ruthlessly dissected by Feinstein[19] and is of interest because it illustrates a number of pitfalls.

3.2. A Representative Population Sample.

Epidemiology is not simply the study of large numbers. It is basically the study of representative samples of subjects. Strictly, the results of a study only apply unequivocally to the patients or subjects studied but results can be extrapolated to a larger group provided the subjects studied were representative of the larger group.

Undoubtedly, a call for volunteers represents one of the worst possible approaches to a study. The reasons why persons may offer themselves may include a belief that they are unusually healthy and that their participation may help others, or a belief that they may be ill and that there may be some benefit in the study

for them.   Even haphazard selection by an investigator can
introduce   all kinds of bias.   An extraordinary example of
selection   bias   is given by Cochrane[20] in commenting on a
comparison of a certain skeletal ratio in cases of Marfans
syndrome and 'unselected normal people'.   The distribution of the
index in the normal subjects shows a remarkable truncation
indicative of a highly biased selection of 'normal controls'.

Clearly, in the case of a possible occupational hazard it is
seldom possible to base a study on all exposed workers, or on a
random sample of the total exposed workforce.   It may only be
possible to study all workers, or a random sample, in one or a few
selected factories.   This represents a compromise, but it may be
acceptable provided there is no reason to suspect that the
factories selected, or the exposure levels *etc.* at the time of the
study, are unusual in any way.   In environmental studies, however,
there is seldom any adequate reason to justify an alternative to
the random sampling of a defined population, for example, a one in
ten sample of all residents within a defined area.   Haphazard
selection or a call for volunteers is utterly unacceptable.

Another aspect of representativeness which is often neglected
is the response rate amongst those selected.   However good the
initial sampling, representativeness can be lost if the proportion
not examined is high.   A remarkable example of this occurred in a
study by Beck *et al*.[21] of byssinosis, a compensatable industrial
disease in cotton workers.   They state that of 377 eligible
workers 217 (58%) were not traceable or had moved away.   Various
indices in these were compared with those in a group of controls
who represented under 60% of the total eligible subjects.   The
reasons for refusal to co-operate are likely to be so very
different in the two groups that comparisons of the small
proportions who did co-operate are likely to be quite misleading.

The selection of subjects for a case-control study involves
certain difficulties and these must be considered in detail.   It
will be remembered that the basis of this strategy is a comparison
of the past exposures of cases with a disease, or an effect, and
of controls who do not have the disease.   The use of hospital
patients as controls has obvious attractions because they are
readily available, they have time to spare and they are usually
co-operative.   However, the opportunities for bias are enormous,
and it may be that the control patients are in hospital for a
condition which shares aetiological features with the disease
under study.   While there is no ideal way to select controls, any
departure from the random selection of individuals within the
defined population from which the case has come represents a
compromise.   It is always possible to draw, for each case, a
control subject selected at random from the electoral list for the
area of residents of the case, or the list of patients held by the
general practitioner of the case.   In the former case the age of
subjects is unknown when selected and it may be necessary to draw
a large number of potential controls, ascertain the ages of these
in a preliminary enquiry, and then select age-matched controls for
the cases.

The matching of controls with cases is a reasonable procedure
because it can increase the efficiency of comparisons but it needs
careful consideration.   Matching for sex and for age are usually
of obvious merit, as is matching for further attributes such as
area of residence or social class.   Matching for other criteria,

such as smoking habit, may have advantages, but is usually very difficult to achieve in practice as such details are not usually known until the control has been interviewed. Furthermore, the more numerous the factors involved in matching, the more difficult it is to find a control for a case. Matching is therefore best kept to a minimum, and allowance made statistically in the analysis for differences between the cases and the controls in factors relevant to the hypothesis under test. Another reason for care in matching is the possibility that differences in a factor of possible relevance are removed by a matching procedure. For example, matching cases and controls for social class may remove differences in environmental or socially determined variables which are of relevance to the disease under consideration.

One aspect of matching seldom receives attention. That is the bias which can be introduced if a control subject refuses and another is substituted. This seems to be common practice and yet it clearly can introduce bias, control subjects who are ill, or who have recently had medical advice or treatment, being more likely to refuse to co-operate. Strictly, if either a case or a control in a matched study has to be omitted because of refusal to co-operate or any other reason, then the case-control pair should be omitted. If this is followed then the approach is likely to be of low efficiency, but if it is not, then the study design is compromised and the conclusions are invalid.

An extraordinary procedure, which is occasionally adopted, is to transfer controls to the nominal list of cases (or *vice versa*) during the conduct of the study. There is a great temptation to do this in drug trials, patients randomized to the drug being counted as controls (or omitted) if they admit to not having taken the drug. Clearly this is a most dangerous procedure because patients whose illness is getting worse are more likely to stop a treatment than those who are improving. The bias is therefore obvious in such a situation but the procedure is just as unacceptable in case-control studies of a hazard. For example, in a study of hair tinters, controls were volunteers and any of these who had ever been exposed to hair dyes were transferred to the 'tinter' group.[18] The bias arises here because, of those who had had past exposure, only those who had had symptoms of sensitivity *etc.* may remember the exposure.

An increase in the sensitivity of a case-control comparison, and a reduction in the bias introduced by the omission of subjects, is achieved if several controls are included for every case.[1] In effect, this reduces the error of estimates of differences but the reduction, and hence the increase in efficiency, is small beyond a ratio of three controls for each case. Again difficulties will arise if there are omissions, and comparisons will be unbalanced if there are different numbers of controls for some of the cases.

3.3. Relevant Outcome Measures. There is no fully adequate substitute for valid clinical estimates of the effect of a hazard and it is important that the limitations in physiological and biochemical variates are appreciated. If a hazard is shown to cause death, disease, or disablement then there is no argument: the effect can be interpreted. However, if raised levels of a substance to which subjects are exposed are found in blood or urine, no conclusions of clinical relevance can be drawn. Furthermore, if changes in biochemical or haematological variates

are found in exposed subjects, it cannot necessarily be assumed
that these are harmful.

3.4. Realistic Estimates of Exposure.

The relevance of
environmental measures to the actual exposure of subjects is a
most difficult aspect of environmental research. Firstly, there
are obvious difficulties in estimating a subject's 'life time'
exposure to any hazard and the best that can often be done is to
group subjects into those who are likely to have had different
levels of exposure. A method of attempting to combine levels and
durations of exposures which involves the construction of an
'exposure matrix' has been suggested.[22] However, further
difficulties arise if a need is seen to discriminate between
low-level chronic exposures and intermittent high-level exposures.
For example, in the rayon industry it has been suggested that
maintenance engineers who had massive exposures to carbon
disulphide on the odd occasion when a machine broke down and
required repair work, would have a greater prevalence of disease
than the workers who ran the machines. Unfortunately adequate
data on exposures are unlikely to be available to enable such a
hypothesis to be tested.

Other difficulties arise when the hazard being studied is
widely distributed. For example, lead occurs in food and water,
in dust, in soil, and in air, and many studies have attempted to
evaluate these sources in terms of their contribution to blood
lead levels. The measurement of each source is difficult and,
although some compromise is usually unavoidable, a number of the
sampling procedures which have been used are likely to be of
little relevance to the exposure of subjects. Dust taken from the
roadside gutter[23,24] or from under a carpet[23] is of unlikely
relevance to human exposure, and while the measurement of dust
transferred from dirty hands to a sticky sweet identifies a
possible route, many assumptions are required if use is to be made
of such data. Strangely, some have been prepared to make such
assumptions, and, for example, an estimated 'sucking rate' has
been applied to lead on hands to give an estimated absorption of
lead.[25] Such procedures are at best useless but may be frankly
misleading.

Particular difficulties arise with regard to smoking, whether
this is of prime interest in a study or examined only as a
possible confounding factor. Particular difficulties arise in
dealing with ex-smokers. The custom in the past has usually been
to amalgamate data for non-smokers and ex-smokers, and to consider
current smokers separately. While the wisdom of this might be
questioned, most subjects who formerly gave up smoking having done
so because of an adverse effect on health, it probably mattered
little what was done because ex-smokers were formerly relatively
few. Now, however, social and economic reasons, together with
fears of later effects, are leading an increasing proportion of
subjects to give up smoking. Not only therefore are there many
diverse reasons for quitting, but the numbers of ex-smokers are
now relatively large and what is done with them is likely to be
important.

On the other hand, measurements of exposure do not have to be
precise to be useful. Grouping of occupationally exposed subjects
by job or department often gives evidence of a graded, or 'dose-
response', relationship. There are numerous examples of this in
the literature and a particularly elegant example showed a marked

gradient in the age-related prevalence of pneumoconiosis in groups of slate workers defined by broad occupational groupings within the industry.[26]

Another very different example of a relatively crude but realistic measure of exposure which has proved useful relates to the exposure of young children to environmental lead. Many studies have failed to detect a relationship between blood lead and lead in the household dust of the child's dwelling.[27] A number of studies, however, have measured lead on the hands of children, either by washing with dilute acid[28] or by removing dirt with a 'wet wipe'.[29] This measure is obviously very crude and yet a significant correlation between hand lead and blood lead has been repeatedly demonstrated[30,31] and, in one study, hand lead measured on a single occasion by a 'wet wipe' technique was found to be the most important determinant of blood lead in young children.[32]

Special problems arise in the estimation of exposure to airborne pollutants. McAulay *et al*.[33] went to enormous lengths to gather realistic data of dust levels in flax mills by sampling in the immediate environment of a random sample of workers in the total employed work force in the whole industry in Northern Ireland. Such efforts are commendable, but resources are seldom likely to be available for such detailed measurements. Furthermore, it must still be admitted that the immediate environment of a worker, as defined by McAulay and his co-workers, represented a compromise regarding what was actually inhaled by the workers; in any case assumptions had to be made about the representativeness of the time period of the sampling, changes in dust levels over time, *etc.* The use of personal sampling devices worn by exposed subjects has much to commend it but these are not without problems.[34]

The sampling of air in relation to the exposure of the general population to pollutants such as lead, sulphur dioxide, *etc.* is especially difficult and whatever technique is adopted questions will arise about the relevance of the measurements made to the true exposure of the subjects. Most people spend much of their time indoors and it is strange therefore that measurements of airborne lead in the USA are made '....2 to 15 m above ground, 20 m from trees, 5-100 m from roadside.... '.[35] Of even less relevance to human exposure are measurements made in petrol filling stations, in a motorway tunnel, and in the central reservation of the M4 motorway.[36] Yet this last is one of the most frequently quoted sources in the UK literature on atmospheric lead pollution. The degree to which outdoor air measurements relate to the true exposure of individuals is unknown but, for a start, indoor measurements would seem to be much more appropriate[37,38] and particularly so in studies of young children.

Water sampling also poses great difficulties. The desire of technicians in the water supply industry seems to be to promote the use of a carefully standardized procedure for water sampling,[39] presumably because this will increase the reproducibility of repeated measurements and assist in the examination of long-term trends, and in comparisons with supplies in other areas. However, subjects in the general population do not draw their water according to any defined procedure.

3.5. Adequate Size of Study.   Studies which are too large represent a waste of resources but, not surprisingly, this is a very rare fault! Studies which are too small, on the other hand, not only waste resources, but may lead to misleading conclusions. This last is particularly likely if the results of a study are used as a basis for a statement that there is 'no effect' or 'no association' between two variables. That is, the 'sensitivity' of a study must be examined in relation to all possible conclusions. It is of course much to be preferred if the likely sensitivity of a proposed study was a major consideration at the time the study was designed.

Sensitivity is one of the most important aspects of biological research both in relation to the design of studies, and in relation to the interpretation of results. To take a very simple example first: suppose one wished to estimate the average exposure of a community, or of a defined group of subjects, to a hazard. One would select a representative (*i.e.* random) group and base an estimate of the overall mean exposure on that of the random sample. The degree of confidence one can put in that sample mean is clearly related to the numbers drawn, but the relationship is with the square root of the number in the sample. That is, in order to double the confidence one can put in a sample mean as an estimate of the population mean, the number in the sample must be increased four times.

Usually, however, the question asked is very much more complicated and this is best illustrated by an example. It might be desired to examine the effect on blood lead levels in the community of the reduction of lead in petrol which took place at the beginning of 1986 in the UK.[40] One might assume that the widespread use of low-lead petrol would lead to a reduction in mean ambient air lead levels in a major city in England from, say, 0.7 μg m^{-3} to 0.3 μg m^{-3}, a reduction of 0.4 μg m^{-3}. A ratio between air lead (in μg m^{-3}) and blood lead (in μg dl^{-1}) has been variously described as one[41] and just under two.[42] One might therefore reasonably predict a change in mean blood lead level following the introduction of low-lead petrol (and assuming no errors in estimation due to laboratory drift *etc.*) of between 0.4 and 0.8 μg dl^{-1} blood. Blood lead levels in a typical urban area in England range between about 5 and 25 μg dl^{-1}, that is, the S.D. of the distribution of blood lead is about 6 μg dl^{-1}. On these assumptions it can be shown that to have a reasonable chance (say, a power of 80%) of detecting a reduction in blood lead levels following the reduction of petrol lead as described, one will need a sample size for surveys before and after the change of between 900 subjects (if the air lead:blood level ratio is two) and 1700 subjects (if the ratio is one). Furthermore, if one is somewhat more cautious and a higher power (90%) and higher levels of statistical significance are chosen then the sample sizes need to be between 3500 and 6700 for a realistic study.

Several important conclusions can be drawn from this example. Firstly, the estimation of a sample size for a study is a most important part of planning. Most of us are repeatedly surprised by the large numbers required for a proposed sensitivity. Secondly, efficient estimations require a number of assumptions; if this were not so and all parameters were known there would have been no need for the study in the first place. It may be that some of these assumptions (for example, the range of a variate in the population) can only be made after a pilot study has been conducted. The most important conclusion, however, is that small and badly planned studies waste valuable resources and are unlikely to yield useful results; indeed they are likely to lead to misleading conclusions and perhaps to a further waste of resources.

One further aspect of this matter of sensitivity deserves consideration. In any report of a study in which the null hypothesis is not rejected, that is a study in which 'no effect' is reported, then it is most important that an estimate of the likely sensitivity of the study is stated. This can always be in terms such as '....had an effect equivalent to such and such occurred, this study would probably have detected it at conventional levels of significance'. Careful consideration of this aspect helps to prevent the dismissing of a possible effect on inadequate evidence.

3.6. The Control of Bias.   The design of research studies should ensure the elimination or the control of bias, and the confidence which can be put in conclusions drawn from a completed study is dependent largely on the degree to which one can be satisfied that bias has been adequately allowed for. Some of the dangers arise from the fact that, while certain sources of bias can be identified, one can never be sure that all sources are known. In clinical trials bias is controlled by a number of measures, notably the use of placebos and the 'double-blind' recording of symptoms and effects. That is, the effect of a drug will be compared with the effect of an identical, but ineffective preparation (such as a dummy tablet) and both patient and doctor will be kept unaware of whether the drug or the placebo is being given. This level of rigour is seldom possible in environmental research, but at the very least the observer who makes measurements of the possible effects of an exposure should be 'blind' with respect to the level of exposure of each subject examined.

In general, researchers are themselves a potent source of bias since they are usually attempting to 'prove' or 'disprove' something. However conscientiously their observations are made, there is always the danger that measurements are unconsciously affected by expectations. Measurement of blood pressure is a good example since there is a substantial subjective element in conventional sphygmomanometry. The assessment of symptoms is even more open to bias, being dependent to some extent on the interest of the interviewer and the concern of the subject interviewed. Recall of past exposures is also open to bias, patients with symptoms or disease being much more likely to recall past events than unaffected subjects.

Thirdly, group bias may arise if the groups being compared are made up in differnt ways. Hill[43] gives an example involving breast-fed and bottle-fed babies. If a breast-fed baby is not thriving, its mode of feeding will probably be changed to the bottle, whereas a bottle-fed baby which ails is likely to get another milk preparation. So when breast-fed and bottle-fed babies are compared, the breast-fed group will always tend to be healthier. This kind of bias arises in any study in which subjects are more likely to drop out of the exposed group than the control group and the reports of such studies should be scrutinized to see what the investigators did with 'drop-outs'. If they were excluded from the analysis, or (worse) added to the control group, the results may be totally misleading.

Most physiological variables fluctuate over time. It follows that if we select those persons who at any moment have the most extreme values of a variate (*e.g.* those with the highest blood pressures) and re-examine them after an interval, their values will tend to be less extreme than they were before. This 'regression to the mean' should be suspected in any reported studies of extreme values which changed after an interval, unless a control group was used or the whole set of subjects (including those with less extreme values) were re-examined.

Perhaps one of the best studied biases is that which has come to be known as the 'healthy worker' effect,[44] though studies other than occupational ones are susceptible to this bias.[45] Persons who are ill or disabled tend not to remain in employment and so 'workers' tend to be healthier and have a better survival than the general population. Clearly the effect is most marked in short-term follow-up studies. For example, in a study of mortality and exposure to flax dust the standardized mortality ratios (SMRs), using the general population as standard, were 56 and 53 in males and females during the first five years of follow-up, 72 and 70 in the second five years, and 87 and 91 in the subsequent seven years.[46] None of these ratios approaches 100, which would be expected if mortality in the workers was the same as in the general population. This 'healthy worker' effect is likely to be inherent in any selected group (*i.e.* blood donors, persons interviewed on the street, samples of friends and neighbours, *etc.*) and illustrates again the inappropriateness of such groups for serious research.

3.7. The Dangers of Confounding. Perhaps, 'confounding' can be said to represent the most basic of all the many sources of uncertainty, and, amongst other limitations, it means that no association, however convincingly described, can ever be equated with causation. Confounding is said to occur when an association is detected between two factors, which is either wholly or partly due to the fact that both these factors are dependent on a third factor.

In the evaluation of an association between two or more factors it is essential that the effects of all other factors relevant to the association are controlled. Thus, in a study of, say, the contribution of air lead to blood lead, other sources of lead may act in a confounding manner; that is, areas of high air lead levels, such as inner cities, are likely to have high levels of lead in other sources, such as water, dust, and peeling lead paint. Furthermore, socially determined factors are likely to be less favourable in inner city areas: diet may be less favourable with regard to factors relevant to lead absorption;[47] smoking and alcohol, which are associated with fairly large differences in blood lead levels,[48] are both social class related.[49,50] Unless these and other relevant sources are measured and controlled, the effect of any single source will be over-estimated.

Control of confounding factors is one of the most difficult aspects of epidemiological research. Not only is it often dificult to measure confounding factors, it is difficult even to identify all the relevant factors with certainty. On the other hand, 'over control' is possible, and the removal of, say, the urban-rural effect, or the social class effect, may remove some of the effect of interest. For example, it is reasonable to control the association between blood lead and intelligence (IQ) in children for the confounding effect of the mother's IQ, because this is relevant both to the IQ of the child and to the cleanliness of the child and of its environment, and hence its blood lead level. However, doing this may remove some of the lead effect itself, as the mother's own IQ may have been reduced by exposure to lead.

Examples of the 'under-control' of confounding factors are legion in the literature. For example, studies of air lead, based on children living around smelters together with other groups of urban, suburban, and even rural children,[28,51,52] are likely to have many uncontrolled sources of confounding. In addition to differences in housing, the children of lead workers have higher blood lead levels than children living in adjacent houses, presumably due to lead-contaminated dust brought home on the clothes and the shoes of workers.[53,54] In fact, some authors[51] comment repeatedly on the difficulties in isolating the likely effect of air due to the high inter-correlation of all the environmental sources, though others[42] have used their data much less critically.

4. Conclusions

Epidemiology is not easy and the perfect study is unattainable. Very careful attention must be paid to every aspect of the design of a study. In particular, the relevance of estimates of exposure to the likely true exposure should be carefully considered, the sensitivity of a proposed study should be estimated, all measurements should be made 'blind' as far as possible, and all likely sources of bias identified and eliminated or controlled.

In fact, however, the greatest abuses of epidemiology are often committed by those who uncritically use evidence from research studies without due regard to their limitations, rather than by those who conducted the studies. As the editor of *Nature*[55] commented with reference to the recent lead debate, '....the case.... has been shamefully and dishonestly exaggerated by (those who) assert that white - or rather grey, is black'.

5. References

1. P. Cole, *J. Chronic Dis.*, 1979, **32**, 15.

2. D. L. Sackett, *J. Chronic Dis.*, 1979, **32**, 51.

3. R. I. Horwitz and A. R. Feinstein, *Clin. Stud.*, 1979, **66**, 556.

4. A. R. Feinstein, *J. Chronic Dis.*, 1985, **38**, 127.

5. S. C. Glauser, C. T. Bello, *et al.*, *Lancet*, 1976, **1**, 717.

6. P. E. Cummins, J. Dutton, *et al.*, *Eur. J. Clin. Invest.*, 1980, **10**, 459.

7. P.O. Wester, *Acta Med. Scand.*, 1973, **194**, 505.

8. J. M. McKenzie and D. L. Kay, *NZ Med. J.*, 1973, **78**, 68.

9. L. K. Atuhaire, M. J. Campbell, *et al.*, *Br. J. Ind. Med.*, 1985, **42**, 741.

10. J. R. Tiller, R. S. F. Schilling, *et al.*, *Br. Med. J.*, 1968, ii, 407.

11. H. F. Thomas, P. C. Elwood, *et al.*, *Nature*, 1979, 282, 712.

12. P. C. Elwood, K. M. Phillips, *et al.*, *Human Toxicol.*, 1983, 2, 645.

13. R. Peto, M. L. Pike, *et al.*, *Br. J. Cancer*, 1976, 34, 585.

14. R. Peto, M. L. Pike, *et al.*, *Br. J. Cancer*, 1977, 35, (1), 1.

15. P. Armitage, *The Statistician*, 1979, 28, 171.

16. M. S. Morton, P. C. Elwood, *et al.*, *Br. J. Prev. Soc. Med.*, 1976, 30, 36.

17. A. S. St. Leger, P. C. Elwood, *et al.*, *J. Epidemiol. Community Health*, 1980, 34, 186.

18. D. J. Kirkland and S. D. Lawler, *Lancet*, 1978, ii, 124.

19. A. R. Feinstein, *Lancet*, 1978, ii, 627.

20. A. L. Cochrane, *Postgrad. Med. J.*, 1965, 41, 440.

21. G. J. Beck, L. R. Maunder, *et al.*, *Am. J. Epidemiol.*, 1984, 119, 33.

22. B. Pannett, D. Coggon, and E. D. Acheson, *Br. J. Ind. Med.*, 1985, 42, 777.

23. R. M. Harrison, *Sci. Total Environ.*, 1979, 11, 89.

24. J. P. Day, M. Hart, and M. S. Robinson, *Nature*, 1975, 253, 343.

25. R. Stephens, *Int. J. Environ. Stud.*, 1981, 17, 73.

26. J. R. Glover, C. Bevan, *et al.*, *Br. J. Ind. Med.*, 1980, 37, 152.

27. P. C. Elwood, *Sci. Total Environ.*, (in press).

28. H. A. Roels, J.-P. Buchet, *et. al.*, *Environ. Res.*, 1980, 22, 81.

29. J. J. Vostal, E. Taves, *et al.*, *Environ. Health Perspectives*, May 1974, p.91.

30. J. W. Sayre, E. Charney, *et al.*, *Am. J. Dis. Child.*, 1974, 127, 167.

31. E. Charney, J. Sayre, and M. Coulter, *Pediatrics*, 1980, 65 (2), 226.

32. J. E. J. Gallacher, P. C. Elwood, *et al.*, *Arch. Dis. Child.*, 1984, 59 (1), 40.

33. I. R. McAulay, G. C. R. Carey, *et al.*, *Br. J. Ind. Med.*, 1965, 22, 305.

34. G. C. Lee, 'Occupational Hygiene', Blackwell Scientific Publications, London, 1980, p.39.

35. C. F. R. 40:58, 'Ambient Air Quality Surveillance', 1982, Quoted in Ref. 2: p.4-2.

36. Transport and Road Research Laboratory, Department of the Environment Report LR 545, 1973.

37. T. D. Tosteson, J. D. Spengler, and R. A. Weker, *Environ. Int.*, 1982, 8, 265.

38. P. C. Elwood, *Environ. Health*, 1984, 92 (1), 12.

39. Water Research Centre, Technical Report No.43, Medmenham Laboratory, Buckinghamshire, 1977.

40. Royal Commission on Environmental Pollution, Ninth Report, Lead in the Environment, HMSO, London, 1983.

41. A. C. Chamberlain, *Atmos. Environ.*, 1983, 17, 677.

42. US Environmental Protection Agency, 'Air Quality Criteria for Lead' (EPA-600/8-83-028A) August 1983.

43. A. B. Hill, 'A Short Textbook of Medical Statistics' 8th Edn., Lancet Ltd., London, 1966, p.31.

44. A. J. Fox and P. F. Collier, *Br. J. Prev. Soc. Med.*, 1976, 30, 225.

45. M. L. Burr and P. M. Sweetnam, *Am. J. Clin. Nutr.*, 1982, 36, 873.

46. P. C. Elwood, H. F. Thomas, *et al.*, *Br. J. Ind. Med.*, 1982, 39, 18.

47. K. R. Mahaffey, *Fed. Proc.*, 1983, 42, 1730.

48. A. G. Shaper, S. J. Pocock, *et al.*, *Br. Med. J.*, 1982, 284, 299.

49. P. N. Lee, 'Statistics of Smoking in the UK', Tobacco Research Council, 1976, Research Paper 1.

50. D. Taylor, 'Alcohol: Reducing the Harm', Office of Health Economics, London, 1981, No.70.

51. C. R. Angle and M. S. McIntire, *J. Toxic Environ. Health*, 1979, 5, 855.

52. A. J. Yankel, I. H. Von Lindern, and S. D. Walter, *J. Air Pollut. Control Assoc.*, 1977, 27, 763.

53. W. N. Watson, L. E. Witherell, and G. C. Giguere, *J. Occup. Med.*, 1978, 20 (11), 759.

54. W. J. Elwood, B. E. Clayton, *et al.*, *Br. J. Prev. Soc. Med.*, 1978, 31, 154.

55. Editorial, *Nature*, 1983, 302, 641.

Section 5: Risk Assessment and Case Histories

16

Risk Assessment - General Principles

By D.P. Lovell

BRITISH INDUSTRIAL BIOLOGICAL RESEARCH ASSOCIATION, WOODMANSTERNE ROAD, CARSHALTON, SURREY SM5 4DS, UK

1. Introduction

The developed world has seen dramatic changes in the hazards faced by its inhabitants during the past century. The old scourges of infectious diseases with their high death toll of infants, children, and adults have been defeated by public health measures, vaccination, and improved medical care. Life expectancy has increased considerably and the main causes of death have changed: circulatory diseases, accidents, cancer, and congenital defects now predominate.

Developed countries are naturally concerned to reduce the rate of premature death still further and are now concentrating on potential hazards in the environment which might lead to accidents or illness. Their worry is in no small part due to the publicity surrounding events such as accidents at nuclear facilities (Chernobyl and Three Mile Island) or chemical plants (such as Flixborough, Seveso, and Bhopal) as well as side effects arising from drug treatment (such as Thalidomide, Practolol, and Opren).

Simplistic paraphrasing of Doll and Peto's finding[1] that 80% of cancers are environmental in origin has also concentrated the public's attention on the chemical additions to their environment. There has also been an increase in the concern that new technologies have risks which were not initially apparent or assessed when they were originally introduced so that public scepticism and questioning of |the benefits of chemicals has increased.

The anxiety about the safety of chemicals in the environment has spread to the use of such chemicals as food additives, to pesticide and herbicide residues contaminating foodstuffs, and to the potential hazards of disposal of various forms of chemical waste.

A consequence has been a demand by some for absolute safety either by the provision of totally safe chemicals or by the removal of hazardous compounds from the environment. Anything less than absolute safety implies a level of risk or the concept that a chemical or procedure is unsafe to some extent. The word unsafe may then become synonymous with dangerous, implying, to some, a degree of recklessness.

One, of course takes risks all the time whether in crossing the road, choosing our occupation, or enjoying our pastimes. Some of these risks are taken by individuals, whereas others are taken by society as a whole.

Some of the choices about taking risks made by individuals are carefully thought through, but others are made without consideration. Similarly at the societal level the degree of accounting for the risks varies from decision to decision.

This chapter is about how to assess and manage the risks arising from hazards associated with chemicals. Chemicals can cause harm in a number of different ways such as by explosion, corrosion, acute or chronic poisoning, or by causing cancers, mutation, or birth defects. They may affect individuals as a result of an accidental release, spillage, explosion, or some other failure of an engineered structure. Alternatively they may act by becoming part of an individual's environment, the food, water, or air supply. Different methods of risk assessment may be appropriate for different hazards. It is assumed here that the hazards associated with a chemical have been identified: see Chapters 8 - 15 and 18.

*Risk assessment** is the process of decision making applied to problems where there are a variety of possible outcomes and it is uncertain which event will happen (Table 1). The problem first has to be defined and the different risks have to be estimated. Then the risks are evaluated in relation to other factors or interests involved and choices made between the possible courses of action. This process of estimation, evaluation, decision making, and implementation is called risk assessment.

Table 1

The stages of risk assessment

(1) Identifying the hazard
(2) Estimating the risk
(3) Evaluating the risk
(4) Deciding on the acceptable risk
(5) Managing the risk

2.   What is Risk?

Dictionaries use risk and hazard as synonyms, interchangeable with words like chance, danger, or peril. In the area of risk assessment they are defined more specifically.

*Hazard** is the circumstance where a particular incident could lead to harm, where harm is death, injury, or loss to an individual or to society.

*Risk** is the chance that a particular adverse event actually occurs in a particular time period or after a specific challenge.

Alternatively, hazard is the potential for harm, risk the chance that the harm will actually happen. A chemical may have the potential to be explosive or toxic because of its inherent properties; risk measures the probability of this chemical realizing its potential harm by reaching an individual or vulnerable parts of an invidivual.

See also Glossary of terms

Risk is a concept based upon probability. Probability is measured on a scale from 0 to 1. Values near 0 imply that an event is unlikely to happen, values near 1 that the event will almost certainly occur. The actual probability of an event requires careful definition of the event being studied. The probability of the reader of this chapter dying in the next 80 years is almost 1 but the probability of that individual dying in the next few minutes is close to 0.

Such a probabilistic definition of risk is closely related to the concept of odds used in gambling or making bets. Waging, betting, gambling, or speculating are all forms of risk taking. The calculations of odds or probabilities are derived essentially from two different approaches: either from past experience of similar events, such as tossing coins and rolling dice, or from estimation of likely outcome from some subjective consideration, as in the case of deriving odds for horse racing. Similar methods will be shown later to be used to estimate risks.

Some risks are obviously high. The chances of surviving acute poisoning by high doses of a toxic chemical or exposure to high doses of radiation are likely to be low. Such risks are relatively easy to evaluate. In other cases the high risks may be related to activities which have long-term effects and are therefore less apparent. It has, for instance, been estimated that a quarter of all young men in Britain will die prematurely from diseases related to cigarette smoking.[2] Conversely, there are other events which are extremely rare: the chance of being killed by lightning in the UK each year is 2 in a million (2×10^{-6}) or the chance of being killed by a tropical cyclone in certain parts of the USA is 5×10^{-7}.

It is useful to put such risks into the context of the risks involved in surviving from year to year. Figure 1 shows the death rates per million by age for males and females in England and Wales in 1980. The death rate falls from 1 in 100 during the first year of life to about $200\text{-}300/10^6$ at about 10 years of age before climbing steeply with increasing age to about a 1 in 10 chance of surviving to the next birthday for people aged about 80. Obviously such population estimates hide variations between individuals and critical groups and are averaged over many events. A severely deformed baby probably has far less chance of surviving the first year of life than a healthy baby; a heavy smoker much less chance of reaching 80 years of age than a non-smoker. All such estimates of risk will represent an amalgamation of sub-groups with different risk profiles.

All activities and even non-activity therefore involve some degree of risk and absolute safety can never be guaranteed. It is therefore necessary to decide what level of risk is acceptable and what priorities should determine how risk is to be managed. Such trade-offs of relative risks become a *risk-benefit analysis*.

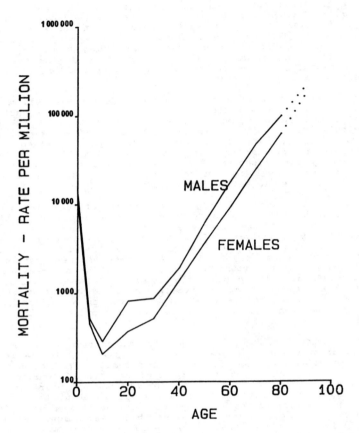

Figure 1 *Mortality rates per million for males and females by age in England and Wales during 1980 (redrawn from Alderson, 1983)*[5]

3.   How Are Risks Measured? - Estimation

We need an approximate estimate of the risks we are likely to take to help make decisions.   In some cases this can be a numerical measure, whereas in others a qualitative statement like big or small may be sufficient.

As individuals it is known how to measure risks subjectively. When we cross a road we estimate the size, the speed, and the distance of an approaching vehicle; we estimate how fast we can walk (or run) across the road, how likely it is that the driver will see us and slow down or swerve to avoid us, how likely we are to slip, and finally how much we *need* to be on the other side of the road.   We have thus made a mental bet about our chances of getting across without being hurt.   The benefit of getting to the other side would then outweigh the risks (hopefully extremely small) of being knocked down. Similar methodologies are used in more formal approaches to risk management.

There are three main approaches to risk estimation (Table 2):

Table 2 *Approaches to risk-estimation*

(1) Estimation of risks from observations on human populations.

    (a)   Actuarial tables
    (b)   Mortality statistics
    (c)   Epidemiological studies

(2) By estimating everything that could go wrong.

    (a)   Event-tree analysis
    (b)   Fault-tree analysis

(3) By expert opinion

    (a)   No observable effect levels for non-carcinogens
    (b)   Extrapolation from dose-response curves for carcinogens

3.1. Estimation of Risks from Observations on Human Populations.

Observations of the risks that the human population is exposed to is the most direct way of estimating risks. Insurance companies have developed large sets of tables for measuring the risks that people are exposed to at different times of their lives. These actuarial tables provide estimates of the risks that different sub-groups of the population are exposed to, and help in the assessment of insurance premiums. Causes of death, for instance suicide, would be classified on the basis of such factors as age, sex, method of suicide, *etc*.

Other sources of estimates of risk are from mortality statistics collected by a range of bodies such as the Office of Population, Census and Statistics (OPCS), Department of Health and Social Security (DHSS), or the Health and Safety Executive (HSE). The Department of Transport, for instance, collects information on road accidents. Table 3 shows some of the risks associated with different occupations. These risks are obviously averaged over many individuals and may overlook sub-groups at greater risk. For instance, the high risk associated with railway staff is probably a reflection of the sub-group of staff who are shunters or workmen on the tracks, whose risks are much higher than those of other rail staff.

Similar tables can be built up for other activities such as risks from various sports or from various medical activities. The risks may be presented either as per million people receiving the treatment or taking part in the activity or per million cases. Table 4 shows some of the risks associated with various medical activities. Such tables of relative risks are useful for identifying sub-groups at risk and helping to determine priorities for action.

There are some difficulties associated with such tabulations. Pochin[3] has called the collection of frequencies 'risk watching' and compared it to collecting stamps or engine numbers, especially when the average frequencies become split into ever finer divisions.

Table 3   *Average annual accidental death rate at work in UK per 10^6 at risk*

Clothing and footwear industries	5
Vehicle manufacture	15
Chemical and allied industries	85
Ship building and marine engineering	105
Agriculture	110
Construction industries	150
Railway	180
Quarries	195
Coal mining	210
Offshore oil and gas (1967-1976)	1650
Deep-sea fishing (accidents at sea 1959-1968)	2800

(from 'Risk Assessment', The Royal Society, 1983)[4]

Table 4   Risks of death from medical procedures (per 10^6 cases)

Vaccination	1
Surgical anaesthesia	40
Childbearing	100
Needle biopsy of liver	200
Immune suppressive therapy for renal transplants	4000
Treatment of ankylosing spondylitis by radiotherapy or radium-224	10000
Former use of 'Thorotrast' as a radiological contrast medium	60000

(from 'Risk Assessment', The Royal Society, 1983)[4]

The nature of the event being studied is also important. The risks arising from various forms of transport depend very much on whether the risks are expressed as per mile travelled, per hour travelled, or per journey. The rankings and the relative risks are altered if expressed on different scales. The death rates per 10^9 km travelled in 1971-1976 for car drivers, pedal cyclists, and motor cyclists were for instance 8, 85, and 165 respectively. The relative risk of cycling would however be much reduced if the relative times to cycle such distances compared with driving were taken into account. Statements such as most pedestrians are killed within 10-15 yards of a zebra crossing, as used in some recent advertisements, do not mean that crossing the road is more dangerous at zebra crossings, rather that more people cross there. Those who cross elsewhere are in fact at far higher risk of being killed. Risk is thus only relevant if compared using sensible bases.

Epidemiology studies are also used for estimating risks to the human population, especially where the cause of the illness or death is more circumstantial and often distant in time, such as smoking and lung cancer. Various epidemiological methods are used to investigate such potential risks. These involve proportional mortality studies where the death rates of an exposed population are compared with a 'standard' population, case-control studies where life histories of people dying or suffering from a disease are compared with people who are not suffering from it, and prospective studies where people exposed to a hazard are followed through subsequent years to investigate future ill-health. Further details of how epidemiological studies are carried out can be found in Alderson[5] and in Chapter 15.

Epidemiological methods are limited to studies of chemicals which have previously been in the environment and where there are sub-groups with different histories of exposure. Novel chemicals or those for which there is widespread general exposure need a more indirect estimate of risk.

3.2. By Estimating Everything that Could Go Wrong.

Epidemiological or actuarial data are not available for many events firstly because there is little previous experience of such circumstances or secondly because the proposed action is novel. An example of the first case is the development of a new chemical plant where little information is available on the safety record of similar such plants (expressed as accidents/plant) as there are few or none of a similar design. An example of the second case would be the development of a new genetically engineered organism and the experiments to be carried out to study its behaviour in the environment.

Estimates of the risks involved in such developments can be obtained by the use of methods which identify all the possible things that could go wrong at each stage, identifying the possible consequences of combinations of such faults, and working out the likelihood of the resulting hazards based upon the probabilities of each individual fault. This is known an *event-tree analysis*. An alternative approach is to determine all the possible hazards and work out how they could have arisen. This approach of working backwards is known as a *fault-tree analysis*. In both cases estimates of the probabilities of the individual faults which might occur are found, from records of how similar components have behaved, from reviews of earlier accidents, or by some form of expert but nevertheless subjective judgement. Such methodology identifies the nature of the potential hazards in complicated plants or processes but mistakes may occur if particular potential accidents are not identified or foreseen or if the estimated probabilities of some of the possible faults are wrong.

3.3. By the Use of Expert Opinions.

Risk estimation of some chemicals may be based upon epidemiological data from exposures of groups such as industrial workers exposed to the compound or alternatively from experimental data derived from studies of the chemical on animals or cell systems to assess somewhat subjectively the safety (or potential hazards) of a chemical.

The purpose of these animal studies is to identify concentrations of a chemical which, within the limitations imposed by the power of the experiment, have no observable adverse effect on the test animals. This level – with the acronym NOEL (no observable effect level) – is by its determination somewhat subjective: what is an observable efect? 'Expert judgement' is used to define levels which are acceptable for human exposure.

In the case of compounds entering food an acceptable (or allowable) daily intake (ADI) is defined by dividing the NOEL from animal studies by a safety factor. This factor is usually in the range of 10 to 1000 depending upon the nature of the chemical, the toxic effects observed in animal studies, the degree of use, and the potential human exposure. The safety factor is often about 100 but the actual value is to some extent arbitrary. Its purpose is to compensate for the low power of animal studies to detect toxic effects, especially at low doses, and differences in sensitivity between the animal species and the exposed human population, and to take into account inter-individual variability within the human population.

In the case of potential cancer-causing agents (carcinogens) an alternative approach to the determination of 'safe' doses is to extrapolate from the results of animal studies.[6,7]   Relatively small numbers of animals are exposed to high doses and the results found are used to estimate the risks at the low doses that the human populations might be exposed to.   Such extrapolation (correctly termed, interpolation) is controversial, as the estimates of the risks at low doses depend considerably on the choice of curve used to fit to the animal data.   Most suggested curves in fact provide 'good fits' to the observed data but the estimates of risk obtained from them may vary considerably. Animal experiments to detect actual effects at low doses which compare with those to which the human population is exposed to would be very large, difficult to carry out, and extremely expensive.

4.   How Do We Evaluate Risk?

Evaluating risk and deciding on 'acceptable' risks is a complex area.   Table 5 lists a number of specific factors which need to be considered before reaching a decision about a specific risk.   Each factor is discussed briefly below.

Table 5 *Evaluating risks and making decisions*

(1) Is there such a thing as absolute safety?
(2) What is an acceptable risk?
(3) The perception of risk
(4) Accepted *versus* acceptable risk
(5) Voluntary *versus* involuntary risk
(6) Are there any offsetting benefits?

4.1. Is There Such a Thing as Absolute Safety?.   On occasions 'absolute safety' has been claimed for a range of chemicals, industrial plants, or processes, or alternatively it has been claimed that no risk is associated with them.

Clearly such claims are inaccurate.   As noted above every activity involves some sort of risk.   This has clearly been recognized by some industries such as the pharmaceutical industry[8] which has been clear in its statement that all drugs have potential risks and side effects.   Industries which claim absolute safety really mean that the risks are very small and are redressed by the resulting benefits.   Unfortunately once the public learn that something which was once presented as 'absolutely safe' now has a risk associated with it they are suspicious.   That distrust grows when further risks, however small, are revealed.

Furthermore, the misplaced pursuit of absolute safety, for example in developing new compounds such as drugs, may cause medically useful chemicals to be rejected.   The existence of unavoidable side-effects may be the other side of those properties of compounds which have pharmaceutical value.

Levels of risk which appear to be acceptable in everyday life were illustrated above.   Can acceptable levels of risk be defined for new activities?

4.2. What Is an Acceptable Risk?. The levels of risk that an individual finds acceptable depend upon a number of factors which will be discussed in more detail later. Society as a whole may determine what level of risk it assumes is acceptable across the population as a whole; this is based upon what risks are routinely accepted. Figure 1 showed the risks of dying between one birthday and the next and Table 3 illustrated some of the risks associated with various occupations. Males over the age of 20 are routinely exposed to risk of death in any one year of >1 in 1000 (10^{-3}). Various occupations and participants in some sports have higher rates in the region of $3-6 \times 10^{-3}$ year^{-1}. One occupation – President of the United States – carries an annual risk of death by assassination of 2%.

At the other end of the scale are risks which are viewed by some as trivial. The Royal Society Study Group on Risk Assessment[4] considered this to be an annual risk of death of 10^{-6} or less (*i.e.* one death per year per million exposed to the hazard or 50-60 annual deaths per year if the whole population of the UK were exposed equally). This level of 10^{-6} annual risk is used almost by convention by many official bodies to represent the level of acceptable risks. Pochin[3] has pointed out that the risk of a child dying from complications arising from common vaccinations is 1 in a million vaccinations, and of complication arising is 1 in 10^{-5}, while the regular train traveller accepts an annual risk of 10^{-5} of death without too much concern.

In selecting an average risk across an exposed group it must be remembered that there may be sub-groups who may be at much greater risk, such as individuals living close to a chemical plant or the young or old exposed to a potential hazard. Individuals may also regard as serious risks which society as a whole considers to be trivial.

4.3. Perception of Risk. Perceptions of the risks involved in an activity vary amongst individuals.

People use different methods to evaluate their own individual risks. In some cases the perception of a group of people may alter the priorities assigned to reducing competing risks. Such 'collective prejudices' depend upon whether for instance the risk is perceived as voluntary or involuntary. Risks which are 'novel' seem to arouse more concern than those that are 'routine'. Hazards which are delayed in their effect such as cancer are considered more worrying than those that happen immediately. Outcomes which are rare, unpredictable, and catastrophic, such as chemical plant explosions, are viewed as more disturbing than those that are common, regular, and small in size, such as road accidents, even if the overall costs in human life and suffering may be similar. There seems to be a 'dread' component to people's perception of certain types of risk.

The failure of people to conform in their perceptions and their choice of actions to reduce risk according to the statistical and mathematical analyses of the problem should not be dismissed. Aspects of human behaviour may be of considerable importance in any evaluation of risks, as the persistence of cigarette smoking illustrates. People also appear to prefer to save identifiable individuals such as named sick children than to respond to appeals to provide resources to save more lives but less identifiable ones by provision of specific services. This relative unwillingness to contribute to the saving of such lives (called 'statistical lives' by Reissland and Harries[9]) is of crucial importance in the implementation of certain types of risk mangement.

There are other aspects of risks which affect individual perception of risks. These are dealt with in the subsequent sections.

4.4. Accepted *versus* Acceptable Risks. Individuals may look more favourably on those risks of which they have experience than on those that seem to be new; the old risks are accepted. In the extreme case the individual accepts that he or she will die but probably cannot view this as acceptable. 'Natural' contaminants and toxic compounds of food may be considered acceptable even though such fungal and bacterial contamination may cause illness, while food additives whose introduction (or identification) in foodstuffs is to assist in preservation may not be acceptable to some people. Like most such issues of risk-perception other dimensions are relevant, such as who obtains the benefit from such 'improved storage' or use of 'previously unusable food'.

4.5. Voluntary *versus* Involuntary Risks. Reactions appear to be different to risks depending on whether they are voluntary or involuntary. A walk through a picturesque but snake-infested area may be viewed differently by the same individual on an adventure holiday or a business trip. Similar factors apply to individuals taking part in relatively dangerous sports like hang-gliding or motor cycle racing. Individuals willing to take such risks may indeed be unwilling to have such activities curtailed in a free society. At the same time they may be unwilling to accept much lower risks from hazards that they are unable to avoid such as low-level radiation from nuclear facilities or pesticide residues in food. A similar dichotomy occurs between risks which are associated with active or passive behaviour. People appear to prefer to expose themselves to risks when they believe they have some control over the events than when they have no control even if the risks are correspondingly reduced.

4.6. Are There Any Offsetting Benefits?. There are costs to reducing risks. They may arise either directly as a consequence of installing safety equipment or removing hazardous components or indirectly by imposing restrictions on people's choice, freedom of action, or movement. *Risk-benefit analysis* is one method suggested for drawing up a balance sheet of the respective risks and benefits.

The immediate problem is how to compare the two sides of the equation. Attempts can be made to relate risk and benefit to some common scale which, if this is a monetary base, converts the procedure into a form of *Cost-benefit analysis*. Assigning monetary values to some procedures may be relatively easy but in others, such as assessing the benefit of a drug, it is difficult. What is the monetary value of reduced mortality, increased life span, and improved quality of life? Similarly what are the costs of adverse effects of chemicals which may result in accidents, injuries, deaths, birth defects, cancer, or mutations?

Estimating the monetary values of a life, for instance, will always be controversial. You and I will perceive our own life as priceless or of inestimable value as it's the only one we've got.

Others may attempt to place a value on it. Various approaches have been taken and there can clearly be no 'correct' approach to this. They include the value of a life as determined by life assurance companies, court awards (which may reflect adversarial litigation), loss of future earnings (which might discriminate against certain groups such as pensioners who might have zero or even negative values), or questionnaire evaluation. In the last case respondents are asked to say how much they would pay to reduce risks, such as how much increase in an air fare in return for improved aircraft safety: the problem is that people do not behave in the way that they indicate in surveys.

An alternative approach has been to estimate what resources are actually assigned to save lives. The estimates vary greatly from £50 inferred from the failure to implement screening to prevent still births to £20 x 10^6 calculated from changes in Building Regulations to prevent further collapses of high-rise flats such as Ronan Point.[10] The Ministry of Transport in 1976 estimated that a life was worth £42,000: £29,000 in loss to the GNP and £13,000 for grief, suffering, and pain.

In some cases the risk-benefit approach is relatively straightforward. An individual suffering from a disease which will prove fatal in two years is offered a drug treatment or operation which has a 10% chance of providing a cure or a 2% chance of immediate death. The risk-benefit equation is clearly laid out; the decision may though still be difficult, as factors such as the quality of any future life after treatment and personal circumstances are of importance.

The risk involved in vaccination of children against whooping cough is a more complex example. Vaccination provides a clear benefit to society as a whole in that fewer children die of whooping cough. Some vaccinated children, though, die or are brain-damaged as a result of the vaccination. The problem arises in that it is not the child who will be vaccinated that is at most risk of whooping cough but younger children who may otherwise be infected. A requirement of altruism by the parents is complicated further in that the vaccine against whooping cough is given to the child and not to the parent who takes the decision.

The questions that arise are: are all the risks and benefits spread evenly through the population or does one group take the risk and another receive the benefits? Food additives, for instance, improve the storage of food, prevent spoilage, improve the appearance, and allow the presentation of previously unused parts of the original foodstuffs. Do the benefits which result from the more efficient use of the food result in lower costs to the consumer or a greater profit to the manufacturer or both? Are the small risks that the consumer may be exposed to compensated by the benefits they recieve? Would the consumer be prepared to pay more or accept food in a different form to remove a perceived risk to themselves from food additives?

5. Managing Risks

The risks having been estimated and evaluated and an acceptable level decided on, there now remains the problem of how to ensure that acceptable risks are achieved. The process of getting and keeping risks below the level society has determined is the area of *risk management*. The concept that society as a whole

determines an acceptable level of risk is of course an over-
simplification. There may not be a consensus of view in the
evaluation of risk but at some point of time a decision will have
been achieved through some process of public opinion, expert
guidance or governmental action. Such processes may result in an
apparent reordering of the priorities for action from those
arising from a strictly mathematically risk-benefit analysis. It
is doubtful though how any other approach can provide outlets for
concerns which to some may appear irrational yet represent genuine
human worries.

Three approaches to risk management can be identified.

5.1. Absolute Safety. In some cases there have been attempts to
develop a strategy which demands absolute safety. Any evidence of
harm from a process or chemical would be sufficient to suspend its
use. The most obvious case is the so-called 'Delaney Amendment'
enacted by the US Congress in 1958 as an amendment to the Food and
Drug Act. This requires that 'no [food] additive shall be deemed
to be safe if it is found .... after tests which are appropriate
for the evaluation of the safety of food additives to induce
cancer in man or animals'.

Such an approach gives no weight to any potential benefit of
the chemical. (In fact one compound, saccharin, which had been
shown at extremely high doses in animal tests to cause bladder
tumours in rats, was eventually allowed limited use. In this case
the risks to groups such as diabetics if saccharin was banned were
taken into account.) It also provides little guidance on how
toxicological studies are to be evaluated to show evidence or not
of carcinogenicity. Recent issues arising from the Delaney
amendment are discussed in more detail by Kessler.[11]

An issue which arises is how does one determine that a food is
free from a particular proscribed constituent or contaminant? The
level of detection of many chemicals such as fungal contaminants
or pesticide residues was until recently at about 1 part in a
million (1×10^{-6}). Technical developments are now making it
possible to find traces of such chemicals at 1 part in a billion
(1×10^{-9}). The costs of preventing such levels of contamination,
particularly by natural components, could well be prohibitive.

5.2. 'As Low as Reasonably Achievable' (ALARA). A second approach
is the development of regulations and procedures which place a
responsibility on the operator, manufacturer, or user to lower the
level of a hazard to as low a value as is reasonably achievable.
This approach, known by its acronym 'ALARA', is dependent on the
definition of the word 'reasonably'. For example, it is applied
in the field of radiological protection. The International
Committee on Radiologial Protection (ICRP) suggests that use of
radiation should be justified by the benefits of any procedure
exceeding the sum of the costs of protective measures and
radiation-induced detriments to health. Furthermore the
protection from the hazard should be increased until the costs of
further improvements clearly exceed the corresponding reduction in
risk. At this point exposures are described as 'ALARA'.

The UK Health and Safety at Work *etc.* Act (1974) has a similar
concept with a requirement that risks are reduced 'so far as is
reasonably practicable'. Legislation on air pollution uses the
phrase 'best practicable means'. The definition of terms like

clearly or reasonably practicable is of course debatable.

'Reasonably practicable' has been interpreted by British Courts as actions which are not 'reasonably practicable' when the costs are seriously out of proportion to the benefits achieved by a reduction in risk. Such a definition is potentially stronger than the ICRP's 'ALARA' as it implies reduction beyond a break-even point. In practice there is so much uncertainty about any estimates of risks or benefits that there is little difference in the approaches.

5.3. Safety Standards. Another approach to risk management is to set fixed safety standards.

Specific upper limits of exposure are set for some chemicals. 'Threshold limit values' (TLV) are set for over 500 common industrial chemicals in the work place.[12,13] These are based upon airborne concentrations of substances and represent concentrations at which it is believed that regular exposure of the workforce will in general have no adverse effects. They are usually derived from experience of exposure in the industries in the past, from volunteer studies, or from extrapolation from experiments using animals. Such TLVs do not provide an assurance of safety; indeed, some workers may suffer illness or have existing conditions aggravated. They represent instead control levels at which exposure is at present considered reasonable. Exceeding such limits is considered reprehensible and may result in legal sanctions.

TLVs are therefore not 'safe' doses and for many compounds no evidence of a 'safe' dose has or can be attained. The use of TLVs is a means of limiting the concentrated exposure to as few individuals as possible using existing procedures without exposing any single individual to disproportionately high concentrations. (see also Chapter 8).

Another example of an upper limit is the acceptable daily intake (ADI) applied to food additives. As explained earlier, these are derived by applying a safety factor to NOELs observed in toxicological studies. ADIs are considered for all practicable purposes to be safe levels. Occasionally exceeding this level is not expected to be harmful and need not necessarily lead to legal action. The ADI is used as a guideline to food manufacturers in their use of ingredients. Manufacturers have to consider the possible daily intakes of an additive from different foodstuffs in calculating how much additive to add to their product.

Both TLVs and ADIs can be altered, usually by lowering them, either as a result of fresh evidence of risks and potential hazards or because technical advances can reduce the exposures.

6. Risk Management in Practice

In practice, managing the risks associated with chemicals uses a mix of safety standards and the 'ALARA' approach. The best example is the control of exposure to radiation other than from natural sources or medical practices. The amount of information on the hazards associated with radiation, the technology for dose assessment, and the considerable public concern about exposure to levels of radiation make radiation a 'model' on which much of the philosophy of controlling chemical exposures can be based.

The principles of risk management of radiation levels to acceptable levels is based upon concepts developed by the ICRP and in Britain by the National Radiation Protection Board (NRPB). The following guidelines came into operation in February 1986.[14] There is a principal limit on the permissible annual dose to members of the public of one mSv (millisievert) and a subsidiary dose limit of 5 mSv in a year for some years provided that the average annual dose over a lifetime does not exceed 1 mSv (a lifetime dose of 70 mSv over 70 years). The 5 mSv upper limit to an individual is an upper limit and should not be exceeded in any circumstances. Rather than being safe it is the lower boundary of what is unacceptable. The overall estimated risk from an exposure to 5 mSv is 1/20,000 of eventual death from radiation-induced cancer. The permitted maximum occupational exposure of 50 mSv in a year carries a corresponding risk of 1 in 2000.

At the same time the ALARA principle is applied so that procedures should reduce individual exposures to well below 5 mSv per year. The NRPB suggests that such procedures should be aimed at reducing exposures to below 1 mSv. Failure to apply ALARA principles can lead to prosecution even when upper control limits are not breached. Improvements in process design, operations, and other factors should then act as a 'ratchet' to lower the risks progressively over time using the ALARA principle.

Similar management pocedures are applicable to potential chemical hazards except that in some cases the risks individuals are exposed to cannot be considered trivial (in these cases the annual risk of death may be considerably greater than 1 in 10^6). In these cases a closer scrutiny of the relative risks and benefits is needed.

7. Concluding Remarks

Risk assessment has an obvious appeal to the user of mathematical methods. It is important though, to recognize both the strength and the weakness of a reliance on reducing such assessments to numerical values alone. Mathematical statements or descriptions initially appear to remove the 'wooliness' which inevitably surrounds terms such as reasonably or acceptable. There are dangers, though, in making the assessment superficially over-precise, incomprehensible to the less numerate, and alienating to those who believe that it is exremely difficult or even impossible to quantify or put in money-terms some of the risks and benefits involved.

Mathematical approaches to risk assessment help to expose a problem to logical analysis, and to identify areas of uncertainty and the issues where relative merits of different actions are clearly seen. Such an approach may not necessarily ease the acceptance of such appraisal by groups perceiving a risk but will provide the intellectual ground work for the appreciation of why such a decision was made. Mathematical analysis can unfortunately be used for hiding inconvenient or muddled thinking behind a facade of apparent technical and scientific expertise. It is important to realize that mathematical assumptions are still assumptions and require estimates of the errors implicit in them. Using ranges of values rather than only a central estimate is a necessary adjunct to risk assessment and forms the basis of a *sensitivity analysis* which tests how general the findings of an assessment may be.

Society is becoming ever more complex and the public is less inclined lightly to accept 'expert' advice without question. It is therefore important that decisions which appear to have a considerable effect on the environment, health, or even life are assessed as openly as possible. Clearly such an approach requires an openness and willingness to accept ciriticism of the methods used. There is a need for wider understanding of the issues involved in risk estimation – particularly in the uses (and abuses) of mathematical methods – among the general public.

Risk assessment will always be a controversial area. There will never be the 'correct' answers and 'absolute' solutions. To assume otherwise would be to accept that there will be no further changes in the views, values, rights, and duties accepted by society and its individual members over time.

8. References

1. R. Doll and R. Peto, 'The Causes of Cancer', Quantitative estimates of avoidable risks of cancer in the United States today, Oxford University Press, 1981.

2. R. Peto, 'Distorting the Epidemiology of Cancer: the Need for a More Balanced Overview', *Nature*, 1980, **284**, 297.

3. E. E. Pochin, 'Risk-benefit in Medicine', in 'Risk-Benefit Analysis in Drug Research', ed. J. F. Cavalla, MTP Press Ltd. Lancaster, 1981.

4. The Royal Society 'Risk Assessment – A Study Group Report', The Royal Society, 1983.

5. M. Alderson, 'An Introduction to Epidemiology', 2nd Edn., Macmillan Press, London, 1983.

6. Scientific Committee of the Food Safety Council, Quantitative Risk Assessment, *Food and Cosmetic Toxicology*, 1980, **18**, 711-734.

7. M. D. Hogan and D. G. Hoel, 'Extrapolation to Man' in 'Principles and Methods of Toxicology', ed. A. W. Hayes, Raven Press, New York, 1982.

8. J. F. Cavalla, 'Risk-Benefit Analysis in Drug Research', ed. J. F. Cavalla, MTP Press Ltd., Lancaster, 1981.

9. J. Reissland and V. Harries, 'A Scale for Measuring Risks', *New Scientist*, 13th Sept. 1979, p.80.

10. D. Wilkinson, 'Nuclear Waste-Perception and Realities', Papers in science, technology, and public policy No.7. Science Policy Research Unit, University of Sussex, 1984.

11. D. Kessler, 'Food Safety: Revising the Statute', *Science*, 1984, **223**, 1034.

12. American Conference of Governmental Industrial Hygienists, 'Documentation of the Threshold Limit Values', 5th edn., ACGIH Inc., Cincinnati, Ohio, 1986.

13. Health and Safety Executive, Occupational Exposure Limits 1985, Guidance Note EH40/85, HSE, 1985.

14. NRPB, 'Dose Limits for Members of the Public', National Radiation Protection Board NRPG-GS4, HMSO, 1986.

17

Risk to Work Forces

By H. Holden

BICC LTD., PRESCOT, LANCS., UK

---

1. Introduction

The management of risk in industry can be influenced in two main ways:

i)   The selection of low-risk options

ii)   Living with risks which are a necessary part of the industry.

The low-risk option can be promising when one considers the chemicals to be used to make a new compound. Sometimes it is possible to replace a hazardous material with a non-hazardous material. This substitution of materials is one of the few ways in which industry can produce a no-risk situation; the most famous of these is the replacement of white phosphorus in matches by red phosphorus. This took place in the 1920s and eliminated the hideous disease known as phossy jaw which affected the predominantly female workforce. The other area where substitution can be considered is in materials with allergenic potential where the choice of material can be used to reduce the incidence of skin or lung sensitivity. For example, consider the three groups of polymers:

i)   Polyesters

ii)   Acrylics

iii)   Epoxy resins

In this case the polyesters are very low allergens and the epoxys the strongest allergens. Thus it is possible when needing to use a resin to choose one which is of low allergenicity although this is only possible when the choice is one which does not require the adhesive properties and aggressive physical nature of the chosen resins.

To live with the risks of an industry should be a last resort, and should be acceptable only when the risk can no longer be reduced. The reduction of chemical risk is primarily the job of the specialist in occupational medicine and the industrial hygienist, with the chemist helping with the substitution of materials. Certain industries are, however, inherently dangerous, such as mining and quarrying, agriculture, forestry, and fishing. The deaths in these industries and in those of a lower order of risk are shown below.

Table 1[1] *Claims for occupational death benefit*

	1981	1982
Agriculture, forestry, and fishing	30	27
Mining and quarrying	56	58
Chemicals and allied industries	11	10
Textiles	3	7

The measurement of potential risk can be assessed by its effects on the human body. The hazardous material may overwhelm the body's defences in a massive way causing death, or in repeated attacks which destroy enough of the cells of the body to create a chronic and continuing loss of function which eventually causes morbidity and finally mortality. Repeated insults which are not of an extensive nature can be dealt with by the body's natural repair systems thus leaving the organism intact with its functions unchanged.

However, there has been a reduction in the incidence of traditional occupational diseases. In 1975 the incidence of notification of prescribed industrial disease was 166 and in 1982 this had fallen to 30.[2] Prescribed occupational diseases are those diseases which are prescribed in respect of defined occupations or conditions at work in which they are recognized to be particular risks. If someone is found to be suffering from a disease and he has recently been employed in an occupation to which that disease is prescribed, then the disease is presumed to be due to the nature of his occupation.

Thus occupational deafness is accepted as a prescribed disease only if a worker has been employed for 10 years in one or more of the specific jobs described in the rules.

This major change in what one could call the old fashioned occupational diseases is due to many factors but primarily due to an increase in knowledge of the risk in these particular occupations and the response of enlightened management to those risks.

However, there has been an increase in the incidence of allergic and stress diseases and an increased perceived risk of carcinogenesis. There has therefore been a major change in the appreciation in the types of illness which could be due to the place of work. This is an area of constant change; the measurement of risk is attempting to hit a moving target.

The management of risk in industry will be discussed under four headings:

i)   Hazard identification

ii)   Risk assessment

iii)   Exposure risk

iv)   Risk monitoring

2. Hazard Identification

The first question one must ask about a hazard is when does it constitute a threat? One answer is when a worker considers it so. His perception of the threat is due to his own assessment of the hazard and his personality. His knowledge is derived from various sources but the major change is in the involvement of the media. Few people who watch television or who read the newspapers misunderstand the association between environmental exposure and many diseases. The influence of the media and politicians is easily seen from the nuclear power story, where the actual threat of death due to radioactive material is very low and yet its association with the atom bomb increases the degree of threat. It is not uncommon to meet people who believe that nuclear power stations can produce nuclear explosions which may wipe out large areas of the countryside. Yet workers in nuclear power stations accept the excellent working conditions compared with those of the coal-fired station. This perception of risk is not always associated with a determination of causal relationships but lies in the ability to decide which of these possible threats to morbidity and mortality are worth considering a risk. The degree of threat is due to the extent of the hazard as seen by the individual.

The level of concern is based on how the individual sees the threat to his or his family's life span. The worker is not able to assess the risk except for the information that he receives both from gossip sources and, more hopefully, from management. If he sees his workmates healthy and well in their 70s, he is unlikely to accept that there is any threat from continuing to work in that part of the plant. He does forget, however, that there may be an increased incidence of a disease which is mathematically estimated, but he does know that the dire consequences do not happen to all the workers in that plant. The incidence of lung cancer among heavy smokers is 1 in 8, which is an extremely high figure, but the level of concern is determined by whether the smoker is addicted and by his age. The very young smoker believes that he may possibly reduce his life span, but at the age of 16 even the age of 25 seems a long way in the future. To the older addicted smoker the belief is that it cannot happen to him and the assessment of the hazard is tempered by the thought that seven out of eight of the heavy smokers will not develop lung cancer.

The threat to the worker and his family from environmental hazards is greater because the worker knows that he has no control over the levels of exposure. The exposure at his workplace is much greater than that of the general population and so determines whether he worries about the specific environmental hazard. The lead worker will not worry about the ambient air pollution because he knows that he is exposed to much higher levels in his workplace and assumes that it is not causing him any harm. It is still common, unfortunately, for workers to accept payment for maintaining employment with some hazardous material. This ignoring of hazards for the sake of money seems to be a very foolish concept, but how many people consider the risk of driving a car to the shops when they know it is more flexible and easier than using public transport?

Most hazards are discovered much later than should be acceptable by society. The research conducted based on epidemiology is far too late for the workers whose deaths indicate a hazard. Research therefore into the early diagnosis of exposure effects is of less importance than the identification of a hazard before workers are exposed to it. Research so far has failed to indicate the hazards of a new material, although markers which identify the possible hazards can readily, albeit at a high cost, be demonstrated. Mutagenic testing was initially thought to be the arbiter of whether a material was a hazard or not but the arguments that range around this particular test showed subsequently that it is not specific. The use of animals such as mice and rats may lead to some degree of objectivity but the cliché 'mice are not little men' explains the difficulty of using a species of animal which is very unlike the human being. The most effective experiments are performed using monkeys, but this is extremely expensive and takes many years to finalize. There are undoubtedly pointers in the animal work to the possibility of hazard and the discovery that rubber antioxidants caused cancer of the bladder in dogs in 1947 was followed by their ban from use in 1948. In industry, the first cases which were seen (Case *et al.*)[3] showed that there was some clustering of bladder papilloma in areas where there were rubber compounding and manufacturing plants. However, even this early discovery of the hazard was not in sufficient time to prevent the large number of deaths which occurred 20 to 30 years later from this disease.

The unhappy attribute of industrial exposure is that there may be a very long latent period between exposure and promotion of the disease. This is particularly true of carcinogens and although it may be possible to monitor the death rates on a continuous basis in order to have early recognition of trends in the development of certain causes of death, they could be entirely unrelated to the work situation and influenced more by the social environment of the individual. Cancer of the lung is acknowledged to be more prevalent in the lower social classes. This may be due to food, housing, pollution, or simply the higher incidence of tobacco addiction. It is this multifactoral causation of disease which produces a great deal of confusion in the assessment of hazards.

If there is a greater awareness in the workforce it is possible to get some early warning of particular diseases occurring in that environment through the knowledge of what is happening to work mates. If there are two or three cases of coronary thrombosis within a short period of time then the workers will need reassurance that this is not due to a workplace hazard, and, although they have little knowledge of the difference between various manifestations of heart disease, they are still aware of the changes which occur in disease patterns.

Cases can sometimes occur of diseases which are unusual in that part of the country. An example of this is the asthma caused by toluene di-isocyanate. Not only is there an unusual incidence of this disease but there is also a particular pattern of symptomatology which leads the worker to question his exposure. These cases commonly present with an asthma which is worse from Monday to Friday; there is recovery at the week-ends, or recovery when on holiday. One difficulty about this is that the acute symptoms may appear whilst at work or in bed at night. The description of the disease and its long-term effects should come from the management of the company rather than uncritical communication from the shop floor.

3. Risk Assessment

In the quantification of risk to people and their environment, the hierarchy of risk consequences can be listed as suggested by Rowe in 1975,[4] and they are:

i)   Premature death

ii)   Illness and disability

iii)   Survival factors

iv)   Exhaustible resources

v)   Security

vi)   Belonging

vii)   Egocentric

viii)   Self actualization

This is based on Maslow's table of human needs, which are listed in such a way that the next highest number becomes the dominant need only when the next lowest number need has been fulfilled. Thus the highest priority need is survival; the body is able to cope and adapt to the most unpleasant conditions simply because all the physiological defence mechanisms have one function only, and that is to prevent death. Adverse consequences are illness and disability, followed by premature death. In the list of risk consequences, under premature death, two areas of risk assessment are of great interest. Firstly, there is the knowledgeable *versus* the unknowledgeable risk; for example, deep-sea fishermen know that there is an increase in the risk of drowning accidents, against the unknowledgeable risk of cancer due to exposure to a material which is used at work. There is also the difference between an expected occurrence and the risk of an unseen hazard such as exposure to ionizing radiations. The worker's acceptance of risk is therefore dependent on knowledge of or understanding of the risk.

The identifiable risk may be accepted by the workforce for reasons which may not strictly be those of objective assessment. The workers in a hot-metal area may have a high morale with low sickness absence merely because they are exposed to a hazardous environment and therefore feel that they are different to those working in the safety of the assembly area. The stigmata of a particular occupation can be an index of the manliness of any worker and there has often been a display of the blackening which occurs on the skin of coal miners following an injury or the full thickness burns which occur from metal splashes. Even though both of these are describing a minor risk, it must mean that there is an increased risk of death in a coal mine and also of severe burning or death in hot-metal exposure. Why then do these men perform these hazardous operations? This can be due to job security, belonging to a group, or the intrinsic needs of wealth or power. It can also further be associated with the kind of family environment, such as coal miners' sons who accept the risks of working in coal mines because their family ethos requires them to ignore or to glory in the risks of injury. Perceived risk is therefore not a mathematical consideration but an emotional consideration and is dependent on many factors.

The identification of risk is a job for the experts and for technology. A mathematical estimation of the odds against having a specific injury requires objective balancing of the risk and the exposure. The acceptable risk may also be set at a level which will not preclude the disablement of those people exposed to a risk. This is seen in the 90 dBA level of exposure for hearing, which level is accepted even though a continuous 8 hour day, 5 day week, 40 years exposure could result in up to 17% of the exposed workers suffering a hearing loss which causes noticeable handicap. [5]

Exposure limits are therefore not determined by a level of no risk but by a level of acceptable risk. The probability would be expressed as the odds of developing an illness for people who are exposed, compared with the odds for a non-exposed person. These odds are defined by using morbidity or mortality statistics.

The risk ratio is determined by the calculation of increased incidence of the illness or death within the exposed group compared with a control group of similar age, ethnic type, and area of domicile.

The simplest method of finding a control group is by using the publication by the Office of Population Censuses and Surveys (OPCS) called 'Mortality Statistics Cause'. This is a review of deaths by cause, sex, and age, and so each exposed person can be matched for these parameters. This provides ready access to a large control group and a standardized mortality ratio can be estimated for each cause of death. The errors can be large because death certification is not an exact science; the control group will include people who are disabled, or unfit to work, who have a different ethnic background, and who live in various parts of the country.

The OPCS also issues classification of deaths coded by occupation. In this way, the quantification of risk can be looked at in very precise groups.

In the field of chronic disease the risk ratio depends on the effects of a material and the exposure in terms of concentration and length of time exposed, *i.e.* quantified exposure.

4. Exposure Risk

The concept of exposure risk is very much a personal one in that the anatomical and environmental factors of an individual must be considered when risk of illness or death is evaluated. The adsorption rate of a chemical material depends on whether it is inhaled or ingested. The percentage of material absorbed by inhalation is somewhere between 25 and 45%, depending on the solubility and activity of the chemical, and by ingestion it is somewhere between 4 and 8%. Even in exposure by inhalation the degree of material inhaled is very much dependent on particle size. It is accepted that the median size of a particle which will be able to reach the alveolar membrane is 7 μm. This means that particles of 15 μm and over are not inhaled and yet they may be ingested by being taken in through the mouth or nasal passages and then swallowed.

Lung damage depends on:

i)   Chemical activity of the substance inhaled

ii)   Particle size

iii)   Concentration in air

iv)   Sensitivity of the patient

The varying rates of absorption, even by ingestion, depend on the hygiene habits of the worker. The smoker ingests much larger amounts of the chemical because of the transfer of the chemical from hands to cigarette and mouth and lips. The retention of inhaled materials depends on the lung function and the ability to clear the particles from the smaller airways, *i.e.* the efficiency of pulmonary clearance. This is reduced by cigarette smoking because of damage to the mechanism whereby the particles are swept up a mucociliary escalator by hair-like cells in the large airways. Eventually the particles, along with mucus, reach the throat where they are coughed up into the area where they can be swallowed. The decreased efficiency of the pulmonary clearance mechanism may cause the increased incidence of lung cancer in asbestos workers who smoke (Table 2), or it may be an actual synergistic effect between the exposure to tobacco smoke and asbestos fibres.

Table 2 *Risk of lung cancer*[6]

Smoking	*Asbestos*	*Risk*
Non-smoker	Non-exposed	1
Non-smoker	Exposed	5
Smoker	Non-exposed	11
Smoker	Exposed	53

It is important to be aware of the target organ because it may lead to screening tests which will determine whether a worker should in fact be exposed to that hazard. It is unwise to expose workers with bronchitis or poor lung function to hazardous materials which may cause lung diseases. Firstly, the incidence of disease will be increased by virtue of the fact that the bronchitic is going to breathe more frequently because of his poor lung function and will inhale more of the hazardous material, thus producing further diminution of lung function and major physical disability. A history of skin rashes or asthma would preclude work with sensitizing agents; thus pre-employment examination is a very effective method of reducing personal risk.

Biochemical determinants commonly rest on the solubility of material in body fluids and it is here that the difference between cadmium oxide and cadmium sulphide can be quoted. The cadmium oxide is readily absorbed *via* the lung tissue and lining of the gut, but cadmium sulphide is extremely insoluble in body fluids, and so has little biochemical effect. Detoxification is also dependent on factors such as liver function, and the effects of chronic alcohol exposure in a person may result in decreased ability to detoxify any tissue poison and so cellular damage and cirrhosis of the liver results.

Apart from these factors consideration must be given to the physical and physiological make-up of the person exposed. The physical appearance of the coal miner of old was that of the rather squat, broad, muscular individual. Now there is a change in the height/weight ratio of the modern worker and one wonders whether their physical attributes would militate against being a good coal miner, particularly in very low seams. Physical disabilities are quite often discussed in terms of worker ability. The employment of a blind person on a standard press, making the same kind of product, produces a worker whose output may be considerably enhanced over that of the sighted worker. This man, however, would be impossible to employ where the assessment of distance or direction is required, for example in handling hot metal. The employment of the overweight, obese individual in compressed air has been said to increase the incidence of the 'bends', and the effect of his physical characteristics would preclude him from being employed in any job which requires mobility and speed of action.

The effect of the psychological make-up of a worker on the risks that he is exposed to begin with his personality and his reaction to the environment. In some cases this may produce stress diseases, such as coronary thrombosis, diabetes, and rheumatoid arthritis. Therefore the problems of exposure risk can be influenced by the physical and mental state of the employee.

Stress is a response to environmental factors and stress diseases are responsible for the majority of morbidity and mortality statistics. In people under 60 years of age it was recently estimated that 45% of the causes of absence from work had a psychological or stress basis. Why these diseases are becoming more common is very dificult to explain but one psychologist made the very apposite remark that the degree of stress was due to the difference between what life is and what the person thinks it should be. It is realized that anyone can become President of the USA, but after all there is only one, and so a great number of people are going to be disappointed. Once again the effect of the media, and in particular the growth of television, gives the impression that beautiful people clean floors in seconds, without dirtying their beautifully designed clothes, and without ruffling their beautiful coiffure. Is it any wonder that our wives are stressed by the simplistic belief in someone's washing powder or floor cleaner?

5. Risk Monitoring

Monitoring involves recurrent observation, recording, and analysis of hazardous events applied to persons in the workplace. This should be tailored to the degree of potential threat, although the possible form of the Control of Substances Hazardous to Health regulations will include the necessity to ensure that *all* occupational hazards are identified and hopefully controlled before causing injury and disease. The onus of demonstrating that risks are under control is firmly laid on the employer who has the need for a formal assessment of risk associated with all workplaces.

Apart from looking at exposures and the toxic effects of substances it will be necessary to institute health surveillance systems to ensure the early recognition of potential risks.

This monitoring of effects should include sickness absence and labour turnover as general indicators of hazardous situations. There should be an emphasis on the workers in environments which are potentially hazardous; absence rates must be assessed in small groups, such as departments or occupations.

The occurrence of specific diseases or even injuries may indicate unsafe systems of work which produce an increase in the hazard rating.

The method of continuous assessment can be important in the assessment of exposure. For example, the methods of *in vitro* or *in vivo* monitoring of body fluids are used. It is relatively simple to measure the amount of metals in blood; this is related to recent exposure. More sophisticated methods of measuring metals in the body, such as neutron activation, can be used to monitor kidney and liver without damage being caused to the person. It is important to look at the individual and also at his work group, because monitoring may give an indication of changes in the environmental conditions. If a group of lead workers has an average blood lead estimation of 400 μg l^{-1} for many months and then this goes up to 600 μg l^{-1}, the ventilation and the system of work should be looked at. If, however, the group has an average level of 400 μg l^{-1} and an individual has a level of 800 μg l^{-1}, then this individual's exposure and his hygiene standards could be the cause.

6. Risk Estimation

The final arbiter is the epidemiology of mortality. The follow-up of all workers exposed to a known chemical hazard is important and one way of determining the hazard is by using life exposure tables which give a history of man years exposure over a working life. It is essential therefore to maintain a record of all the chemicals used in various departments, along with an occupational history, this being an important profile for the assessment of the degree of risk and one which may demonstrate an association between exposure and cause of death. It is too late to intercede on behalf of those who have died, but continuous epidemiological monitoring can indicate trends in causes which may lead to some timely preventive action being taken.

Excess cancer risk can be estimated by using a model such as the Environmental Protection Agency's human exposure model.[7] This can also be used to estimate the exposure levels which would produce an effect on the population outside the factory. The general population risk can be expressed as an expected annual cancer incidence, local interpretation being based on an appreciation of the geographical area rate and possible clustering which occurs owing to point sources of exposure. Alternatively the excess cancer risk can be expressed as the lifetime probability of cancer for the exposed population. It is possible to estimate the risk numerically if the degree of exposure is known. This probability of development of a specific cancer is expressed as the probability per microgram of chemical in each cubic metre of inspired air during a lifetime exposure.

7. Conclusions

Risk estimation and hazard identification in industry are a cost-effective part of management techniques, and large companies may possess the expertise and motivation to tackle these problems, particularly if financial savings can be shown. The difficulty arises when the strictures of society are encountered; this is not a random meeting of event and consequence, but the ground is prepared by industry itself.

Every cost-benefit analysis is frozen in time and few would deny that the perceived threats to the worker change as each month passes. The hazards of ten years ago would not now be tolerated by the enforcing authorities. There is no turning back and, although risks can be quantified by the use of more sophisticated methods, those who ask for risk assessment will have to understand the answers. The quantification of probabilities, the limits of assumptions, and the identification of consequences all make prognostication difficult but, as the hazards in industry change from the gross pathology of the past to the more esoteric diseases of the future, there will be a greater need for social evaluation and acceptance of the risk produced by industry. The good communicators will encourage their understanding and acceptance by society.

8. References

1. Health and Safety Executive Statistics 1981/82, ISBN 0 11 883816 4.

2. Health and Safety Executive Statistics 1981/82, ISBN 0 11 883816 4.

3. R. A. M. Case and J. T. Pearson, *Br. J. Ind. Med.*, 1954, 11, 213.

4. W. D. Rowe, 'An Anatomy of Risk', US Environmental Protection Agency, Washington DC, 1975.

5. BS 5330; 1976, Method of test for estimating the risk of hearing handicap due to noise exposure.

6. F. D. K. Liddle, and J. A. Hanley, *Br. J. Ind. Med.*, 1985 (June), **42**, (6), 389-396.

7. Environment Protection Agency, Human Exposure Model (HEM).

18

Use of Toxicity Data - A Case Study of Di-(2-ethylhexyl) Phthalate

By J.W. Bridges

ROBENS INSTITUTE OF INDUSTRIAL AND ENVIRONMENTAL HEALTH AND SAFETY,
UNIVERSITY OF SURREY, GUILDFORD, GU2 5XH, UK

1. Approaches to the Assessment of Human Hazard from a Toxic Chemical

Depending on the availability of information a range of approaches may be employed to assess the hazard to man from a chemical.

The most valuable is good-quality data on the effects of the chemical in man; however, information in man where available is frequently insensitive or unreliable, or refers solely to short-term exposure. Normally risk assessment has therefore to be based only on toxicity data obtained in animals. Where only animal data are available, it is important to establish as far as possible:

That the form of animal exposure is relevant to the likely human exposure situation;

That studies have been performed to acceptable standards, in particular that a comprehensive thorough investigation of possible adverse effects has been conducted;

The findings are reproducible;

The precise nature of the adverse effects, their dose dependency, and possible reversibility;

Whether or not the effect is species specific, *i.e.* whether the effect observed is likely to occur in man. This is usually assessed by utilizing knowledge of comparative anatomy, physiology, and biochemistry of test species and man.

In an attempt to bridge the gap between the animal findings and the human exposure situation, several philosophies may be adopted:

1. Use of a range of test species, the belief being that if findings in several species agree, similar effects will be seen in man. A 'no effect level' in each species may be identified by direct study or by mathematical extrapolation of the dose-response findings.[1] In the absence of evidence to the contrary, findings in the most sensitive species are used for exrapolation purposes. To allow for imprecision in extrapolation and the fact that no studies are made in the test procedures of disease, age, other chemical effects, *etc.*, an arbitrary safety factor (*e.g.* 10, 100, 1000) is usually applied to define the 'safe level' for man.[2]

2. Extrapolation on the basis of drug metabolism and pharmacokinetic investigations in the toxicity test species and in man, particular emphasis be given to matching of biologically active metabolites, to the blood levels and clearance rates of chemical and its active metabolites.[3]

3. Investigations of mechanisms of toxicity.[4] Such an approach is of particular use in establishing possible threshold doses for toxicity. A major question is the degree of detail needed to provide confidence in assessing human hazard.

4. Comparison with closely related chemical structures for which there has been considerable human exposure.[5]

In the following discussion, data on di-(2-ethylhexyl) phthalate (DEHP) are used to illustrate some of these approaches to toxicity data evaluation and hazard assessment.

2. Chemical and Biological Properties of DEHP[6]

Di-(2-ethylhexyl) phthalate (DEHP) is a very widely used plasticizer in polyvinyl chloride (PVC) products. It is also used in a number of cosmetics and fragrances, as a pesticide carrier, and as a constituent of certain insect repellants, industrial oils, and munitions. DEHP and its hydrolysis product, the mono-ester mono-2-ethylhexyl phthalate (MEHP) are very widely distributed in the environment:[7] few, if any, humans are unexposed to the chemical, albeit the levels of exposure are typically very low. Higher exposure groups include workers involved in DEHP manufacture and in production of products incorporating DEHP, and individuals undergoing frequent blood transfusions (the source of DEHP being the plastic bags in which the blood is stored).

In the following discussion the importance of using information from a variety of types of study for hazard assessment purposes is illustrated.

3. Information from Toxicity Studies

3.1. Sub-acute Studies. Very large daily oral doses of DEHP, in excess of 3 g kg^{-1} body weight per day, are required to produce a significant increase in mortality in rats or mice.[8]

Sub-lethal effects of note in the rat at high doses include some pathological changes in the liver and testes,[9] Some effects on red blood cells, and loss of body weight. At lower doses only minimal changes in tissue pathology are observed.[10]

Sub-lethal effects in mice at high doses include pathological changes in the liver, kidney, intestine, and spleen and some loss of body weight.[11]

Sub-lethal effects in dogs at high doses [intravenous (i.v.) route] include pulmonary changes and some body weight loss.[12]

Sub-lethal effects in monkeys at high dose levels (i.v. route) include changes in liver histology and liver function.[13]

DEHP is not significantly teratogenic in rats, but high doses (2% in the diet) cause an increase in resorptions.[14] Mice are far more sensitive: a dose of 0.1% in the diet may cause embryotoxicity and some skeletal malformations. By direct extrapolation of dose-response data a 'no effect level' of 0.05% in the diet has been identified in the species.[15]

3.2. Lifetime Studies. The results of the National Cancer Institute (NCI) study[11,16] on the effects of lifetime ingestion of high doses of DEHP on rats and mice are summarized in Table 1. It should be noted that dose-related increases in testicular atrophy and in hepatocellular carcinoma were observed in both rats and mice. Other dose-related adverse effects in the rats included reduction in body weight and neoplastic liver nodules. Interestingly a dose-related decrease in interstitial cell tumours of the testes and in pituitary carcinomas was also found.

Table 1 *Summary of toxic effects of DEHP in lifetime rodent studies*

Rats (2 years oral (p.o.)

Dose (p.p.m.)	*Testicular atrophy (%)*	*Hepatocellular carcinoma (%)*	
		Male	*Female*
0	2	2	0
6000	5	2	4
12000	90	10	16

Mice (2 years oral (p.o.))

0	2	10	0
3000	4	29	14
6000	14	38	34

In mice other dose-related effects included an increase in chronic renal inflammation and in hepatocellular adenoma. It is unfortunate that only very high dose levels were used in these lifetime studies and therefore it is not possible to deduce from the data alone whether the effects observed are confined to high doses or would also occur at lower exposure levels.

3.3. Mutagenicity Studies. 'Classical' chemical carcinogens are believed to initiate their effects by interaction with DNA, either directly or through the aegis of a chemically reactive metabolite(s). Such effects on DNA may be identified by measurement of induced mutations, increased levels of DNA repair, or chromosome damage. Separate test systems have been devised to measure these individual endpoints. DEHP and its hydrolysis product MEHP typically give negative results in these test systems (see Table 2).

3.4. Conclusions from Toxicity Studies. From the findings in sub-acute, chronic, lifetime, and mutagenicity tests it may be concluded:

1. DEHP has a low general toxicity.

2. At high doses, DEHP is a hepatocarcinogen in rats and mice; it also produces testicular atrophy but not testicular tumours.

3. DEHP and MEHP are non-mutagenic. They are therefore not typical hepatocarcinogens.

4. DEHP is teratogenic to mice, but not significantly so to rats.

The above findings are very useful in identifying the toxic properties of DEHP. However, they do not allow identification of either a 'no effect level' in rodents for DEHP-induced hepatocarcinoma or testicular atrophy, nor do they establish whether findings in rats and mice with DEHP are indicative of possible human hazard. To gain information on these crucial questions data from drug metabolism,[4] pharmacokinetic, and mode of toxic action studies are necessary.

Table 2 *Summary of effects of MEHP and DEHP in mutagenicity tests*

	MEHP	DEHP	Ref.
Ames test	Negative	Negative	17
Mammalian cell mutations			
In vitro	Not studied	Negative	18
In vivo	Positive	Positive	19
Chromosome changes			
In vitro	Positive	Negative	20
In vivo	Negative	Negative	21
Dominant lethal test	Negative	Negative	22
Unscheduled DNA synthesis test *in vitro*	Negative	Negative	23
DNA binding *in vivo*	not studied	very low	24

4. Information from Metabolism and Pharmacokinetics Studies

Following oral administration, DEHP is extensively hydrolysed in the gut lumen and/or wall to the monoester (MEHP).[25] In addition a variety of tissue esterases are able to hydrolyse DEHP to MEHP.[26] In the rat MEHP is well absorbed from the gut and undergoes extensive metabolism to form a number of products (see Figure 1).[27] Good absorption of DEHP is also likely following its inhalation. In contrast to the situation for most chemical carcinogens, there is no evidence that DEHP is converted into any reactive intermediate metabolites. However, two chemically stable biologically active oxidized metabolites (ω-1)-hydroxy-MEHP and (ω-1)-keto-MEHP are produced. Considerable species differences have been observed in the metabolism and pharmacokinetics of DEHP (see Table 3). In the rat MEHP and its oxidation products are excreted largely in the unconjugated form, whereas in hamster, guinea pig, marmoset, and man conjugation with glucuronic acid is a major metabolic pathway. In the guinea pig oxidation of MEHP is a minor pathway, whereas in all other species, including man, this is the major fate of absorbed MEHP. In the dog, marmoset, and man absorption of an oral dose of DEHP is poor, whereas in other species the majority of an ingested dose is absorbed.

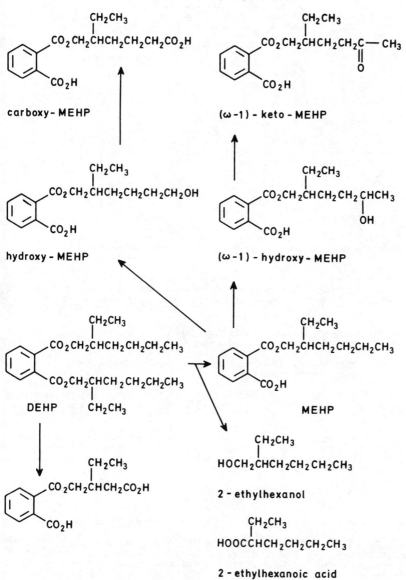

Figure 1 *Major metabolic pathways of DEHP in the rat*

Table 3 *Species differences in the metabolic fate of DEHP*[28] *50 mg kg*[-1] *oral (p.o.) per day.*

Metabolite	% Metabolites in 24 h urine			
	Rat	Hamster	Guinea Pig	Marmoset
MEHP	0	2	24	20
ω-Oxidation products	49	25	2	8
$(\omega-1)$-Oxidation products	36	65	4	65
% Conjugation of $(\omega-1)$-oxidation products	0	42	2	24
% Conjugation of MEHP	0	trace	50	trace
% Dose excreted in urine	48	83	-	22*

*Dose 2 g per animal. Figure is for percentage absorption

In the rat DEHP related material shows little tendency to accumulate in body tissues other than fat, and covalent binding to liver DNA is minimal.

4.1. Conclusions from Metabolism and Pharmacokinetics Studies. The findings on metabolism and pharmacokinetics indicate:

1. Man is probably less well able to absorb an ingested dose of DEHP/MEHP than the rat.

2. MEHP and the active $(\omega-1)$-oxidation (unconjugated) metabolites might be expected to attain lower blood and tissue levels in man than in the rat.

3. Because the rat, in contrast to man, does not form conjugated metabolites, the relevance of this animal species for assessing the possible toxic hazard to man from exposure to DEHP is somewhat questionable.

In the following, only mechanisms involved in the hepato-carcinogenicity of DEHP are discussed.

5.   Information from Mode of Action (Mechanism) Studies

5.1. *In vivo* Studies with DEHP in the Rat (Dose-Response and Time Sequence).[29]   DEHP causes hypotriglyceridaemia and hypo-cholesterolaemia in rats with alterations in hepatic phospholipid levels and composition and in triglyceride distribution. In the rat many, if not all, of these lipid changes appear to be associated with the proliferation of both the endoplasmic reticulum and peroxisomes in the liver. In order to ascertain the mechanism underlying these effects of DEHP and their relevance to the subsequent development of liver tumours, it is necessary first to study the dose-response relationship and the time sequence of the various hepatic changes.

<u>Initial but transitory changes</u> include increases in DNA synthesis and in the depositon of fine lipid droplets, at first in all hepatocytes but subsequently mainly in periportal hepatocytes.

<u>Early persistent changes</u> include proliferation of endoplasmic reticulum and peroxisomes together with some mitochondrial changes. The increase in endoplasmic reticulum membrane is associated with enhanced activity of the P450-dependent enzyme laurate hydroxylase which catalyses the ω-hydroxylation of fatty acids and of DEHP. The effect of DEHP on peroxisomes is of particular interest because not only are the peroxisomes much more numerous, but they also differ from those of control rats in a number of respects, *i.e.* they lack certain enzymes (*e.g.* urate oxidase); they have very high levels of fatty acid β-oxidation enzymes but rather low levels of catalase (*N.B.* catalase in peroxisomes is responsible for the detoxication of hydrogen peroxide, an obligatory by-product of fatty acid β-oxidation) and they are much more fragile than normal peroxisomes.

<u>Changes of slower onset</u> include liver enlargement, glycogen loss, and reduction in microsomal glucose-6-phosphatase activity (a useful index of subtle liver damage)

<u>Later changes</u> include a considerable increase in autophagic lysosomes and deposition of lipofuscin particles (a complex of peroxidized lipids and proteins). Both these changes imply that damage to the membranes of one or more organelle types has occurred. Whereas the 'early' and 'slow onset' changes appear to occur over a considerable dose range, these 'later changes' appear to be associated with high doses only.

<u>Long-term changes</u>. After a year, small hepatic foci appear which stain differently from other parts of the liver. By eighteen months nodules may be found on the liver surface and by two years some well differentiated non-invasive carcinomas may be seen.

<u>5.2. Species Comparisons *in vivo*</u>. In hamsters, rats, and marmosets sub-acute exposure to DEHP and MEHP causes minor, if any, changes in hepatic peroxisome proliferation,[30] suggesting that rats and mice may be uniquely sensitive to these chemicals. However, it is dificult to be certain whether this species difference has drug metabolic and pharmacokinetic origins, or is due to the inherent differences in responsiveness of hepatocytes from different species.

<u>5.3. Structure-Activity Studies *in vivo*</u>. A considerable number of chemicals have been found to produce peroxisome proliferation in rodents *in vivo* (see Table 4). Many show no obvious structural relationship to DEHP. All the potent peroxisome proliferating agents which have been examined in rodent lifetime studies have produced hepatocarcinomas.[31] The hypolipidaemic drugs (*e.g.* clofibrate and fenofibrate) are of particular interest in this regard, because they possess little structural similarity to DEHP but produce the same sequence of histological and biochemical changes as DEHP[32] (see above); they are non-mutagenic and do not bind to DNA and there is considerable information on their acute and chronic use in man at relatively high dose levels.

Valid carcinogenicity studies have not been conducted on most commercially used phthalates. Short-term *in vivo* studies in rats have shown that di-(n-octyl) phthalate and di-(n-hexyl) phthalate

produce minimal efects on peroxisomal numbers and that the profile
of other hepatic effects differs in a number of significant ways
from that produced by DEHP.[33]

Table 4 *Examples of chemicals causing peroxisomes proliferation in rat/mouse liver*[31]

A   *Hypolipidaemic drugs*

   (i) Clofibrate analogues

Clofibrate	Gemfibrozil
Fenofibrate	Benzafibrate
Nafenopin	Ciprofibrate

   (ii) Structurally unrelated to clofibrate

Tibric acid	Wy 14,643
Clorocyclizine	

B   *Plasticizers and related compounds*

Di-(2-ethylhexyl) phthalate	2-Ethylhexanol
Di-(2-ethylhexyl) adipate	2-Ethylhexanoic acid
Di-(2-ethylhexyl) sebacate	

C   *Other chemicals*

   Polychlorinated aliphatic hydrocarbons
   High-fat diet
   Various pyrethroids and biphenyl ether herbicides

5.4. *In vitro* Studies. To establish the mechanism(s) responsible
for hepatic peroxisome proliferation it is necessary to develop an
in vitro model which mirrors the early events seen *in vivo*. The
obvious model is a primary maintenance culture of rat hepatocytes.
This preparation produces the same metabolites as those formed *in
vivo* and shows very similar endoplasmic reticulum and peroxisomal
changes.[34−36]

Dose-response studies in rat hepatocytes have shown that DEHP,
MEHP, and the (ω-1)-hydroxy and (ω-1)-keto metabolities of MEHP
are each potent peroxisome and endoplasmic reticulum proliferating
agents.[36] Similar effects are observed with hypolipidaemic drugs
but, as *in vivo*, not with mono-(n-octyl) phthalate and
mono-(n-hexyl) phthalate.[37]

Primary hepatocyte maintenance cultures may be prepared from
many mammalian species. It has been found that hepatocytes from
marmosets, guinea pigs, and man do not respond significantly to
DEHP, MEHP, or (ω-1)-oxidation products of MEHP.[38] Thus it would
appear that the species differences in the *in vitro* response to
MEHP have pharmacodynamic rather than pharmacokinetic causes.

5.5. Proposed Mechanism of Action. Early events. The earliest
identifiable hepatic event following dosing of rodents with DEHP,
or addition of MEHP to cultured rat hepatocytes, is inhibition of
fatty acid oxidation. Within 24 hours fine droplets of lipid are
seen to accumulate within the hepatocytes. (These are reminiscent
of the effects of administering high fat diets to rodents.)
Within 48 hours peroxisomal and endoplasmic reticulum

proliferation can be seen.   This sequence of changes may be
interpreted as shown in Figure 2.

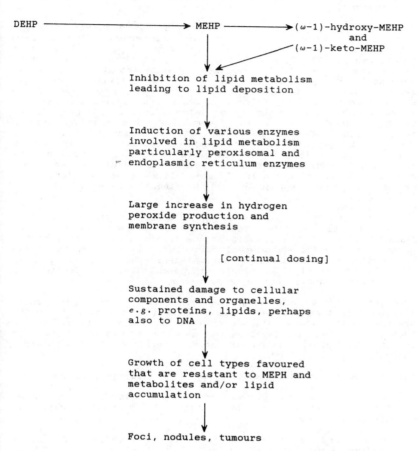

Figure 2 *Proposed mechanism of DEHP Hepatocarcinogenicity*

MEHP and/or its (ω−1)-oxidation products bind to a 'receptor'
site (perhaps on one or more of the enzymes responsible for lipid
oxidation).   As a result fatty acid oxidation is inhibited and
levels of fatty acid and triglyceride build up.   Because MEHP has
a branched chain it is resistant to metabolic degradation and
hence its effects probably persist (in contrast to the situation
for many straight-chain phthalates).   Either the accumulating
endogenous lipid or the retention of MEHP-related material causes
the stimulation of the expression of genes for lipid oxidation
enzymes, in particular peroxisomal enzymes and endoplasmic
reticulum enzymes.   (*N.B.* This response is presumably the normal
physiological mechanism in rodents for handling excess lipid and
metabolism-resistant lipids.) Species differences are presumed to
arise both because different levels of lipid are required to

trigger the expression of the genes for fatty acid oxidation enzymes between species and because the profile of enzymes induced is different.

Medium-term events. The question remains as to the relationship between peroxisome proliferation and subsequent liver cancer. Several hypotheses have been propounded to explain this relationship. The first and most widely supported is the active oxygen hypothesis. This hypothesis is dependent on the known generation of hydrogen peroxide during peroxisomal oxidation of fatty acids and P450-dependent ω-oxidation of fatty acids by the endoplasmic reticulum. Following organelle proliferation there is a great increase in oxidation of fatty acids, and hence hydrogen peroxide production which overwhelms the active oxygen detoxication systems. One consequence of this is an increased fragility of peroxisomes. The enhanced release of active oxygen will continue as long as DEHP exposure continues. If exposure is prolonged a number of sustained changes to cellular constituents may occur (*e.g.* DNA oxidation, changes in membrane functions), one or more of which may produce a permanently altered cell (see Figure 2).

An alternative mechanism by which continual exposure to DEHP may produce permanently altered cells is through the continuous turnover of subcellular membranes due to the proliferation of the peroxisomes and the endoplasmic reticulum. In this regard it may be apposite that a number of non-mutagenic chemicals which cause hepatocarcinogenicity in rodents are also potent proliferators of subcellular organelles (*e.g.* metapyrilene, DDT).

Long-term effects. The stages between formation of a permanently altered cell type and development of foci, nodules, and tumours are poorly understood: one scenario is that such cells may thrive because they have an impaired capacity to metabolize DEHP or to respond by proliferation of peroxisomes and/or endoplasmic reticulum.

5.6. Conclusions on Mechanisms Studies. The findings indicate that:

1. Peroxisomal and endoplasmic reticulum proliferation may be an essential step in the chain of events leading to hepatocarcinomas.

2. Rodents are far more sensitive to these events than other species, including man.

3. The proposed mechanisms that suggest hepatocarcinogenicity is solely a high-dose phenomenon.

6. Overall Assessment of Human Hazard from DEHP Exposure

Dose-response findings in rat and mouse lifetime studies do not allow direct identification of a 'no effect level' for hepato-carcinogenicity. However, results from mutagenicity tests indicate that DEHP is not a typical chemical carcinogen. Thus it is inappropriate to use the mathematical models devised to assess human hazard from genotoxic carcinogens.[39] Drug metabolism and pharmacokinetic data indicate that man may be less able to absorb DEHP/MEHP and may produce lower levels of the unconjugated active metabolites than rats. Mechanism studies imply that rodents are uniquely susceptible to the hepatocarcinogenicity effects of

DEHP/MEHP and that the effect may be purely a high-dose phenomenon. Information on the hepatic effects of the hypolipidaemic drugs in man is important to the risk assessment of DEHP. Although these drugs are administered chronically to man in high doses there is no evidence of significant hepatotoxicity to man despite careful investigation. Because their mode of action is similar to that of DEHP in rodents this is reassuring evidence that DEHP is unlikely to produce hepatocarcinogenicity in man. Bearing in mind the low doses of DEHP to which man is exposed, DEHP and its metabolite MEHP are unlikely to prove a carcinogenic hazard to man. [*N.B.* Conclusions regarding possible testicular or teratogenic effects in man require a separate analysis].

7. Future Prospects

Too often in assessing hazard to man, data for effects of chemicals in standard animal toxicity tests are used without recourse to any other form of information. The case study described above for DEHP illustrates the value of bringing together information from a range of disciplines in identifying toxic hazard to man. Such a comprehensive investigation of a chemical is of course easy to justify where widespread human exposure to a chemical is, or is likely to be, involved. However, if human exposure to significant levels of a chemical is unavoidable some understanding on how the chemical causes toxicity is likely to be necessary if confidence is to be placed in the extrapolation of animal data to the human exposure situation.

8. References

1. J. K. Haseman, 'Statistical Issues in the Design, Analysis and Interpretation of Animal Carcinogenicity Studies', *Environ. Health Perspect.*, 1984, **58**, 385.

2. Casarett and Doull, 'Toxicology – the Basic Science of Poisons', 2nd Edn., ed. J. Doull, C. D. Klaassen, and M. O. Amdur, Macmillan, New York, 1980.

3. J. W. Bridges, L. F. Chasseaud, G. M. Cohen, and C. H. Walker, 'Applications of Pharmacokinetics in Animals and Alternatives in Toxicity Testing', ed. M. Balls, R. J. Riddell, and A. N. Worden, Academic Press, London, 1983, pp.31-75.

4. D. E. Hathway, 'Molecular Aspects of Toxicology', Royal Society of Chemistry, London, 1984.

5. ECETOC Monograph No.8, 'Structure-Activity Relationships in Toxicology and Ecotoxicology: an Assessment', Brussels, 1986.

6. General Review of Biological Properties, Proceedings of the Conference on Phthalates, held at Washington D.C. June 9-11, 1981, *Environ. Health Persp.*, 1982, **45**, 1-153.

7. ECETOC Technical Report No.19, 'An assessment of the Occurrence and Effects of Dialkyl ortho-Phthalates in the Environment, Brussels, 1985.

8. H. C. Hodge, 'Acute Toxicity of Rats and Mice of 2-Ethyl-hexanol and 2-Ethylhexyl Phthalate', *Proc. Soc. Exp. Biol. Med.*, 1943, **53**, 20.

9. T. J. B. Gray, K. R. Butterworth, I. F. Gaunt, P. Grasso, and S. D. Gangolli, 'Short-term Toxicity Study of Di-(2-ethylhexyl) Phthalate in Rats', *Food Cosmet.Toxicol.*, 1977, **15**, 389.

10. M. Nikonorov, H. Mazui, and M. Piekacz, 'Effect of Orally Administered Plasticizers and Polyvinyl Chloride Stabilizers in the Rat', *Toxicol. Appl. Pharmacol.*, 1973, **26**, 253.

11. NTP Carcinogenesis Bioassay of Di-(2-ethylhexyl) phthalate (CAS No. 117-81-7) in F344 Rats and B6C3F1 Mice (feed study), NIH publication No. 82-1773, March 1982.

12. H. A. Rutter, 'Toxicity of Plastic Devices Having Contact with Blood. Acute and Sub-acute Toxicity of Di-(2-ethylhexyl) phthalate in Dogs', Contract No. NIH NHLI-72-2991B, Animal Report, June 1972-October 1973.

13. M. S. Jacobson, S. V. Kevy, and R. J. Grand, 'Effect of a Plasticizer Leached from Polyvinyl Chloride on the Sub-human Primate: a Consequence of Chronic Transfusion Therapy, *J. Lab. Clin. Med.*, 1977, **89**, 1066.

14. R. Wolkovski-Tyl, C. Jones-Price, M. Marr, and C. A. Kimmell, Teratologic Evaluation of Di-(2-ethylhexyl) phthalate (DEHP) in Fischer 344 Rats', *Teratology*, 1983, **27**, 85A.

15. Y. Yaki, Y. Nakamura, I. Tomita, K. Tsuchikawa, and N. Schimio, 'Teratogenic Potential of Di- and Mono-(2-ethylhexyl) phthalate (DEHP), in Mice', *J. Environ. Path. Toxicol.*, 1980, **3**, 33.

16. W. M. Kluwe, J. K. Haseman, J. F. Douglas, and J. E. Huff, 'The Carcinogenicity of Dietary Di-(2-ethylhexyl) phthalate (DEHP) in Fisher 344 Rats and B6C3F1 Mice', *J. Toxicol. Environ. Health*, 1982, **10**, 797.

17. E. Zeiger, S. Haworth, W. Speck, and K. Mortelmans, 'Phthalate Ester Testing in the NTP Program's Environmental Mutagenicity Test Development Program', *Environ. Health Perspect.*, 1982, **45**, 99.

18. P. E. Kirby, R. F. Pizzarello, T. E. Lawlor, S. R. Haworth, and J. R. Hodgson, 'Evaluation of Di-(2-ethylhexyl) phthalate (DEHP) and its Major Metabolites in the Ames Test and L5178Y Mouse Lymphoma Mutagenicity Test', *Environ.-Mutagen*, 1983, **5**, 657.

19. I. Tomita, T. Aoki, and N. Inui, 'Mutagenicity/Carcinogenicity Potential of DEHP and MEHP', *Environ. Health Perspect.*, 1982, **45**, 119.

20. M. Ishidate, and S. Odashima, 'Chromosome Tests with 134 Compounds on Chinese Hamster Cells *in vitro* - A Screening for Chemical Carcinogens', *Mutation Res.*, 1977, **48**, 337.

21. D. L. Putnam, W. A. Moore, L. M. Schectman, and J. R. Hodgson, 'Cytogenetic Evaluation of Di-(2-ethylhexyl) phthalate and its Metabolites in Fischer 344 Rats', *Environ. Mutagen.*, 1983, **5**, 227.

22. C.J. Rushbrook, T. A. Jorgenson, and J. R. Hodgson, 'Dominant Lethal Study of Di-(2-ethylhexyl) phthalate and its Major Metabolites in ICR/SIM Mice', *Environ. Mutagen.* 1982, **4**, 387.

23. J. R. Hodgson, B. C. Myhr, M. McKeon, and D. J. Brusick, 'Evaluation of Di-(2-ethylhexyl) phthalate and its Major Metabolites in the Primary Rat Hepatocyte Unscheduled DNA Synthesis Assay', *Environ. Mutagen.*, 1982, **413**, 388.

24. P. W. Albro, J. T. Corbett, J. L. Schroeder, S. Jordan, and H. R. Matthews, 'Pharmacokinetics, Interactions with Macromolecules and Species Differences in Metabolism of DEHP', *Environ. Health Perspect.*, 1982, **45**, 19.

25. I. R. Rowland, R. C. Cottrell, and J. C. Phillips, 'Hydrolysis of Phthalate Esters by the Gastro-intestinal Contents of the Rat', *Food Cosmet. Toxicol.*, 1977, **15**, 17.

26. B. G. Lake, J. C. Phillips, J. C. Linnell, and S. D. Gangolli, 'The *in vitro* Hydrolysis of Some Phthalate Diesters by Hepatic and Intestinal Preparations from Various Species', *Toxicol. Appl. Pharmacol.*, 1977, **39**, 239.

27. P. W. Albro, R. Thomas, and L. Fishbone, 'Metabolism of Di(2-ethylhexyl) Phthalate in Rats, Isolation and Characterization of the Urinary Metabolites', *J. Chromatog.*, 1973, **76**, 321.

28. J. C. L'Huguenot, Ph.D. Thesis, University of Dijon, 1984.

29. F. E. Mitchell, S. C. Price, R. H. Hinton, P. Grasso, and J. W. Bridges, 'Time and Dose Response Study of the Effects on Rats of the Plasticizer Di-(2-ethylhexyl) phthalate', *Toxicol. Appl. Pharmacol.*, 1985, **81**, 371.

30. C. Rhodes, C. R. Elcombe, P. L. Batten, H. Bratt, S. J. Jackson, I. S. Pratt and T. C. Orten, ''Comparative Toxicity of 2-(Diethylhexyl) phthalate (DEHP) in Rats and Marmosets', *Toxicol. Lett.*, 1983, **18**, 63.

31. J. W. Reddy and N. D. Lalwai, 'Carcinogenesis By Hepatic Peroxisome Proliferators. Evaluation of the Risk of Hypolipidaemic Drugs and Industrial Plasticizers to Man', CRC *Critical Rev. Toxicol.*, 1983, **12**, 1.

32. S. C. Price, R. H. Hinton, F. E. Mitchell, D. E. Hall, P. Grasso, G. P. Blane, and J. W. Bridges, 'Time and Dose Study on the Response of Rats to the Hypolipidaemic Drug Fenofibrate *Toxicology*, 1986 (in the press).

33. A. H. Mann, S. C. Price, F. E. Mitchell, P. Grasso, R. H. Hinton, and J. W. Bridges, 'Comparison of the Short-term Effects of Di-(2-ethylhexyl) phthalate, Di-(n-hexyl) phthalate and Di-(n-octyl) phthalate in Rats', *Toxicol. Appl. Pharmacol.*, 1985, **77**, 116.

34. T. J. B. Gray, J. A. Beamand, B. G. Lake, J. R. Foster and S. D. Gangolli, 'Peroxisome Proliferation in Cultured Rat Hepatocytes Produced by Clofibrate and Phthalate Ester Metabolites', *Toxicol. Lett.*, 1982, **10**, 273.

35. T. J. B. Gray, J. A. Beamand, B. G. Lake, J. R. Foster and S. D. Gangolli, 'Peroxisome Proliferation in Primary Cultures of Rat Hepatocytes, *Toxicol. Appl. Pharmacol*, 1983, **67**, 15–25.

36. A. M. Mitchell, J. C. L'Huguenot, J. W. Bridges, and C. E. Elcombe, 'Identification of the Proximate Proliferators Derived from Di-(2-ethylhexyl) phthalate', *Toxicol. Appl. Pharmacol.*, 1985, **80**, 23.

37. D. J. Benford, S. Patel, H. J. Reavy, A. Mitchell and N. J. Sarginson, 'Species Differences in Response of Cultured Hepatocytes to Phthalate Esters', *Food Chem. Toxicol.*, 1986 (in the press).

38. A. Mitchell, 'Hepatic Peroxisome Proliferation: Mechanisms and Species Differences, Ph.D. Thesis, University of Surrey, 1985.

39. R. Peto, M. C. Pike, L. Bernstein, L. S. Gold and B. N. Ames, 'The TD50: a Proposed General Convention for the Numerical Description of the Carcinogenic Potency of Chemicals in Chronic Exposure Animal Experiments', *Environ. Health Perspect.*, 1984, **58**, 1.

19

Effects of Pesticides on Non-target Organisms

By D. Osborn

INSTITUTE OF TERRESTRIAL ECOLOGY, MONKS WOOD EXPERIMENTAL STATION,
ABBOTS RIPTON, HUNTINGDON, PE17 2LS, UK

1. Introduction

In the United Kingdom real worries about the unexpected effects of
pesticides began with deaths of foxes in Northamptonshire and the
failure of Peregrine falcons to recover from the persecution they
suffered during World War II. Sheail[1] gives an illuminating
account of the growth of concern. The peregrine's problems are
described by Ratcliffe.[2] Elsewhere in the world, environmental
problems arising from pesticide use in particular, and chemical
use in general, were becoming apparent during the 1960s and 1970s,
and Carson's widely read book 'Silent Spring'[3] made the general
public aware of the possibility that pesticides could, by having
effects on non-target organisms as well as pests, fundamentally
upset 'the balance of nature'. This chapter outlines some of the
environmental and research problems encountered in studying the
effects of chemicals on organisms living in the natural
environment, and specifically concentrates on the effects of
pesticides on non-target organisms.

A working definition of pesticide and 'non-target organisms'
may be helpful as different organizations and authorities give
these words meanings which differ considerably. Pesticides are
often defined as being those chemicals used in agriculture to
control the severity and incidence of pests and diseases which
reduce agricultural yields; in addition, they have a number of
non-agricultural uses. There are some grey areas: for example,
are agricultural veterinary products pesticides, and are chemicals
indentical to those used in agriculture still pesticides when used
as wood preservatives *etc.*? For present purposes the broadest
view will be taken of a pesticide, *i.e.* any chemical applied to
land or materials to prevent economic losses resulting from the
activities of living organisms. This is broadly in line with the
definition given by Mellanby.[4] It should be remembered that
pesticides are designed to kill pests. Pesticides includes
herbicides and hence effects on plants are also described.

Non-target organisms are all those organisms which are not the
intended specific targets of a particular use of a pesticide. The
term, therefore, includes animals living in and around farmland
such as beneficial insects (*i.e.* those insects which provide a
'natural' means of limiting pest damage), and also animals living
at some distance from the point of application of a pesticide
which, by some route, accumulate harmful residues of the chemical
or its breakdown products in their tissues. The definition also
encompasses those organisms which are not directly affected by the
toxic action of the pesticide, but also those which are indirectly

affected. This latter group of organisms may suffer for a number of reasons; for example, pesticide use may have removed a food source, or more drastically altered the ecological structure of the habitat.

2. General Points

1. In studies of the effects of pesticides it is important to remember that these chemicals are designed to kill living organisms. They must have this capacity in order to be effective. There is a chance that they will not only kill the target pest organism but other organisms as well. There is an even greater chance that they will have a recognizable adverse effect on an organism and a greater chance still that they will have some sub-lethal effect, though this will not necessarily be a deleterious effect of physiological significance.

2. To be effective, pesticides must persist for long enough in the environment to make contact with and have an effect on the target, *i.e.* they must persist at an effective concentration for periods of many days to many decades (in the case of timber treatments). Considerable effort must be devoted to determining persistence times, and the consequences of persistence in assessing the likely effects of pesticides on non-target organisms.

3. A persistent chemical may have its adverse effect far from its site of use, and may be passed from one animal to another. In certain circumstances, and this is by no means a general rule, pesticides may accumulate either in those organisms which are higher in the food chain or in older organisms.

4. No amount of laboratory testing will effectively deal with difficulties arising from the unexpected effects of pesticides: the only real solution to this problem is to conduct close examination of a product once it is in use to ensure that no unexpected efects are occurring as a result.

5. The mechanisms by which many pesticides exert these effects on non-target organisms are unknown. This makes it extremely difficult to make any attempts to prevent such problems arising in future.

6. Attempts should not be made to hide the fact that we are using dangerous chemicals in a situation (the environment) that is by no means understood. It is not even known how many of what type of organism should occur in a particular type of environment, although some progress along these lines is being made in the case of the invertebrate fauna of British streams and rivers.[5] Equally, knowledge of such vital resources as energy and the nutrients cycle in many ecosystems is so rudimentary that there is little understanding of the basic processes that control the function of natural and semi-natural ecosystems (see also Chapter 10). Sight must never be lost of these facts. In simple terms, it is not known how the balance of nature is being maintained or what that balance is or should be.

It is vital that pesticides are used judiciously and that those having unacceptable adverse effects are speedily removed

from the market.   Long-term damage to man and the environment,
could result if this cautious approach is not adopted..

The remainder of this chapter briefly describes *some* of the
known effects of pesticides on wild organisms.   Most of the
examples are drawn from the UK, North-West Europe, USA, and
Canada, where many problems have been well documented.   The
effects on non-target organisms are perhaps largely unknown in
many parts of the world.

This last paragraph carries with it an assumption that the
effects of pesticides can be detected.   In reality, detection and
evaluation of the effects of chemicals on organisms living in the
natural environment are extremely difficult.   There are very many
reasons for this.   Two of the main problems faced by research
workers studying such matters are:

(i)   A change or effect can only be adequately documented if
           there is a baseline (or at least a reference point)
           against which the change or effect can be measured.   For
           most species on Earth such baseline data do not exist.
           On very rare occasions data have been gathered retro-
           spectively.

(ii)   Once a change or effect has been noticed, it can be
           difficult to pinpoint the cause.   The change may be due
           to a single chemical, to a mixture of chemicals, or to a
           mixture of interacting effects which may, in addition to
           chemicals, include long-term changes in land use and
           shorter-term changes in specific industrial or
           agricultural practices.

3.   Types of Living Organisms Affected by Pesticides

There is much evidence detailing the effects that pesticides have
on birds, mammals, fish, amphibia, invertebrates, and soil
micro-organisms.   The invertebrates studied range from bees to
soil arthropods and earthworms.   There is much less information
about the effects of pesticides on non-target plants than there is
on animals, and for some groups of organisms *e.g.* bryophytes,
lichens; few, if any, studies have been done.   There is very much
less certainty on the importance of the recorded effects.

It is impossible to give a complete account of all the known
effects of pesticides in this Chapter.   Consequently examples are
given which illustrate the current difficulties in attempting to
prevent the undesirable effects of pesticides, and point out where
further information is required.

It must also be recognized that, despite all the efforts to
prevent the marketing of harmful products being made by industry
and regulatory agencies, the detection of the adverse effects of
pesticides is still largely a retrospective activity.   A chemical
comes into use but unknown adverse effects are noted and remedial
action is initiated.   Predicting the adverse effect is difficult
for most types of organism largely because it is not known how
environmental variables influence the availability of pesticide
formulations and their active ingredients to wildlife.   From a
scientific research point of view this makes studies on effects of
pesticides on wild organisms appear to have a very fragmentary
nature.   This is also because most of the known effects have been

noticed fortuitously. There have been few instances where a research programme has been mounted to look for particular effects on non-target organisms in the wild. Some such studies are now being done involving careful field observations and of changes that occur when organisms are exposed to different pesticide application regimes in real agricultural situations.

4. Birds

4.1. Effects on Birds. Effects on this group of animals, along with those on fish, are those ones which have given rise to the most public concern. Fortunately, the long interest that biologists have had in wild birds has led to the development of an extensive database about their breeding biology and ecology. Important aspects of the physiology of wild birds and their behavioural characteristics (e.g. migration patterns) are also well documented. This means that, for birds, the normal state is relatively well known, so deviations from that state can be detected with comparative ease. Also, large-scale mortalities of birds, even of the smaller ones, are obvious and such wildlife incidents are reported regularly. Large-scale mortalities are easily seen almost as soon as they occur. This greatly increases the chances of detecting the cause and preventing repeated mortalities.

Despite the relative ease with which problems can be identified in birds the types of effect detected are generally limited to cases of mortality or gross effects on population dynamics which lead to declines in population size. A further limitation is that the latter effect can only be detected and accurately quantified for those species whose population size is accurately known in a given geographical area. So, effects on birds of prey are well documented, whereas evidence of effects on passerines is largely limited to statistics for mortality incidents.

More subtle effects on birds (including indirect ones) of the kind that might alter social interactions and might change breeders into non-breeders and *vice versa*, or of the kind that might change the survival or breeding prospects of individuals with different genetic characteristics cannot, at present, be detected.

Work in the wild is badly needed but the necessary co-operation between different scientific disciplines has yet to come about. The best methods available to date involve such schemes as the British Trust for Ornithology's Common Bird Census, whose effectiveness and validity is perhaps best demonstrated in the breeding season on the farmland of lowland Britain.

4.2. Mortalities of Birds. Tables 1-4 give some information relating to mortalities of birds that have occurred in several parts of the world. These tables are based on mortality incidents thought to be of an unusual nature where groups of birds have been found dead in close geographical or temporal proximity. Typical incidents might be of these types: (i) several hundred starlings found dead floating on a reservoir, or (ii) unusually high frequency of mortality of individual herons in an area of a few square miles. The first incident had a natural cause, but the second was caused by toxic chemicals - including some persistent organochlorine pesticides.

Table 1   *Some examples of effects on birds*

Insecticides

Bird type/species	Pesticide/Use	Effects noted
Birds of prey	Indirect exposure DDT (insecticide in agriculture, sheep dip, *etc*.)	Reduced breeding success. Eggshell thinning
Birds of prey	Indirect exposure to Dieldrin, Aldrin, Endrin (seed dressing)	Deaths of individuals
Geese	Organophosphorus compounds (seed dressing)	Deaths of groups of birds
Song birds	Indirect exposure to DDT (insecticide for fruit trees *etc*.)	Deaths and local populations reduced or eliminated.
Forest birds (Warblers)	Organophosphorus compounds	Populations reduced temporarily
Various species	Heptachlor (against fire-ant)	Populations temporarily reduced
Pigeons, Doves, Corvids, Game birds	Dieldrin/Aldrin, Heptachlor (seed dressing)	Death
Ducks	Carbofuran (granular formulation)	Death
Grebes	DDD (Gnat control on clear lake)	Population almost eliminated
Gulls, Herons	Indirect exposure to organochlorine insecticides	Shell thinning. Reduced breeding success in some species
Brown pelican	Indirect exposure to organochlorine insecticides	Shell thinning and decline of population in some areas
Various species	Parathion	Deaths of individuals

Other Agricultural Chemicals

Bird type/species	Pesticide/Use	Effects noted
Game birds, Birds of prey	Mercury (seed dressing fungicides)	Deaths of birds (groups and individuals)
Snail kites	Molluscicide	Deaths of individuals
Raptors and Rooks	Thallium (rodenticide)	Deaths of individuals and decline of population
Raptors	Fluoracetamide (*vs*. jackals)	Deaths of individuals
Raptors	Endrin	Deaths of individuals

Information in these tables is drawn from Refs. 6 and 7.

Table 2 *Causes of death amongst a sample of bald eagles found dead in the USA* (reclassified from Ref.6).

Main cause	*%*
Direct persecution by man[a]	43
Accidental death	21
Natural causes[b]	12
Pesticides[c]	10
Uncertain	14
Total attributable to human activity[d]	53

[a]Includes some cases of 'primary' exposure

[b]Includes diseases

[c]Refers to 'secondary' exposure to organochlorines; mostly HEOD residues were detected.

[d]This is a minimum figure; some of the accidental deaths involved collision with power lines.

Table 3 *Classes of chemicals killing wildlife in England and Wales in the 1970s and early 1980s*

	Approx. % of total numbers of incidents
Insecticides	50
Rodenticides	25
Herbicides	10
Molluscicides	10
Nematicides	5
Mammal poison	5

The table does not reflect the number of incidents which actually occurred. For example, the mammal poison, strychnine, was implicated in a large number of incidents.

Data abstracted from the series of reports such as in Ref.8.

Table 4 *Classes of the causes of wildlife incidents investigated partly or wholly by staff at ITE Monks Wood 1979-86*

Incident type	Approx %
Industrial	25
Agrochemicals[a,c]	20
Natural causes[b]	15
Diseases	10
Amenity/recreational activities[c]	5
Unknown	20
Total attributable to human activity	50

[a]These studies are often conducted in conjunction with MAFF (ADAS) Tolworth Laboratory who conduct a national scheme designed specifically to detect the effects of pesticides, mortality being the main effect monitored (see Ref. 8).

[b]Includes the largest single incident where over 30,000 sea-birds were washed ashore on the east coast of the UK.

[c]Does not include the very large number of cases of swan mortality caused by lead poisoning, nor the deaths of individual predatory birds killed by organochlorine insecticides. These later deaths are monitored in a study of pesticides residues in predatory birds run by staff at ITE Monks Wood funded by the Nature Conservancy Council.

Tables 1-4 draw attention to a number of matters concerning incidents of wildlife mortality. First, mortality incidents involving large, obvious species of high wildlife conservation interest tend to predominate. This suggests a considerable bias in the manner in which these incidents are reported. Presumably, numbers of smaller, less obtrusive birds also die in unusual circumstances, but are rarely found or not thought to be of great concern by the people that find them.

Secondly, it is quite clear that pesticides are not the only cause of wildlife mortalities. Many incidents appear to be related to industrial activities or to other human influences. The Tables also suggest that, of those pesticides that do cause wildlife incidents, insecticides are, by far, the group that causes most problems. However, many of these incidents involve the mis-use of pesticides[9] and, perhaps, should be set to one side in the scientific arguments about the general environmental safety of pesticide use.

An interesting example of the unexpected effects of chemicals used in agriculture is provided by the chemical famphur. This material can be applied orally, topically, or intramuscularly, and is used against warble fly larvae in cattle. Its use in this way makes it a veterinary product, and in its effect, it could be considered to be an organophosphorus insecticide. Magpies have been killed both in the UK and in the USA by this chemical, and some reports have been published which detail both the 'primary' poisoning of magpies and the 'secondary' poisoning of hawks eating affected prey.[8,10]

4.3. Sub-lethal Effects of Pesticides on Birds. The first example of a sub-lethal effect of a pesticide on a bird was probably the egg-shell thinning phenomenon first described by Ratcliffe[11] and recently re-examined by Moriarty *et al*.[12] A detailed review of the phenomenon is given by Cooke,[13] and its significance for the breeding biology of raptors is discussed by Newton.[6]

DDE, the stable metabolite of the insecticide DDT, is said to be the cause of egg-shell thinning and consequent declines in the breeding success of birds of prey. The phenomenon has been studied in the laboratory[14] as well as in the field.[15]

Egg-shell thinning is a good example of how difficult it is to identify an effect in the wild, to evaluate its significance for the species suffering the effect, and also to identify the mechanism by which the suspect chemical causes its effect. It is a testimony to the magnitude of these difficulties that 20 years after the phenomenon was first discovered arguments can still be raised which question not only the significance of the effect, but whether the effect is really present in some species or not.[12]

Other sub-lethal effects on birds have been noted from time to time, as a result of exposure to pesticides (*e.g.* Ref. 16) but the general significance of the effects which have been identified is as yet impossible to assess.

4.4. The Role of Physiology and Biochemistry in Understanding the Effects of Pesticides on Birds. An understanding of the physiology of the breeding cycles of wild animals in any detail is only just beginning and studies of aspects of their biochemistry which are concerned with the pharmacology of pesticides in birds (*e.g.* metabolism of xenobiotics, identification and characterization of the sites of action) are still in their infancy, although significant progress is being made in the area of xenobiotic metabolism.[17] There is little doubt that, if we are to understand the full potential impact of pesticides on such central processes as breeding biology, more work on the sub-lethal effects of chemicals on the physiology and biochemistry of breeding biology will be needed. This will be particularly productive as it seems likely to help greatly in explaining the inter-species differences in sensitivity to the effects of pollutants which often undermine our efforts to predict environmental hazards from tests on a few 'representative' laboratory species.[18]

4.5. Indirect Effects of Pesticides on Birds. There is little hard evidence to support the idea that pesticide use has had an indirect effect on birds, either by reducing food supply or causing habitat changes. However, results emerging from the Game Conservancy's 'Cereals and Game Birds' project[19] strongly suggest that the general use of pesticides has impaired the reproduction of partridges on the farmland of lowland Britain, by removing the insect fauna that the young chick feeds on immediately on hatching. The results of this study will be followed very closely as it is one of the few pieces of work in the world attempting to gain some measure of the impact of overall pesticide and agrochemical use.

5. Effects on Mammals

Although mammals were amongst the first groups of animals to be identified as suffering the adverse effects of pesticides,[1] there has been relatively little work on the effects of pesticides on wild mammals, although, of course, a great deal of mammalian toxicology studies are conducted in the course of developing new pesticides.

Effects on mammals have proved very difficult to quantify. As already mentioned, Sheail[1] gives details of the investigations into 'fox-death', a phenomena perhaps first noticed as long ago as 1959. The first detailed accounts were reported from the Fitzwilliam Hunt, near Oundle, Northants. Many animals other than foxes were involved, and although there were some early indications that small quantities of organochlorine compounds were present in affected animals, other causes, *e.g.* diseases like acute hepatitis, were also suggested. By early 1960 it was clear that the phenomenon, which had increasingly disturbed hunting, conservation, and agricultural interests was largely confined to arable farming areas where large numbers of birds (on which foxes might have preyed) had died in wildlife mortality incidents. Eventually the speculation and gradually accumulating circumstantial evidence that implicated dieldrin and heptachlor seed dressings as the base cause of 'fox-death' were confirmed by experimental feeding of foxes. This work led to the conclusion that liver residues as low as 1 mg kg^{-1} dieldrin or 4 mg kg^{-1} heptachlor could be indicative of the death of a fox being caused by these pesticides.[20]

Some other work on mammals has been less conclusive. However, the sensitivity of some bats to applications of organochlorines such as DDT and lindane has been demonstrated.[7] In addition there is some, albeit circumstantial, evidence to link declines in populations of aquatic mammals (*e.g.* otters) with organo-chlorines.[21]

At present there is little evidence to suggest that mammals are affected by modern pesticides, apart from some evidence of sub-lethal effects.

However, the study of the effects of pesticides on small mammals involves a number of special difficulties, and it seems likely that some effects on small mammals will only be quantified if intensive trapping studies are conducted in carefully selected habitats. Some work of this type is being conducted, and publication of the results of this work are awaited with interest.

6. Effects on Invertebrates

6.1. Bees and Butterflies.

These are two groups of invertebrates about which there is much concern.

Bees are important economically, and butterflies are of great conservation value. Because bees visit some agricultural crops, notably oilseed rape, there is a likelihood that they will come into contact with chemicals that will kill them in large numbers. Incidents such as this now occur annually and MAFF has established a scheme to monitor these events.[6] Some changes have been made to the active ingredients used on oilseed rape in an attempt to minimize this problem.

Concern about the effects of pesticides on butterfly populations remains unestablished. A recently published atlas of British butterflies[23] makes it plain that there have been marked declines in the numbers of sites at which many butterflies are found. However, as very small changes in the fine structure of the habitat can have important influences on butterfly breeding, it is not possible to ascertain the effects of pesticides on butterflies as distinct from the many other changes that have taken place in the environment.

6.2. Other Invertebrates. A considerable quantity of data has been gathered about effects of pesticides on non-pest invertebrates because it seems obvious that chemicals applied to kill invertebrate pests might also kill their non-target relations.

The greatest agricultural impact of pesticides on non-target invertebrates may have been to reduce the numbers of pest predators.[18] This effect on predators may be a long-term effect, and some have argued that more predatory species have suffered from pesticide applications than have pest species. Two accounts, one describing a specific case[24] and the other[25] the currently perceived general overviews of this topic, provide more information. In short, it is possible to identify many effects of pesticides on populations of, say, soil invertebrates, but equally possible to describe many effects as being short-term. An exception might be the influence of aldicarb on earthworm populations.[26]

6.3. Soil Micro-organisms (Including Fungi). Short of the use of complete soil sterilants, the effects of pesticides on the microflora and microfauna of soils seem small in so far as it has been possible to measure these effects.

However, it is clear from a recent conference[27] that much needs to be done before we can say categorically that pesticides have no important effect on soil micro-organisms. Many of the tests of soil micro-organism effects are effectively tests of ecosystems function, and it is widely acknowledged amongst ecologists that many of the species normally present in an ecosystem may be removed without many of the functions of the ecosystem being adversely affected: *i.e.*, simplistically, a marsh remains a marsh in the absence of cuckoos and reed warblers.

7. Effects on Plants

As most pesticides are used on agricultural crops, chemicals that have a high phytotoxicity to plants in general are useless as pesticides, with the obvious exception of the total weedkillers such as paraquat and glyphosate.

However, some crops can be inadvertently affected by applications of pesticides to nearby fields. For example, lettuce, tomato, and beans may be crops that are particularly sensitive to suffering damage from herbicide application in those environmental conditions when spray drifts from one field to neighbouring ones.

Crop plants can also be damaged by the application of fungicides, and there is even evidence that, under certain conditions, herbicides may cause changes in plant-fungi relationships that increase the chances that the plant will succumb to a stress of one kind or another (*e.g.* drought or disease[27]).

Effects of pesticides on non-target plants other than those that are crops are not well documented, as far as adverse effects in the real world are concerned. Once again it is difficult to distinguish between effects that are due to pesticides (except for *gross* damage caused by spray drift, mis-use of herbicide, or careless application), and those effects which have other causes. In particular, for plants, the use of increasing amounts of fertilizer could be an important confounding variable, especially as pesticide and fertilizer use have increased over the same period of time.

There is no doubt that several plants are much more sensitive to herbicides than are others, and a recent study on herbicides for use on nature reserves provides a convenient source of information on this topic.[28]

8. Conclusion

This chapter describes some of the effects of pesticides on non-target organisms, but points out that our knowledge is fragmentary in many respects. For too long a retrospective approach which deals with problems as they arise has been relied upon. Our ability to predict possible adverse effects is slowly improving, but more progress with predicting effects on aquatic organisms is being made than is being made for adverse effects on terrestrial wildlife. This state of affairs will persist until the results of physiological and biochemical research to explain interspecies differences in both lethal and sub-lethal sensitivity to pesticides can be used. Once this information is available it may be possible to integrate this knowledge with the retrospective ecological evaluations that have to be used in isolation at present, and to have proved so difficult to generalize and apply.

9. References

1. J. Sheail, 'Pesticides and Nature Conservation: the British Experience 1950-75', Oxford University Press, Oxford, 1985.

2. D. A. Ratcliffe, 'The Peregrine Falcon', T. and A. D. Poyser, Calton, 1980.

3. R. Carson, 'Silent Spring', Riverside Press, Cambridge, USA, 1962.

4. K. Mellanby, 'Farming and Wildlife', Collins, London, 1981.

5. D. Moss, M. T. Furse, J. F. Wright, and P. D. Armitage, *Freshwater Biol.* (in press).

6. I. Newton, 'Population Ecology of Raptors', T. and A. D. Poyser, Berkhamsted, 1979.

7. F. L. McKewan and G. R. Stephenson, 'The Use and Significance of Pesticides in the Environment', John Wiley & Sons, New York, 1979.

8. Haffiadas, 'Pesticide Science 1980', Reference Book 252(80).

9. C. J. Cadbury, 'Silent Death; the Destruction of Birds and Mammals through the Deliberate Mis-use of Poisons', Royal Society for the Protection of Birds, Sandy, 1980.

10. C. J. Henry, L. J. Blus, E. J. Volbe, and R. E. Fitzner, *J. Wildl. Manage.*, 1985, **49**, 648.

11. D. A. Ratcliffe, *J. Appl. Ecol.*, 1970, **7**, 67.

12. F. Moriarty, H. M. Hanson, and A. A. Bell, *Environ. Pollut.*, A, 1986, **40**, 257.

13. A. S. Cooke, *Environ. Pollut.*, 1973 **4**, 85.

14. J. L. Lincer, *J. Appl. Ecol.*, 1975, **12**, 781.

15. I. Newton and J. A. Bogan, *J. Appl. Ecol.*, 1978, **15**, 105.

16. C. E. Grue, G. V. N. Powell, and M. J. Chesney, *J. Appl. Ecol.*, 1982, **19**, 327.

17. C. H. Walker, *Drug Metals Rev.*, 1978, **7**, 295.

18. P. J. Bunyan, in 'Environment and Chemicals in Agriculture', ed. F. P. W. Winteringham, Elsevier Applied Science, London, 1985, p.307.

19. G. R. Potts, in 'Origins of Pest, Parasite, Disease, and Weed Problems', 18th Symposium of the British Ecological Society, ed. J. M. Cherret, and G. R. Sagar, Blackwell, Oxford, 1977, p.183

20. D. K. Blackmore, *J. Comp. Path. Ther.*, 1963, **73**, 391.

21. P. R. F. Chanin, and D. J. Jefferies, *Biol. J. Linn. Soc.*, 1978, **10**, 305-328.

22. D. A. Jett, *Environ. Toxicol. Chem.*, 1986, **4**, 225.

23. J. Heath, E. Pollard, and J. A. Thomas, 'Atlas of Butteflies in Britain and Ireland', Viking, Harmondsworth, 1984.

24. J. P. Dempster, *J. Appl. Ecol.*, 1969, **6**, 339.

25. C. A. Edwards in 'Comportment et Effects Secondaires des Pesticides dans le Sol', ed. M. Hascoet and E. Steen, INRA, Paris, 1985, p.27.

26. Ph. Lebrun in 'Environment and Chemicals in Agriculture', ed. F. P. W. Winteringham, Elsevier Applied Science, London, 1985, p.105.

27. 'Comportment et Effects Secondaires des Pesticides dans le Sol', ed. M. Hascoet and E. Steen, INRA, Paris, 1985.

28. 'The Use of Herbicides on Nature Reserves', ed. A. S. Cooke, (Focus on Nature Conservation No.14), Nature Conservancy Council, Peterborough.

20

Comparative Risk Assessment - The Lessons of Cultural Variation

By S. Jasanoff

PROGRAM ON SCIENCE, TECHNOLOGY AND SOCIETY, CORNELL UNIVERSITY, ITHACA, NEW YORK 14853, USA

1. Introduction

In the proliferating literature on risk assessment and risk management, there are two schools of thought about the role that science can play in guiding national policy. One view, particularly favoured by the scientific community,[1] holds that public concern about many technological hazards, including toxic chemicals, is based on misunderstanding and misinformation which science has a duty to correct. According to this analysis, greater expert involvement is needed in order to offset public hysteria about risk. It has been argued that scientists should play a prominent part not only in developing objective measures of risk but also in explaining their analytical methods to the public, and in placing risks from different sources in perspective through explicit calculations of 'relative risk'. Scientists believe that they have a further responsiblity to ensure that risk control policy is grounded in sound technical data and that political considerations do not affect the interpretation of evidence. Such initiatives by the expert community, it is suggested, can steer policy-making in more rational directions, leading to results that are consistent with the actual, rather than the perceived, significance of technological risks.

A contrasting viewpoint, emerging primarily from the social science literature on risk, has tended to emphasize the political and cultural dimensions of risk decisions. Anthropological research, for example, suggests that both primitive and advanced societies make essentially arbitrary choices in treating some risks as important and others as insignificant.[2] The implications of scientific uncertainty have been studied by political scientists, who note that, in areas of genuine disagreement, interest group affiliations frequently colour the way experts themselves look at evidence.[3] Cross-cultural comparisons of risk management point to a similar conclusion by illustrating how experts in different countries vary in their evaluations of the same risk or its acceptability.[4] All of these analyses stress the subjective element in judgements about risk and call into question the expert's capacity to place risk policies on a rational, value-neutral footing.

This chapter seeks to clarify the role of science in risk decision-making by exploring how far cultural predispositions actually influence expert assessments of risk. To what extent, for example, are decisions which are conventionally regarded as 'scientific', such as the weighting of competing studies and methodologies, affected by national attitudes towards risk? Are there systematic variations in the way scientific knowledge is produced, tested, and validated in the risk management systems of different countries? How do experts in different national settings behave when confronted by gaps in the scientific record? Are there detectable differences in the credibility they accord to indirect evidence of harm or in their propensity to use predictions and extrapolations when hard data are not available?

In the interests of economy and clarity, this chapter limits itself to a comparison of chemical risk assessment in two countries: the United Kingdom and the United States. The first part of the chapter reviews the major reasons for expecting divergences between the processes and results of risk assessment in these two countries. The second section compares the assessment of two toxic substances, asbestos and lead, in the UK and the USA in order to demonstrate how variations have arisen in the expert evaluation of risk. Drawing on these case studies, the chapter's concluding section re-evaluates the role of science in public decision-making about risk.

2. The Causes of Divergence

It is frequently assumed that the scientific component of risk decisions can be separated from and insulated against the social and political considerations that also enter into such policy-making. This, however, is an overly simplified picture. There are at least three reasons, each explored more fully below, for expecting that even the most technical aspects of risk decisions cannot be wholly immune to socio-cultural influences. First, scientists themselves are members of national and cultural groups and, as such, are likely to share in their culture's underlying assumptions about risk. Secondly, many questions about risk can be framed in scientific terms, but cannot be answered by science alone. Expert responses to these 'trans-scientific' issues may therefore reflect personal or institutional values in addition to technical understanding. Thirdly, the scientific assessment of risk information almost invariably takes place within a framework of laws and institutions that impose their own constraints on the interpretation of evidence. A law to regulate environmental chemicals may require, for example, that doubtful scientific issues should always be resolved so as to provide greatest protection for persons at risk. Such legal requirements reflect a pervasive political consensus that cannot help influencing the appraisal of specific technological risks.

2.1. The Impact of Culture. Mary Douglas's work[5] on beliefs about pollution and the environment has convincingly established that the concept of risk varies from culture to culture. Seemingly objective assertions about the physical environment are rooted in social practices that affect the perception of nature. Thus, two tribes living on the banks of the same river disagree on which season is hot and which cold. Neither is clearly 'right', since the temperature range by day and night is roughly the same in both seasons. The difference seems to lie in the agricultural systems of the two tribes. The group that has to complete all its work in the short dry season fears the heat of this period much more than the group that distributes its outdoor work evenly through the year. More recently, Douglas and Wildavsky have argued that primitive peoples are not alone in holding culturally conditioned views about risk.[6] Their work suggests cultural explanations for the elevation of statistically minor risks, such as the risk of cancer from exposure to environmental chemicals, to the forefront of social and political conern in advanced Western societies.

Recent work on the psychology of risk perception provides additional evidence that public attitudes to risk are shaped by factors other than simple statistical probability. A number of groups have carried out interesting empirical research in this area, including Slovic, Fischhoff, and their colleagues in the USA,[7] and Green and Brown in the UK.[8] Their findings suggest that public perceptions of risk are influenced both by the qualitative characteristics of the hazard and by the degree of control individuals can expect to exercise against it. For example, the public in both the UK and the USA appears more concerned about involuntary than voluntary risks, and at least the US public tends to exaggerate the probability of risks that are either unfamiliar or have a catastrophic potential (see also Chapter 16).

To what extent are scientists subject to the cultural and psychological influences that affect the lay public's perception of risk? The accepted dogma within science has long been that such external factors have little or no bearing on the way science is done. There is a body of literature in the sociology of science, however, which sugests that scientific concepts and principles can be developed and sustained because of their relevance to a wider social context.[9] One might expect to detect such influences most clearly in the area of policy-relevant science, where the connections between scientific inquiry and social concerns are most pronounced. Indeed, cross-national studies of regulatory policy have produced some evidence, albeit scattered and unsystematic, that alternative approaches to risk analysis are evaluated differently by scientists in different countries.[10] The case studies in this chapter make a small additional contribution to this literature.

2.2. Trans-Science.

Alvin Weinberg, who popularized the concept of trans-science in the early 1970s, was among the first scientists to point out that questions about risk often lie outside the bounds of both current knowledge and feasible scientific inquiry.[11] Weinberg defined trans-science as consisting of those questions which can be asked of science and yet which cannot be answered by science. They thus 'transcend science', although epistemologically they are questions of fact. In a recent paper, Weinberg has argued that regulators should not present such unanswerable questions to the scientific community at all, but should proceed on the assumption that certain kinds of uncertainty simply cannot be resolved.[12] Alternatively, Weinberg suggests that we should recognize a new area of inquiry called 'regulatory science', where the uncertainties are so great that ordinary standards of scientific proof cannot be required to apply. For Weinberg, much of the discipline of probabilistic risk assessment - the prediction of extremely rare events - would fall into the domain of regulatory science.

American lawyers and administrators dealing with the problem of trans-science have adopted a somewhat different perspective. Recognizing that questions about risk often have no definitive scientific answer, they have suggested that uncertain or disputed technical issues should be resolved as a matter of policy by the appropriate regulatory authority.[13] The need for such 'science policy' decisions has been widely accepted in connection with regulating chemical carcinogens in the USA. Several federal agencies, including the Environmental Protection Agency (EPA) and the Occupational Safety and Health Administration (OSHA), have developed explicit principles for evaluating epidemiological and toxicological data on chemical carcinogenesis.[14] The subject of considerable scientific and political controversy, these US 'cancer policies' never had a counterpart in British regulatory practice, which prefers not to articulate the principles of scientific interpretation in explicit, generic form.

The relationship between risk assessment and science policy has been a matter of concern to US regulators for some years, especially in the context of controlling toxic chemicals. The basic question is this: Are there any aspects of evaluating chemical risk that can be treated as wholly scientific and can therefore be resolved independently of all legal or policy considerations? In 1981 this question was referred to a special committee of the National Academy of Sciences (NAS) and in early 1983 the committee published its completed report on risk assessment by the federal government.[15] The report drew a fundamental conceptual distinction between *risk assessment* ('the use of the factual base to define the health effects of exposure... to hazardous materials') and *risk management* ('the process of weighing policy alternatives and selecting the most appropriate policy response').[16] The committee left no doubt that these two elements of risk decision-making should be kept distinct as far as possible. Yet the report recognized that even the risk assessment component is not purely technical, but involves judgements based on both scientific and policy considerations. Accordingly, the committee recommended that risk assessment should not be delegated to an expert body standing outside the federal regulatory structure, but should rather be carried out by each agency in line with its legal and policy objectives.

In the UK, a Royal Society study group prepared an almost contemporaneous report on risk assessment and risk perception.[17] Differences in the terminology used by the British and American reports point to the absence of a convenient uniform vocabulary for speaking about risk. The Royal Society report conceptualized risk policy-making as a three-part process consisting of *risk estimation* ('the identification of outcomes and estimation of their probability and magnitude'), *risk evaluation* ('the determination of the significance of risk to those affected'), and *risk management* ('judgements concerning the acceptability of risk').[18] In this conceptual scheme, the term 'risk estimation' comes closest in meaning to the NAS committee's 'risk assessment'.

The Royal Society study group was concerned more with identifying valid analytical approaches than with ways of structuring the process of decision-making. As a result, the group's final report did not directly address questions as to whether risk estimation should be performed by an independent scientific committee or by a regulatory agency using in-house expertise. Like the NAS report, however, the Royal Society report stressed the uncertainties involved in risk estimation, particularly in the case of exposure to toxic chemicals, and noted the need for consultation with exposed groups or for other forms of public accountability.[19]

Both British and American experts thus agree that the relatively technical exercise of risk assessment (or estimation) is fraught with uncertainties and demands considerable judgement. The NAS committee was forthright in admitting that judgement in this context involves not only scientific but also policy considerations. The important point, however, is that even experts do not believe that the results of risk assessment are uniquely determined by science. Consequently, it is not unreasonable to expect external factors, such as cultural differences between countries, to lead to discrepancies in the assessment of risks.

<u>2.3. The Legal and Institutional Framework.</u> The third and most significant source of cultural influences on risk assessment is the legal and institutional framework within which decisions about risk are made. Statutes and implementing organizations reflect broad public understandings about how governmental power should be exercised and to what ends. Even among very similar Western societies, there are subtle differences of doctrine and tradition about the appropriate functions of government and the role of citizens in public decision-making. In recent years, a number of cross-national studies have traced the impact of these fundamental political divergences on public policies concerning the environment and health and safety. Brickman, Jasanoff, and Ilgen, in particular, looked in detail at similarities and differences in the chemical control laws and implementing institutions of four countries: the United Kingdom, France, West Germany, and the United States.[20] The study found significant variations in the process of decision-making in all four countries, with the UK and the USA differing most sharply overall. Of the numerous striking differences between these two nations, perhaps most significant for purposes of this chapter are those relating to the use of scientific expertise in decisions about toxic chemicals.

Three features of the British and American regulatory systems carry particularly important consequences for risk assessment: the formulation of legislative standards and obligations, the procedures for obtaining and analysing technical information, and the provisions for challenging governmental decisions about risk. As noted in several recent studies, the American approach to all three phases of decision-making tends to be formal, adversarial, and rule-bound. First, US risk management laws generally contain detailed standards telling regulators when, and how, to assess risks, and they frequently specify strict timetables for planning and sequencing regulatory action. Secondly, American legislation permits private parties to participate actively in technical decision-making, not only by presenting their own analyses of the relevant science, but by questioning the contrary opinions of their adversaries. Thirdly, American legislation provides unusually generous opportunities for dissatisfied parties to seek judicial review of regulatory decisions, and courts are quite prepared to scrutinize the basis for challenged risk determinations.

British environmental and health and safety regulation, by contrast, is usually carried out under broad enabling laws that leave regulators considerable discretion to decide when and how to assess risks. Public participation usually takes the form of informal consultation and generally occurs *after* the regulatory agency has developed the scientific information base with the aid of its expert advisers. Neither health and safety standards nor the underlying risk assessments are ordinarily subject to judicial review. In any case, British courts have traditionally deferred to the government on matters of policy and expert judgement, and they would be reluctant to look too critically at an agency's scientific and technical reasons for taking regulatory action.[21]

These general characteristics of the British and American policy systems have had a substantial impact on the way science is incorporated into risk decisions. In the UK, scientific advice is developed with little interference from administrative or judicial circles. By contrast, US agencies, always concerned about reversals in court, have been driven to justify their risk determinations in terms that judges are likely to approve. Rule-making procedures dictated by Congress are already remarkably formal by British standards. In many cases, however, agencies have supplemented these with further procedures in order to convince the courts that they have given a fair hearing to all points of view. In the area of carcinogen regulation, the federal agencies have taken care to develop explicit policies aimed at assuring judges that they have resolved issues of scientific uncertainty in a rational, non-arbitrary manner. Finally, judicial review has been instrumental in promoting the use of quantitative risk assessment. The Supreme Court held in 1980 that OSHA's proposed workplace standard for benzene was invalid because the agency had failed to produce a persuasive quantitative estimate of the risk of cancer.[22] The decision considerably narrowed the scope of an agency's discretion to decide whether or not it is appropriate to carry out a formal risk assessment.

Because of these procedural divergences, the 'standard' approach to building the scientific basis for chemical regulation looks rather different in the United Kingdom and the United States. In the UK most of the scientific advice required by the regulatory agencies is drawn from a network of technical advisory committees. Major regulatory institutions, such as the Health and Safety Commission (HSC) or the Ministry of Agriculture, Fisheries and Food (MAFF), rely on standing committees for reviews of the scientific literature and evaluations of the risk presented by toxic chemicals. Two such bodies have acquired special prominence in recent regulatory controversies: HSC's Advisory Committee on Toxic Substances (ACTS) and MAFF's Advisory Committee on Pesticides (ACP). Established under authority of the Health and Safety at work *etc.* Act, the UK's worker safety statute, ACTS, is a 'tripartite' committee composed of representatives from industry, unions, and government. ACP's membership, by contrast, is limited to academic scientists. However, both ACTS and ACP delegate the responsibility for assessing specific chemicals to *ad hoc* working parties of scientific experts. This practice undercuts the potential for systematic differences in the assessment of toxic substances by the two advisory committees. In particular, reliance on technical working groups reduces the possibility of political influence on the evaluation of occupational health hazards by ACTS.

British advisory committees have no legal duty to conduct open meetings or to explain their scientific determinations to the public. In the absence of statutory requirements, committee practices vary widely with respect to publishing the reasons for particular policy recommendations. On the whole, the advisory system is weighted towards confidentiality, particularly when proprietary information is considered. In a large majority of proceedings involving toxic chemicals the public never has an opportunity to review or criticize the basis for committee recommendations. But in a growing number of cases, including both asbestos and lead, public concern has outweighed the government's traditional preference for confidentiality and has prompted the disclosure of advisory committee reports. The British rule-making process, however, provides no systematic opportunity for public discussion of such documents.

In reviewing toxic chemicals, British advisory committees are expected not only to evaluate the available scientific information, but also to recommend control measures to the relevant government department. Though conceived as purely advisory, these recommendations carry great weight with political decision-makers. Instances of government officials over-ruling their technical advisors are extremely rare. For all practical purposes, therefore, the expert committee is a powerful source of policy guidance as well as of scientific expertise in the British system of chemical regulation.

Chemical regulation in the United States begins with steps that are quite comparable to those followed in the United Kingdom. Agencies generally compile and review scientific information produced by research scientists in academia, in industry, or in independent public institutions such as the National Institutes of Health (NIH) and the National Institute for Occupational Safety and Health (NIOSH). Supervision by independent scientific panels is sometimes required by law, as in the pesticide regulation and air pollution control programmes of the Environmental Protection Agency (EPA).[23] However, even without such statutory requirements, US agencies are accustomed to seeking advice from a variety of standing and *ad hoc* advisory bodies. These are ordinarily composed of academic scientists, although pluralistic committees like ACTS are not unknown in the USA.

US committees, however, exercise less influence on risk assessment than their British counterparts. Agencies in the USA are usually well equipped with in-house expertise. As a result, their advisory committees are relegated to a clearly subordinate role, seldom extending beyond the review of studies and analytical documents prepared by or for the administrative agencies. The preparation of risk assessments, for example, is generally delegated to the agency's own staff scientists or to consultants selected by the agency. The development of actual policy recommendations, such as emission standards, residue limits, or threshold limit values (TLVs), falls outside the mandate of the advisory committees. Similarly, the agencies are in no way constrained to follow the recommendations of such private standard-setting organizations as the American Conference of Governmental Industrial Hygienists (ACGIH). These arrangements are consistent with the prevailing view that the scientific input to risk decisions (risk assessment) should be separated from the consideration of policy (risk management), and that enforceable regulatory policies should be made only by agencies that are fully accountable to the courts and the public.

Later stages in the American regulatory process dilute the influence of advisory committees still further. After being screened by the appropriate committee, the scientific case for action (or inaction) developed by the agencies is generally subjected to a public hearing, whose form varies according to the interests at stake. The results of such hearings must be incorporated into the final decision, and judicial review is available as a check to ensure that agencies do not wilfully disregard significant arguments raised by the affected parties. The ultimate controls on the agencies are legal and political rather than scientific. Administrators are answerable to Congress and the courts for acting in the public interest. Support from the expert community helps administrators make their case in these fora, but science seldom binds the agencies to particular courses of action, and administrators can deviate from the recommendations of their expert advisers if there are persuasive grounds for selecting a different regulatory option.

3. Risk Assessment in Operation

The regulation of both asbestos and lead provides interesting opportunities for examining the role of culture in risk assessment. The health effects of both substances are relatively well understood. Asbestos is a known human carcinogen and lead is a neurotoxin capable of causing brain damage and death. Given the large amount of information available about the two substances, one would expect a high level of expert agreement about the risks they present to the public. Cross-cultural differences in risk assessment are therefore particularly suggestive and offer a promising starting point for looking at the impact of external social factors on the scientific evaluation of risk.

4. Asbestos

For policy-makers seeking undisputed evidence of a hazard to health, asbestos comes close to being the proverbial 'smoking gun'. The substance has been commercially exploited on a large scale since the end of the nineteenth century, with demand increasing rapidly after the first world war. Because of its binding and fire-retardant properties, asbestos has been used in a wide variety of consumer products, including insulation materials, brake linings and other friction products, textiles, and cement. As its beneficial uses multiplied, asbestos revealed itself as a potent threat to human health. Some of its disastrous effects – irreversible lung disease and cancer – were recognized more than seventy years ago. By the 1970s, asbestos was identified not merely as a hazardous substance, but as one of the most serious public health problems of the decade.

In many respects asbestos presents a unique case for risk management policy. Not only are its effects on human health empirically demonstrable, but there is a substantial body of epidemiological evidence from which conclusions can be drawn about the degree of risk at different exposure levels (see also Chapter 15). The problem for regulators in both the UK and the USA has been to design reasonable public health measures based on this large mass of evidence. This exercise clearly includes both a scientific and a socio-economic component. Considerations such as the impact on employment and the cost and availability of substitute products have played an unavoidable role in both British and American strategies for regulating asbestos. Indeed, regulatory agencies in both countries are required to take economic factors into account in controlling toxic chemicals, and in the USA the apparently overzealous implementation of this requirement by the Office of Management and Budget has caused considerable political scandal.[24] But although social and economic considerations have influenced the pace and stringency of regulation, their impact on the assessment of health risks should presumably be slight, since the evidence of health damage is so compelling. This case study focuses mainly on the relatively apolitical phase of scientific assessment.

Positive results in well designed and well conducted epidemiological studies provide regulators with the strongest possible ammunition for controlling substances of proven benefit to society. In the case of asbestos, government agencies have had a wealth of such evidence at their disposal. Dozens of studies link the substance to asbestosis, lung cancer, and mesothelioma (a rare, asbestos-induced form of cancer) in exposed groups from countries in Europe, North America, and South Africa. This supply of information is 'universal' in the sense that the same studies are available for analysis by scientists and government officials in any country concerned about the health effects of asbestos. A comparison of asbestos risk assessments in the UK and the USA, however, shows that marked divergences can appear in the characterizations of risk that scientists in different countries construct from the same studies.

In the UK, the task of reviewing the health risks of asbestos and recommending control strategies was delegated in 1976 to a specially appointed Advisory Committee on Asbestos (ACA), composed of independent scientists as well as experts nominated by labour, industry, consumer groups, and local government.[25] This committee in turn established five working groups to provide detailed information on medical issues, environmental monitoring, legal and administrative controls, production, and substitutes. ACA also requested a Toxicity Review from an *ad hoc* two-member panel and commissioned additional papers from outside experts on the 'ill effects of asbestos on health' and on 'asbestos control limits'. The paper on ill effects proved especially influential in guiding government policy. When the expert reports had been completed the committee solicited further information from the public and held three days of open meetings at which twelve outside contributors presented additional comments. All these submissions were incorporated into the committee's overall assessment of asbestos risks and into its policy recommendations. ACA's final report was published in 1979 by the Health and Safety Executive (HSE), the principal implementing agency for worker protection policies in the UK.

In the USA, two agencies have been largely responsible for assessing the health risks of asbestos: EPA and NIOSH, a research agency located in the Department of Health and Human Services. One of NIOSH's most important functions is to prepare risk assessments and recommend exposure standards for toxic substances in the workplace. These 'criteria documents' play a major role in guiding OSHA's priority-setting and standard development. The 1972 and 1976 criteria documents for asbestos substantially influenced OSHA's policies on the substance. Consistent with its more general regulatory mandate, EPA is concerned with most non-occupational exposures to toxic substances and has taken the lead in assessing risks to the general public from asbestos in public buildings and the environment. The following paragraphs compare the analysis by ACA in the UK of the scientific data on asbestos with comparable analyses by NIOSH and EPA in the USA in the mid-1970s and early 1980s.

In reviewing the epidemiological data on asbestos, ACA came to the significant conclusion that different fibre types differ in their capacity to induce human disease. The committee based these findings on the toxicity review carried out by its two independent consultants, E.D. Acheson and M.D. Gardner. These experts read the epidemiological studies as supporting quite different hazard assessments for the three major fibre classes: crocidolite, amosite, and chrysotile.[26] The strongest indications of hazard were found for crocidolite and the weakest for chrysotile, although no fibre type was completely exonerated. The argument that disease incidence varies with fibre type seemed most convincing in the case of mesothelioma. Several of the studies reviewed by Acheson and Gardner appeared to confirm these patterns of variation unequivocally, and the findings seemed all the more conclusive because they recurred in four different exposure contexts (miners, process workers, neighbourhood and domestic exposure, and the geographical distribution of mesothelioma).

In the field of health hazard assessment, however, the evidence is seldom one hundred percent consistent, and British cancer experts had to deal with numerous conflicts in the data on asbestos. Their handling of one disputed study indicates how the interpretation of troublesome individual items was reconciled with the preferrred 'reading' of the larger body of data. A study by Enterline and Henderson found only one case of mesothelioma in a cohort of asbestos production and maintenance workers who had lived to retirement age.[25] Since this group was primarily exposed to chrysotile, the study appeared to support Acheson and Gardner's general thesis that chrysotile presents a relatively low risk of mesothelioma. The Enterline and Henderson data, however, were challenged by Borow, who claimed to have found many cases of mesothelioma in younger members of the same study population.[25] While not dismissing Borow's observations out of hand, Acheson and Gardner discounted them in various ways. They noted, for example, that Borow gave 'no data on the histories of exposure of the individual cases'.[25] The authors also advanced an alternative explanation of Borow's findings, which accorded reasonably well with their own theories about the linkage between mesothelioma and crocidolite. Specifically, they noted that cases of mesothelioma appeared in the study population only after the introduction of crocidolite into the factory under investigation, and even then only after the passage of the predictable latency period.[27] The factory opened in 1917, but crocidolite was not used until 1929, since when it has been in continuous use by men making pipes. Almost all the cases reported by Borow occurred since 1960 and in men. In the large workforce of women employed making textiles from chrysotile only one case of mesothelioma was found.

ACA was persuaded by Acheson and Gardner's analysis and endorsed the view that the risks of asbestos-related disease vary with the fibre type. For mesothelioma, in particular, ACA summarized the case as follows: 'evidence from miners; from process workers exposed to a single fibre type; from the distribution of neighbourhood and domestic cases; and from the geography of mesothelioma, when combined, presents a powerful case from four different sources that crocidolite has been more dangerous than chrysotile and anthophyllite. The position of amosite may be intermediate between crocidolite and chrysotile.[28] These conclusions, in turn, provided the basis for control measures directed at specific fibre types, as can be seen below.

In the USA, by contrast, NIOSH's 1976 assessment of asbestos did not identify fibre type as a factor of particular relevance to the characterization of risk. The agency's position was summed up as follows: 'Although some differences in both the fibrotic and the carcinogenic responses to asbestos fibres may depend on the type of fibre administered, all types have definitely shown both these kinds of action.[29]   In other words, NIOSH was more concerned about the fact that *any* fibre type can produce a pathological response than about risk differentials associated with different fibre types.   The agency's interpretation of the Enterline and Henderson study and of Borow's subsequent findings was consistent with this position.   Both studies were reported in NIOSH's brief summary of epidemiological evidence on individual types of fibres, but, unlike Acheson and Gardner, NIOSH scientists were not interested in rebutting Borow.   Instead, the NIOSH appraisal attributed the difference between the two studies to 'methodologic variations', specifically, to the fact that 'Enterline and Henderson had limited their investigation to men age 65 or over, while many of the mesothelioma cases reported by Borow *et al*. had died before that age'.[30]   This explanation preserved the credibility of the Borow study, which happened to agree with NIOSH's basic determination that chrysotile, along with crocidolite, is definitely implicated in the etiology of mesothelioma, and should be treated with equal caution.

The 1979 ACA report and the 1976 NIOSH report also differ in their approach to the issue of a threshold for asbestos.   Both reports acknowledge that there is no safe level of exposure to asbestos; asbestosis, respiratory cancer, and mesothelioma can all occur even at extremely low exposures.   But British and American experts arrived at this conclusion along somewhat different paths.   In its first criteria document for asbestos, NIOSH focused primarily on the risk of asbestosis and recommended a standard aimed at preventing this disease.   The hope was that such a standard would also provide adequate protection against asbestos-induced cancer.   By 1976, however, both NIOSH and OSHA were more concerned about the risk of cancer, and the evidence of risk at very low exposure levels supported the position taken by both agencies that it is practically impossible to establish safe thresholds of exposure to carcinogens.   Thus, NIOSH recommended in 1976 that the asbestos standard should be set at the lowest detectable level, an approach consistent with the agency's general principles for controlling carcinogens.[31]   In a roughly contemporaneous document on asbestos, OSHA stated the emerging federal position on the threshold issue more explicitly: 'While some level, below which exposure to a carcinogen does not cause cancer, may conceivably exist for any one individual, other individuals in the working population may have cancer induced by doses so low as to be effectively zero.   This is not to say that researchers will never find a threshold for a carcinogenic substance, but it does mean that the threshold concept for carcinogens is, at present, more a matter of responsible regulatory policy than a precise scientific determination'.[32]

ACA, on the other hand, seemed to base its acceptance of the no-threshold model for asbestos on empirical evidence about this particular substance rather than on a general policy concerning carcinogens.   The British committee attached great weight to the fact that existing epidemiological studies showed increasing risk with increasing exposure, and that there were well attested cases of cancer following limited contact with asbestos in homes and around mines, factories, or docks.   The analysis of data bearing on risks at low exposure levels suggests that British scientists were not prepared to accept the no-threshold hypothesis as a matter of principle (or policy).   For example, in evaluating the evidence relating to cancers of the gastrointestinal tract, the Acheson and Gardner study reported that 'there is insufficient data to determine whether or not a threshold exists'.[33] The statement implies a readiness on the authors' part to admit that there could be a threshold for some forms of asbestos-induced cancer, a position that NIOSH appeared less inclined to accept after 1976.

Scientific opinion in the USA and the UK also diverged in assessing the risks of asbestos exposure to the public at large. ACA took up the question in a limited form in its 1979 report.[34] The committee reviewed the studies pertinent to exposure assessment and identified some problems that might be encountered in estimating the risk to the general public. For example, ACA noted that discrepancies in measurement techniques make it difficult to compare data from environmental monitoring to data gathered from studies of industrial exposure. A more recent HSC report on asbestos by R. Doll and J. Peto again emphasizes the uncertainties involved in measuring asbestos exposure in buildings, although the authors do provide some risk estimates based on data from the Ontario Royal Commission on Matters of Health and Safety Arising from the Use of Asbestos.[35] American scientists, however, have accepted the idea that a conversion factor may be used in order to compare the data from environmental and industrial studies. For example, William Nicholson of Mount Sinai School of Medicine, a leading US centre for asbestos research, testified on the use of this technique at a Congressional hearing in March 1982.[36]

ACA made no attempt to develop a quantitative estimate of risk to exposed populations even in contaminated buildings, where the risk to the general public is believed to be greatest. The committee made a number of proposals for reducing public exposure to asbestos, but these were based on a general philosophy of keeping dust emissions under control.[37] Although the British government instituted monitoring programmes to measure asbestos fibres in the environment and in buildings, these actions have not led to the establishment of exposure standards or directly to asbestos stripping programmes in schools and other buildings. In the USA, on the other hand, EPA, under pressure to do something about asbestos in schools, ventured to prepare a precise quantitative estimate of the lifetime risk and the number of premature deaths for students and adult employees.[38] In so doing, the agency was clearly prepared to bridge some of the methodological uncertainties noted by ACA, suggesting a readiness to use and accept predictive models that was not matched in the UK at that time.

While it is difficult to be more than impressionistic on this point, the asbestos case suggests that there may be a predisposition in each country to rely on domestic rather than foreign studies for purposes of health risk assessment. In estimating the environmental risk from asbestos, for example, Nicholson presented data on dose-response relationships for cancer from several studies of American industrial workers.[39] Acheson and Gardner's report also addressed the question of dose-response relationships,[40] but, at least in the case of asbestosis, apparently preferred data from British studies at Rochdale to comparable work done in the United States and Canada under somewhat different exposure conditions. In the area of exposure assessment, both the Mount Sinai group and EPA have relied on air sampling in American homes and buildings, including schools with damaged asbestos surfaces.[41] Acheson and Gardner's account of exposure in the non-occupational environment mentioned studies by Nicholson and other foreign researchers, but attached more significance to a study done by Byrom and his colleagues on asbestos-contaminated buildings in the UK.[42]

One possible reason for cross-national differences like these is that the phenomena being studied in the area of health risk assessment - disease incidence, mortality rates, building contamination - are so dependent on local conditions that study results cannot easily be transferred across national boundaries. Another explanation is the so-called 'not invented here' syndrome,[43] which allegedly makes scientists more receptive to protocols and methodologies developed in their own country than to those from abroad. If this is a genuinely important factor in the evaluation of scientific data, then a national bias is bound to creep into health risk determinations until the methods of data collection and risk assessment are fully standardized across national lines. This point has obvious implications for the role of science in public decisions about risk.

Differences between the asbestos control policies adopted in the UK and the USA reflect, to some extent, the divergences identified above in the two countries' scientific assessments of risk. Following ACA's recommendation, for example, British policy has distinguished among different fibre types in prescribing exposure limits. Crocidolite is most stringently controlled. ACA recommended a statutory ban on the use or import of this form of asbestos and a control limit in the workplace of 0.2 fibres ml^{-1}, and for chrysotile a still higher control limit of 1 fibre ml^{-1}.[44] ACA had already recommended a ban on the use of crocidolite in manufacturing industrial products. Import of this fibre type ended in 1970 through voluntary action by the asbestos industry. Action on these proposals was delayed for a time while Britain participated in negotiations at the European Economic Community (EEC) on an asbestos directive that would apply to all member countries. But under public pressure the Health and Safety Commission (HSC) eventually decided to implement the ACA proposal, even before the EEC negotiations were completed. The asbestos control limits were tightened further in 1983, not in response to new evidence, but because industry agreed to accept stricter standards.[45] Although HSC has expressed concern about risks to the general public, there has been no attempt by the British central government to regulate asbestos levels in the environment, though there is a requirement to monitor after asbestos stripping were necessary.

These policy outcomes contrast markedly with regulatory approaches pursued in the USA. The begin with, US agencies have not attempted to link their asbestos control measures to specific fibre types. The pace of standard-setting, however, has been slow, even in the workplace, where asbestos has long been recognized as a top priority. The substance was included in OSHA's first promulgation of standards in May 1971, at which time the workplace exposure limit was set at 12 fibres ml^{-1}. In 1972, NIOSH recommended a 2 fibres ml^{-1} exposure limit, following the lead set by the British Occupational Hygiene Society in 1968 after its survey of asbestosis among textile workers. Pressured by the unions, OSHA first issued an emergency temporary standard (ETS) of 5 fibres ml^{-1} in late 1971. A permanent standard of 5 fibres ml^{-1} was promulgated in 1972 and was lowered to 2 ml^{-1} four years later.[46] The current US occupational standard for asbestos thus continues the 1972 NIOSH recommendation, and does not reflect any more recent scientific knowledge about the risks of cancer from low-dose exposures to the substance. Despite an interim setback in court,[47] OSHA was preparing to propose a new asbestos standard in late 1985 which would finally bring regulatory policy more in line with available knowledge.

In recent years, EPA has become a second major arena for regulatory controversies over asbestos. One reason is the mounting public concern about environmental exposure, especially in schools. Congress and a number of environmental groups have turned their attention to EPA because of the agency's authority under the Toxic Substances Control Act (TSCA) to regulate substances that cannot be adequately controlled by any other federal agency. In the past few years, EPA has not only carried out a quantitative risk assessment for asbestos in schools, but has worked on a variety of proposals to collect information about the uses of asbestos, to warn certain user groups, to inspect buildings and require removal of deteriorating asbestos, and to ban or phase out some uses of the substance. Although the agency has moved on many fronts, its initiatives, as in OSHA's case, have been slow to produce tangible results. Frustrated by the pace of policy-making and by interference from the Office of Management and Budget, EPA announced in early 1985 that it lacked authority to ban asbestos under TSCA, and moved instead to have OSHA and the Consumer Product Safety Commission (CPSC) take over further rule-making.[48] However, this decision was rescinded following a public outcry, and in early 1986 EPA proposed a ban on five major uses of asbestos and a phase out of all remaining uses over a ten-year period.[49]

5. Lead in Petrol

The assessment and regulation of lead in petrol offers another interesting case study of the impact of culture, particularly law and politics, on the interpretation of scientific evidence. For ease of comparison, consideration will be given only to the regulatory history of leaded fuels, even though lead has been controlled in other environments, such as the workplace and the paint and food processing industries, in both the United Kingdom and the United States. The comparison begins with the USA, since leaded fuels were regulated there some years before comparable action was undertaken in the UK.

Once lead enters the bloodstream, following its uptake from inhaled air and from food and water contamination, its effects are relatively easy to monitor, although numerous controversies about behavioural and other sub-clinical impacts still remain to be resolved. But the uncertainties are much greater when it comes to determining the precise amount of risk contributed by different sources of pollution. Prior to 1970, for example, there was general agreement that automobile emissions accounted for as much as 90% of airborne lead. Yet, from the standpoint of human health, there were many unanswered questions about the significance of airborne lead in relation to other sources such as diet. Accordingly, the major problem for regulators in both countries has been to decide how stringently to control different sources of exposure to lead. Given the ubiquity of lead and the multiple pathways of human exposure, controversies have inevitably arisen over whether lead in fuels is worth regulating, and if so, to what degree.

EPA's consideration of this policy question was moulded from the start by its organic statute, the Clean Air Act of 1970, which framed the agency's authority to regulate hazardous fuel additives in the following terms:

'The Administrator may, from time to time on the basis of information obtained under subsection (b) of this section or other information available to him, by regulation, control or prohibit the manufacture, introduction into commerce, offering for sale, or sale of any fuel or fuel additive for use in a motor vehicle or motor vehicle engine (A) if any emission products of such fuel or fuel additive will endanger the public health or welfare....'[50]

EPA took the position that it was not necessary to show that lead from fuels alone would harm public health in order to meet the 'will endanger' standard of the law. Rather, EPA argued that it was enough to show that such pollution could 'present a significant risk of harm to the health of urban populations, particularly to the health of city children'.[51] The petrochemical industry challenged this interpretation as a misreading of the risk standard actually intended by Congress. In industry's view, the 'will endanger' formulation of the Clean Air Act required 'a high quantum of factual proof, proof of actual harm rather than of a significant risk of harm'.[52] But in the end it was the EPA administrator's reading of the law that prevailed, since the US Court of Appeals for the District of Columbia Circuit agreed with the agency that 'endanger means something less than actual harm'.[53]

EPA's compilation and analysis of scientific evidence were consistent with its legal position. The agency used data from animal, clinical, and epidemiological studies to establish the existence of a risk to health from exposure to lead, and to identify automobile emissions as a substantial causative factor. EPA argued, in particular, that blood lead levels of 40 μg 100 g^{-1} should be taken as dangerous even though clinical evidence of harm to health was found only at much higher concentrations. EPA also concluded that airborne lead presents a direct threat to public health through inhalation, and that urban children are subject to a further risk through the deposit of airborne lead and its mixing with dust and dirt. The last point was supported by a chain of inferences showing, first, that dust in urban areas, especially near highways, is unusually contaminated with lead, and, second, that a high percentage of children are prone to pica, or dust eating, which renders them particularly susceptible to harm from such surface deposits of lead (see also Chapter 15).

Almost every point in the agency's reasoning came under attack from industry on the general ground that there was no *evidence* for proceeding from one step to the next. Industry claimed for instance, that EPA had not shown any adverse health effects at the 40 μg 100 g^{-1} blood lead level, that EPA's estimates of direct absorption of lead from the ambient air were flawed, and that there was no evidence that children with pica in fact swallow dust containing excessive concentrations of lead. The reviewing court, however, interpreted the agency's burden of proof in less rigid terms, requiring from EPA only enough evidence to provide a reasonable basis for each of its major conclusions. Applying this standard of proof, the court was satisfied that the agency had properly met its combined burden of fact-finding and policy-making in deciding to phase out the use of lead in petrol.

In the UK, by contrast, a permissible limit of 0.4 g 1^{-1} was still in effect for lead in petrol in early 1981.[54] At that time, the Department of Health and Social Security (DHSS) commissioned a new study of lead-related risks to health from a working party of scientists chaired by Patrick Lawther.[55] The Lawther Report found causes for concern and recommended additional reductions in public exposure to lead, but it stopped short of calling for the complete elimination of lead from petrol. On the basis of this report, the government announced in mid-1981 a plan to reduce the allowable lead level in fuel to 0.15 g 1^{-1}. This decision provoked a public campaign for non-leaded petrol by a newly organized activist group called CLEAR (Campaign for Lead-free Air).

Unlike most small, single-issue campaign groups, CLEAR possessed two advantages that greatly strengthened its position. The first was the political sophistication of its organizer, Des Wilson, who brought to the task several years of experience forming and directing campaigns on housing and environmental issues. The second advantage CLEAR enjoyed was sufficient financial support from private sources to wage a more polished and visible campaign than would normally have been possible for a group of this size. Wilson recognized that the principal challenge was to present his organization's risk assessment to the public so convincingly as to undercut the scientific judgement of the government's own expert advisers, the Lawther Committee. CLEAR decided that it could attain this objective only through a campaign to educate the public about the risks of exposure to lead. Such a direct appeal to the public was seen as the most effective means of building pressure for speedier governmental action on lead.

While the organizers of CLEAR intended to wage a political campaign, they admitted from the start that the campaign would eventually have to achieve enough scientific credibility to change the minds of powerful actors within the political establishment. The effort was especially daunting in view of the fact that the government had just rejected the possibility of banning lead in petrol on the basis of the Lawther Committee's report. The campaign also had to defuse opposition from the gasoline industry, which characterized CLEAR's positions as emotional and insisted that there was no evidence linking low-level lead exposure to health. To build an effective rebuttal, CLEAR needed not only strong scientific support for the existence of a risk, but the means to bring this information dramatically before governmental decision-makers and the public.

By good fortune, international science helped CLEAR's campaign strategy in two ways. First, influential studies published in Italy and the United States suggested that eliminating lead in petrol did indeed cause a drop in blood lead levels. Establishing this linkage, however, was not enough to overcome the reluctance of some experts to acknowledge risk without definite evidence of harm to human health. To convince the sceptics, CLEAR needed to show, if possible, that lead found in the ambient air can cause measurable adverse effects on health. Here, the campaign received vital support from epidemiological studies in the USA suggesting that children with relatively high levels of lead in the blood display a lower IQ and other adverse behavioural symptoms. To publicize these studies, which established the desired causal connection between lead and health, CLEAR organized an international scientific symposium on lead and health.

In advertising the risks of lead to the public, CLEAR's strategy was to borrow the forms and procedures of official scientific investigations while pursuing an anti-establishment policy. The model chosen by the campaign was the public inquiry, the UK's equivalent of the American administrative hearing. As in official inquiries, contributions were invited from experts on the issue, in this case the scientists who had produced major studies on the effects of low-level exposure to lead. The campaign also followed official practice in selecting as chair for the symposium an individual whose credentials were beyond reproach: Michael Rutter, a leading British child psychiatrist, a member of the Lawther Committee, and a scientist enjoying high credibility within government. At the end of the three-day symposium, Rutter summarized the proceedings and appended his own conclusion: 'The evidence suggests that the removal of lead (from fuel) would have a quite substantial effect of reducing lead pollution, and the costs are quite modest by any reasonable standard'.[56] Since Rutter had previously joined in the Lawther Committee's recommendation against a ban, this statement marked an important change of position, and dealt a serious blow to the earlier report. With Rutter's support assured, CLEAR achieved the scientific credibility which was one of its major goals.

CLEAR's public presentation of the scientific data on lead and health was not the only factor that ultimately prompted a change of policy on leaded petrol. It was extremely significant for example, that the Royal Commission on Environmental Pollution (9th Report) decided after its own review of the lead data that a ban on leaded fuels was advisable, since it was prudent from a public health standpoint to reduce the dispersal of lead from all possible sources. Nevertheless, CLEAR's scientific campaign must be credited for part of the momentum that made the government announce in April 1983 that it would press for a Europe-wide ban on lead in petrol. Des Wilson persuasively describes his organization's role in changing the public's – and indirectly the government's – perception of the risks of lead:

'The other side, particularly the industries, have their own version of events. Above all, they argued and would still argue that we exaggerated the health case and deliberately frightened people for no reason. It is, of course, true that we aroused concern all over the country about the health risk. But that was our task. It was precisely because the authorities refused to acknowledge it that we had to do this. I do not believe that people should be denied the knowledge of research and studies that indicate health problems. The people of Britain are adult and mature enough to consider the facts objectively and reach their own conclusions. Their instinctive human response was that they did not care whether the evidence was conclusive or not, for they could see that there were sufficient organizations and people of substance who were prepared to state that it was a risk, and they wanted action on the basis of risk. Not for them conclusive proof, achievable only by what would in effect have been experiment with their children'.[57]

Wilson's suggestion that British officialdom would have continued to wait for conclusive evidence of a risk to health unless CLEAR had politicized the issue is partly supported by events following the international symposium. Shortly before the government announced its policy change, the Department of Environment (DOE) released a study on the effects of low-level lead exposure on children. Although the study found a correlation between low IQs and high blood-lead levels, it showed that other variables, such as social class, would adequately account for the findings.[58] DOE, in other words, seemed still to be looking for evidence that would establish the risk of lead beyond a reasonable doubt. It is questionable whether this standard of proof could have been satisfied within the foreseeable future. Significantly, the Royal Commission was reluctant to draw any firm conclusions about the effects of low concentrations of lead on children's behaviour, and a more recent report by the Medical Research Council found that an elevation in the body lead burden of some British children had no measurable effects on IQ.[59]

6. Science and Risk Assessment

The two case studies presented here illustrate basic differences in the assessment of toxic hazards by experts in the UK and the USA. In both, the tendency in the UK was to read the evidence of risk to human health more narrowly than in the USA.

   Thus, in the case of asbestos, British experts associated different degrees of risk with different fibre types, whereas their counterparts in NIOSH and OSHA viewed all forms of the substance as equally hazardous. Faced with equivocal information about asbestos in the general environment, EPA concluded that the risk was sufficient to justify a programme of asbestos removal. British authorities, however, withheld action on asbestos in public buildings pending further scientific investigation of the environmental risk.

   For lead, EPA constructed a complex chain of inferences to argue that the substance should be banned from petrol. The UK's Lawther Committee, by contrast, found the evidence insufficiently persuasive to justify an immediate ban. Official policy in the UK ultimately followed the American approach, but only in response to political pressure and to more evidence suggesting a demonstrably adverse effect on the health of exposed children.

   The absestos case exemplifies a methodological difference that has been noted in other comparative studies of chemical risk assessment in the UK and the USA.[60] American agencies seem prepared to use mathematical models to estimate risk in situations where British experts either view the risk as insignificant or the techniques of extrapolation as unproven. From the standpoint of the British regulatory process, the discussion of quantitative risk assessment remains largely academic, mainly limited to the deliberations of expert groups such as the Royal Society. In the USA, however, the government's internal and external scientific advisers are centrally involved in developing and validating methods of estimating risk from limited and indirect evidence. Indeed, the use of these predictive techniques has become at least as important a source of controversy in American regulatory proceedings as the initial qualitative interpretation of hazard information.[61]

   The asbestos case study most clearly illustrates the way interpretations of individual pieces of evidence can be influenced by culturally conditioned philosophies of assessment. The available studies do not indisputably prove whether the health effects of exposure to asbestos vary in relation to fibre type. No matter how one reads the evidence, there are some studies, such as the Enterline and Henderson data, that either remain ambiguous or point in the wrong direction. Both British and American experts had to find *ad hoc* explanations to reconcile such 'truant' data with their preferred scientific theories. The British approach, however, led to a more narrowly formulated hypothesis about asbestos-induced disease and to a lower projection of the total risk than in the USA. This result seems consistent with a greater inclination in the USA to act on the basis of probable rather than provable harm.

It would be foolish to speculate on the basis of two brief case studies whether such differences in the assessment of toxic substances reflect genuine attitudinal differences between British and American scientific experts. The most one can say is that certain judgements commonly regarded as 'scientific' – the weighting of competing studies, the validation of predictive models – do indeed vary from one country to another. To the extent that such cross-national differences exist, they can be bolstered, and even magnified, by the 'not invented here' syndrome, which allows scientists to be more sceptical of research in other countries that does not fit in with their own methodological presuppositions.

Perhaps the clearest lesson to be drawn from the two case studies is that the assessment of toxic hazards is subject to diferent kinds and degrees of external (*i.e.* non-scientific) influences in the UK and the USA. All decisions about risk acceptability involve a fundamental value judgement as to how risk-averse society should be. Though this judgement is usually viewed as a part of risk management, it also enters into the interpretation of scientific data. When an expert rejects a methodologically imperfect or irrelevant study or concludes that more evidence is needed to complete an assessment, he is implicitly deciding that the level of risk is not high enough to overcome his professional scepticism about the quality or amount of the data. In the British regulatory process, scientific experts generally have the last word in making such decisions, even though they contain a value component. The American regulatory system, however, often specifies by law or administrative regulation how risk-averse the decision-maker should be. Such prescriptions can profoundly affect the interpretation of risk data, as seen in EPA's assessment of lead in petrol. Here, the statutory phrase 'will endanger' was used by the agency and the courts to justify a highly precautionary risk assessment. In the UK, there was no comparable law dictating whether the authorities should act even in the absence of actual harm. As a result, the Lawther Committee's conclusion that there was no need for an immediate ban did not have to be modified to fit a broader societal concept of acceptable risk.

In the USA, then, the existence of precautionary legislation and judicial decisions mitigates the danger that experts will assess risks according to standards not sanctioned by society as a whole. Administrative regulations designed to bridge gaps in the scientific data, such as the 'cancer policies' used by US agencies, serve a similar purpose, since such rules also permit regulators to act without evidence of actual harm. The potential divergence between lay and expert assessments of risk, and the acceptability of risk, is greater in the UK, where scientists are not subject to as many externally imposed standards of risk-aversiveness. Ordinarily, such disagreements do not arise, perhaps because public trust in expertise remains higher than in the USA. As the lead case illustrates, however, the gap between the scientist's and the public's perceptions of risk may occasionally grow large enough to assume political significance. In such situations, emotional, media-oriented debates over scientific issues may be difficult to avoid, even though experts are likely to find this approach to assessing risks and risk acceptability extremely unpalatable.

7. Conclusion

Cross-cultural comparisons of chemical regulation illuminate the extent to which risk assessment and acceptability are influenced by wider social and political considerations. Uncertainties in the information base require experts to exercise judgement that is not merely scientific, but is also based on a more subjective appraisal of the acceptability of risk. The exercise of such subjective judgement can be influenced by a variety of factors, including underlying cultural views about pollution and the environment, and, more immediately, by the benefits of the substance being evaluated, and by legal requirements that specify how much evidence is needed in order to establish the existence of risk. Of course, such legal prescriptions themselves reflect a broad public consensus about what risks are worth regulating and how stringently they should be regulated.

Differences in the assessment of risks from asbestos and lead in the United Kingdom and the United States underscore the futility of seeking exclusively scientific solutions to risk controversies. Certainly, scientific judgement has to play an important part in risk determinations, and continuing international dialogue is needed in order to ensure that the data and methodologies used in risk assessment enjoy broad support from the scientific community. At the end of the day, however, each national regulatory system must decide on its own terms how to deal with the uncertainties that remain after all the available evidence has been put together. This largely political decision will be made differently in different countries, consistent not only with their legal and bureaucratic traditions, but with prevailing cultural attitudes towards risk.

8. References

1. 'Risk: Man-Made Hazards to Man', ed. M. G. Cooper, Clarendon Press, Oxford, 1985.

2. M. Douglas, 'Purity and Danger', Routledge and Kegan Paul, London, 1966.

3. 'Controversy', 2nd Edn., ed. D. Nelkin, Sage, Beverly Hills, CA, 1984.

4. B. Gillespie, D. Eva, and R. Johnston, 'Carcinogenic Risk Assessment in the United States and Great Britain: The Case of Aldrin/Dieldrin', *Social Stud. Sci.*, 1979, **9**, 265.

5.　M. Douglas, 'Environments at Risk', in 'Risk and Chance', ed. J. Dowrie and P. Lefrere, Open University Press, Milton Keynes, 1980.

6.　M. Douglas and A. Wildavsky, 'Risk and Culture', University of California Press, Berkeley, 1982.

7.　P. Slovic, B. Fischhoff, and S. Lichtenstein, 'Perceived Risk', in 'Societal Risk Assessment: How Safe is Safe Enough?', ed. R. S. Schwing and W. A. Albaro, Plenum Press, New York, 1980.

8.　C. H. Green and R. A. Brown, 'Counting Lives', *J. Occupational Accidents*, 1978, **2**, 55; 'It all Depends What You Mean by "Acceptable" and "Risk"', School of Architecture Research Unit, Dundee, 1981.

9.　'Science in Context', ed. B. Barnes and D. Edge, Open University Press, Milton Keynes, 1982, pp.187-231.

10.　R. Brickman, S. Jasanoff, and T. Ilgen, 'Controlling Chemicals: The Politics of Regulation in Europe and the U.S.', Cornell University Press, Ithaca, NY, 1985.

11.　A. Weinberg, 'Science and Trans-Science', *Minerva*, 1972, **10**, 209.

12.　A. Weinberg, 'Science and Its Limits: The Regulator's Dilemma', *Issues in Science and Technology*, Fall, 1985, pp.59-72.

13.　T. McGarity, 'Substantive and Procedural Discretion in Administrative Resolution of Science Policy Questions: Regulating Carcinogens in EPA and OSHA', *The Georgetown Law Journal*, 1979, **67**, 729.

14.　S. Jasanoff, 'Risk Management and Political Culture', Russell Sage Foundation, New York, 1986 (in the press).

15.　National Academy of Sciences, 'Risk Assessment in the Federal Government: Managing the Process', National Academy Press, Washington, DC, 1983.

16.　Ref. 15, p.3.

17.　Royal Society, 'Risk Assessment', Royal Society, London, 1983.

18.　Ref. 17, p.23.

19.　Ref. 17, pp.72, 90-91.

20.　R. Brickman, S. Jasanoff and T. Ilgen, Ref. 10, pp.28-53.

21.　*Bushell v. Secretary of State for the Environment*, 3 W.L.R. 22 (1980).

22.　*Industrial Union Department, AFL-CIO v. American Petroleum Institute*, 448 U.S. 607 (1980).

23. N. Ashford, 'Advisory Committees in OSHA and EPA: Their Use in Regulatory Decision-making', *Sci. Technol. Human Values*, 1984, 9, 72.

24. 'Proposed Asbestos Ban: OMB Accused of Illegal Interference', *Chemical and Engineering News*, October 14, 1985, pp.6-7.

25. Health and Safety Commission, 'Asbestos', Final Report of the Advisory Committee, HMSO, London, 1979.

26. Ref. 25, Vol. 2, pp.47-49.

27. Ref. 25, p.29.

28. Ref. 25, Vol. 1, p.53.

29. NIOSH, 'Revised Recommended Asbestos Standard', Washington, DC, December 1976, pp.41-42.

30. Ref. 29, p.35.

31. Ref. 29, p.93.

32. OSHA, 'Occupational Exposure to Asbestos', *Federal Register*, 1975, **40**, October 9, p.47656.

33. Ref. 25, Vol. 2, p.41.

34. Ref. 25, Vol. 1, pp.35-45.

35. R. Doll and J. Peto, 'Asbestos: Effects on Health of Exposure to Asbestos', HMSO, London, 1985.

36. US House of Representatives, Committee on Energy and Commerce, Sub-committee on Commerce, Transportation, and Tourism, 'EPA's Failure to Regulate Asbestos Exposure', 97th Congress, 2nd Session, pp. 64-65.

37. Ref. 25, Vol. 1, pp.89-92.

38. Bureau of National Affairs, *Chemical Regulation Reporter*, Current Report, April 19, 1985, p.69.

39. Ref. 36, pp.57-60.

40. Ref. 25, Vol. 2, pp.34-41.

41. Ref. 36, pp.51-52.

42. Ref. 25, Vol. 2, p.42.

43. F. McCrea and G. Markle, 'The Estrogen Replacement Controversy in the USA and UK: Different Answers to the Same Question?', *Social Stud. Sci.*, 1984, **14**, 15.

44. Ref. 25, Vol. 1, pp.74-78.

45. J. Locke, 'Fixing Exposure Limits for Toxic Chemicals in the UK', unpublished paper presented at WHO Symposium on Risk Assessment and Risk Management, 1985, pp.12-17.

46. OSHA, Occupational Exposure to Asbestos, *Federal Register*, 1975, **40**, October 9, p.47653.

47. *Asbestos Information Association v. OSHA*, 727 F.2d 1137 (5th Cir. 1983).

48. Bureau of National Affairs, *Environmental Reporter*, Current Developments, February 1, 1985, p.1315. For an announcement of EPA's reversal of the referral decision, see *ibid.*, March 15, 1985, p.1443.

49. 'EPA Proposed Ban on Asbestos Use', *Chemical and Engineering News*, January 27, 1986, p.6.

50. 42. U.S.C. 1857f-6c(c)(1)(A).

51. *Ethyl Corp. v. EPA*, 541, F.2d 1 (D.C. Cir. 1976), p.12.

52. Ref. 51.

53. Ref. 51, p.13.

54. D. Wilson, 'Pressure: The A to Z of Campaigning in Britain', Heinemann, London, 1984, pp.156-80.

55. 'Lead and Health', Report of a DHSS Working Party on Lead in the Environment, HMSO, London, 1980.

56. Ref. 54, p.170.

57. Ref. 54, p.180.

58. Ref. 54, p.176.

59. Medical Research Council, 'The Neurophysiological Effects of Lead in Children: A Review of Recent Research: 1979-1983', MRC, London, 1984.

60. S. Jasanoff, Ref. 14.

61. *Gulf South Insulation v. Consumer Product Safety Commission*, 701 F.2d 1137 (5th Cir. 1983).

Section 6: Legislation on Chemicals

21

European Community Regulation of 'New' and 'Old' Chemicals

By C. Whitehead

COMMISSION OF THE EUROPEAN COMMUNITY, RUE DE LA LOI 2000, B-1049 BRUSSELS, BELGIUM

1. Introduction

European Community environmental policy started in 1973 with the adoption of the First 5 year European Community Environmental Action.[1] A main focus of that Programme was on repairing the damage to the environment caused by what may be termed 'old' chemicals – that is, certain industrial chemical products developed during particularly the 1950s and 1960s. Insofar as European Community environmental laws deal with the control of environmental pollution, they may be viewed as part of the policy for the control of risks from existing chemicals.

The Second and Third Environmental Action Programmes[2,3] added a further emphasis to EC chemicals control policies – the principle of 'prevention rather than cure'. A number of EC environmental directives adopted since 1977 are based on this principle, including the directives on the notification of new chemical substances, the prevention of major industrial accidents and environmental impact assessment. This anticipatory, preventive approach was laid out in the European Commission's recent Communication to the Council on New Directions in Environment Policy:

'Prevention should remain the key objective of environment policy. To this end, environment requirements should be integrated into legislation and decision-making'.[4]

This second generation of EC environment directives takes a broader approach to chemicals regulation by looking at the life cycle of potentially hazardous chemicals from research to industrial production, to use, to ultimate disposal, by tailoring the law to identify potential hazards throughout this life cycle, and by defining obligations for the producer, national governments, and European Commission to identify potential hazards and take certain measures to monitor and control such hazards. Thus, regulations on old as well as new chemicals are being gradually integrated into a single system of environmental protection.

This chapter reviews the main elements of the European Community's policies to control the risks to health and the environment from hazardous chemicals. It gives particular attention to the directives on the notification of new chemicals, the labelling of new and old chemicals, the prevention and control of industrial accidents, restrictions on marketing and use, and future initiatives on EC chemicals control policies in the 1980s.

2. The European Community's Approach to Chemicals Control

In signing the Treaty of Rome and subsequent treaties of accession, the 12 member states of the European Community created a common government and pledged to work through it to improve the quality of life of all their citizens.[5] As this commitment is understood today, 'improving the quality of life' means not only protecting the physical environment against degradation that would harm human health, but also ensuring the rational management of the physical resources of Europe, and, as the European Council acknowledged at its meeting of 29-30th March 1985, promoting economic growth by making environmental policy an essential component of the economic, industrial, agricultural, and social policies of the European Community.[6]

The more than 100 EC environmental laws adopted in the past 12 years are based on the realization that many environmental protection issues cannot be solved by national legislation alone. This is particularly true for chemicals control policies. The burgeoning international movement of industrial investments, chemical products, and wastes means that the world is facing a new, more complicated generation of potential threats to the environment that must be solved through international action if they are to be solved at all.

There are two kinds of international problem from hazardous chemicals. First, local problems from the use of hazardous chemicals in industrial plants or waste disposal facilities which, because of international trade, may arise in many different countries. Secondly, the transnational or even global impacts of some hazardous chemicals, such as the acidification of the soil and plant life from power plants and automobiles, changes in global climate, or the rapidly increasing rate of extinction of plant and animal life caused by large-scale industrialized methods of agriculture or changing land-use patterns.

Thus, by acting in concert through the European Community, the member states can adopt and implement policies that, if taken by a single national government alone, would be ineffective, handicap domestic industries *vis-à-vis* foreign competitors, or create trade barriers. Furthermore, EC action can strengthen and give added momentum to national actions by fostering co-ordination of research, exchange of information, and reducing costs. In many cases national initiatives have been the source of EC legislation, thereby extending progress in national environmental protection policy to all the citizens of the European Community.

3. The Regulation of 'New' Chemicals

To control the health and environmental risks from new chemicals the EC has adopted a preventative approach, requiring manufacturers to carry out a defined 'Base set' of tests and a risk assessment of these chemicals before marketing. New chemicals are regulated by the Council Directive amending, for the sixth time, Directive 67/548/EEC on the approximation of the laws, regulations, and administrative provisions relating to the classification, packaging, and labelling of dangerous substances (79/831/EEC), which is known as the '6th Amendment'.[7]

The 6th Amendment was developed through more than three years of close consultation with all interested parties within the EC.

Other major industrialized countries were informed of the progress
of the Directive through the meetings of the Organization for
Economic Co-operation and Development (OECD) Environment Committee
and the work of the OECD Chemicals Programme. Thus, the Community
endeavoured to achieve international agreement on the elements of
a system of preventive control of new chemicals not only
internally for nine member states but also externally for the 24
industrialized member countries of the OECD.

The 6th Amendment and, indeed, most EC environmental
legislation has two fundamental goals:

To create harmonized conditions of marketing, thereby ensuring
the free movement of products in trade within the EC.

To protect public health and the environment against the
potential hazards of chemicals.

It establishes a single testing and notification system for
new chemical substances marketed in the EC. It creates a common
information base about the potential hazards and risks of new
chemicals which begins with the pre-market 'notification' - a
single doorway through which domestic manufacturers or importers
can bring new chemicals to the markets of all the member states.
By filing a proper notification for a new chemical with one member
state the manufacturer or importer receives the right to market
the chemical throughout the entire EC. Thus, the notification is
a legal innovation that is considerably less burdensome to
industry than the national permit procedures that apply to
pharmaceuticals or pesticides.

3.1. Information and Testing Requirements. In brief, the Directive
requires that a new chemical substance - that is, one placed on
the EC market (see also Section 3.2.) for the first time after
18th September 1981 - must be notified to a national competent
authority at least 45 days in advance (Articles 5,6). The *ca*.
90,000 existing chemicals which were already on the market between
1st January 1971 and 18th September 1981 are not subject to
notification, and will be listed in the European Inventory of
Existing Commercial Chemical Substances (EINICS), which will be
published in English later in 1986.

Substances already subject to Community regulations, such as
pharmaceuticals, narcotics, food and feeding-stuffs, and
radioactive substances, are excluded from the scope of the
Directive (Article 1.2). Pesticides and fertilizers are exempt
from notification insofar as they are subject to equally strict
national or community requirements, or pocedures that are not yet
harmonized (Article 1.4). Low-volume (under 1 tonne year^{-1}) and
research chemicals and polymers containing less than 2% of a new
monomer are regarded as having been notified (Article 8.1). A
limited 'announcement' must still be submitted to the national
competent authority in each country in which it is placed on the
market, but the announcements need not be transmitted to the
European Commission or the other member states.

Each manufacturer or importer of a new chemical substance must
notify it. Thus an extensive information system of producers and
users is created that can be used to monitor hazards, exposures,
and implementation of the Directive.

The notification must contain four items of information: (see also Chapters 11, 12, and 22).

A technical dossier (the 'Base set') supplying the information necessary for evaluating foreseeable risks, whether immediate or delayed, that the substance may entail for people or the environment (Annex VII).

A declaration concerning the unfavourable effects of the substance in terms of the various uses envisaged (*i.e.* the notifier's risk assessment of the chemical).

The proposed classification and labelling.

Proposals for recommended precautions for safe use (Article 6.1).

The information in the technical dossier - or 'Base set' - is a screen for health and environmental hazards. Annex VII of the Directive contains a detailed description of what is required, including physico-chemical data, toxicological and ecotoxicological tests, production quantities, uses, and safety considerations. These tests are predictive in nature; they were chosen to provide indications of potential adverse effects in biological systems at a reasonable cost to the manufacturer.[8] For example, information about physico-chemical properties is relevant not only to the identification of the substance, but also to the evaluation of the potential exposure of people and the environment. This information then allows the prediction of the distribution of a substance in the different sectors of the environment through accumulation, degradation, and other processes.

The Directive takes the principle of a common set of test data several steps further by requiring more thorough toxicological and ecological testing when marketing levels reach 100 tonnes year^{-1} or 500 tonnes total, and again when marketing levels reach 1000 tonnes year^{-1} or 5000 tonnes total (Annex VIII). The totals refer to the total quantities on the EC market, and not to the total quantity marketed by a single manufacturer or importer. The Annex VIII tests are designed to provide concrete information about actual as well as potential health and environmental hazards. Thus, they are proportionately more time-consuming and costly to perform than those in the 'Base set'. These testing programmes will be tailored to each individual chemical by the competent authority in consultation with the notifier and the other member states.

Follow-up information (Article 6.4) may be required from the notifiers before these production levels are reached, for example, in response to:

A new knowledge of the effects of the substance

New uses

Changes in the properties of the chemical caused by a modification of the substance.

Annual marketing levels of 10 tonnes, or a total of 50 tonnes (Annex VIII) (see also Chapter 11).

The test methods (Annex V)[9] were promulgated in 1984 and follow closely the test methods adopted by the OECD Council in 1981, so that EC producers will also be able to use most or all of their test data generated for the 6th Amendment to meet the data requirements of other OECD countries (see also Chapter 22).

In July 1985, the Commission proposed a Council Directive on the standards and enforcement of good laboratory practice for tests on chemical substances.[10] It will implement the OECD Council actions of 1981 and 1983 on Good Laboratory Practice. The draft Directive simply requires the member states to 'take all measures necessary to ensure that laboratories apply the principles of good laboratory practice (GLP) specified in Annex 2 to the Decision of 12 May 1981' of the Council of the OECD on the mutual acceptance of data for chemical products (Article 1). Laboratories must certify that the tests have been carried out in compliance with GLP and the member states must carry out on-the-spot inspections and verifications of studies (Article 2). The member states must inform the European Commission of the procedures they have taken to verify compliance with the Directive, and notify the names of the responsible authorities (Article 3).

Lest the data requirements be too rigid, the Directive provides for flexibility where circumstances warrant it. For example, information may be omitted from the 'Base set' if it is not technically possible to provide it or if it does not appear necessary (Annex VII) (see also Chapter 11 for full details).

3.2. Implementation. All member states (except Spain and Portugal) have adopted legislation or regulations intended to implement the 6th Amendment. Broadly speaking, the system is clearly one of the success stories of European environmental policy, and a notable example of successful international law and implementation. Nevertheless, one of the Commission's highest priorities at the moment is to ensure complete substantive as well as formal compliance with the Directive by all the member states. Only in this way can it fulfil its function as a unified European system of regulation. The Commission has initiated infringement procedures against a number of member states for failure to adopt national provisions that conform in detail to the Directive.

The Commission and the national competent authorities meet four times a year in Brussels under the chairmanship of the Commission to discuss and facilitate the substantive implementation of the Directive. In the early months of implementation during 1981-82, the group took a number of decisions relating to the treatment and transmission of the information in the notification. First, the group agreed on a standard form for the summary of the notification dossier to be circulated by the Commission to the member states. After discussion in the group, in 1984, the Commision adopted a Decision on the publication of the list of notified new chemical substances.[11] The list is to be published in the Official Journal annually, beginning in the year of publication of EINECS, the inventory of existing substances, probably 1986.

At the same time, the group was faced with several legal questions regarding implementation. For instance, in consultation with the Commission, a decision could be taken on the definition of the term 'placing on the market' and on the scope of the

exemptions from notification defined in Article 8.1. Still under
discussion is the question of which categories of substances, such
as those used in pesticides *etc.*, would have to be notified
because EC and national regulation on these substances is not as
strict as the 6th Amendment.

Discussion with European experts on the testing methods for
Annex VIII is already well underway. The EC is again working
closely with the OECD Chemicals Programme on this topic.

There will be repeated notifications of new chemicals because
each new chemical must be notified each time a different producer
or importer places it on the market. To alleviate the notifier's
burden and avoid duplication of testing and administration the
Directive permits the competent authority to agree that the
notifier refer to the results of the studies carried out by one or
more previous notifiers, provided the latter have given their
agreement in writing (Article 6.2). This applies only to the
tests in the technical dossier, not the other three categories of
information, and is intended to ensure that all notifiers of the
same chemical are treated equally. The Commission and the
competent authorities are discussing possible means of simplifying
the procedure for repeat notifications.

3.3. Labelling. (See also Chapter 23). A major achievement was the
adoption on 29th July 1983 of the Labelling Guide and the criteria
for irritation and corrosion, which completes Annex VI of the
Directive.[12] Before that time, the Directive contained criteria
only for the labelling of very toxic, toxic, and harmful
substances. The Labelling Guide came into effect on 1st January
1986. By this date, all existing dangerous substances which had
not yet been given an EC label and listed on Annex 1 of the
Directive were to be provisionally labelled by industry, on the
basis of available knowledge. Annex 1 currently contains about
1000 substances for which a uniform EC label has been adopted.
Naturally, there is great interest in extending the EC label to
other important dangerous substances. To this end, the
Commission has set the following three priorities for action:

Substances for which the provisional labelling is different
between firms.

Substances which have carcinogenic, mutagenic, or teratogenic
properties.

All notified new dangerous substances.

The member states have used their right to propose specific
dangerous substances for addition to Annex 1. Currently 520
pesticides, 52 solvents, 35 possible carcinogens, and a number of
other substances have been proposed. A committee of special
experts has been established to recommend labelling for
carcinogenic, mutagenic, or teratogenic substances on the basis of
the criteria in the labelling guide.

4. The Regulation of 'Old' Chemicals

'Old' or 'existing' chemicals have long been the subject of
environmental regulation for the protection of specific
environmental media - air, water, soil, or for the control of
industrial emissions and operations. In addition, beginning in

1980, there has been considerable international discussion about the development and interpretation of data related to existing hazardous substances and rational procedures by means of which governments could generate lists of 'priority chemicals' for testing and regulation.[19]

4.1. The Regulation of Chemical Products.

In addition to specific sectors of chemical products such as pharmaceuticals, agricultural chemicals, or cosmetics, or restrictions on the use of specific substances in specific products, the European Community may regulate chemicals in products generally under the 1976 directive on marketing and use.[14] This directive sets out a general scheme for restrictions on the marketing and use of specific products by means of Council Directives designating specific chemicals for addition to the Annex. Thus far, the directive has been used to regulate PCBs and PCTs, trichloroethylene, tetrachloroethylene, and carbon tetrachloride in ornamental objects, *tris*-(2,3-dibromopropyl) phosphate in clothing, benzene in toys, fire retardants, sneezing powders, and asbestos.

4.2. Regulation of Industrial Operations for the Prevention of Accidents.

The 1982 Council Directive on the Major-Accident Hazards of Certain Industrial Activities,[15] known as the 'Seveso Directive' is aimed at controlling major accidents by carefully allocating new responsibilities among industrial plant operators, governments, and the Commission of the European Communities. It came into force on 8th January 1984.

The Seveso Directive sets up a legislative framework based on the principle that the manufacturer bears the main responsibility for taking 'all the measures necessary to prevent major accidents and to limit their consequences for man and the environment' (Article 3). In the design and operation of an industrial plant, the manufacturer must consider the possible causes of accidents, monitor critical points in production processes, anticipate the combinations of events that might give rise to an accident, and introduce and maintain stringent safety measures. Should a major accident nevertheless occur, manufacturers must have adopted plans and procedures for limiting the impacts of the accident. The member states must ensure that manufacturers meet these responsibilities (Articles 3,4).

The Directive imposes general or specific responsibilities, depending on the hazardousness of the chemical involved. In the case of less dangerous chemical processes, manufacturers must be able to prove to the national competent authority at any time that they have 'identified existing major-accident hazards, adopted the necessary safety measures, and provided the persons working on the site with information, training and equipment in order to ensure their safety' (Article 4).

In the second case, certain industrial activities are made subject to a notification procedure, *e.g.* if the chemical is listed in Annex III and may be present in the threshold quantity or if the chemical is listed in Annex II and simply stored in the threshold quantity (Article 5). Annex III lists 178 dangerous chemical substances, among which are more than 50 toxic or highly toxic substances, about 60 explosive substances, and about 20 carcinogens. For the most dangerous substances, such as TCDD, the quantity threshold is only 1 kg, while for some flammable liquids the threshold rises to 50,000 metric tonnes.

The notification must contain detailed information about:

Substances and manufacturing processes; including physico-chemical information, forms in which the substances may occur or be transformed under foreseeable abnormal conditions, hazards and risks, safety precautions and emergency procedures.

The industrial plant; including siting, exposed groups and environment, sources of danger arising from the location of the plant, preventive measures, and technical controls.

Possible major-accident situations; including emergency plans, safety equipment, alarms, and resources.

The Directive covers both new and existing industrial activities. It permits manufacturers to submit initially a short declaration for existing industrial activities with the full notification due by 8th July 1989 (Articles 9.3, 4). The notification for new industrial activities must be submitted before the activity begins.

In case of an accident, the manufacturer must immediately inform the national competent authority about all aspects of the accident, emergency measures, and steps foreseen to limit the effects of the accident to human health and the environment. The competent authority must see that the necessary emergency measures are taken and that a full investigation of the accident is carried out (Article 10). In addition, the member state must immediately inform the European Commission of the accident and provide a complete report on the origin of the accident and the emergency, and short- and long-term measures taken to control the effects of the accident and to prevent a future occurrence. The European Commission is setting up a register of information about major accidents.

The Seveso Directive is a breakthrough in European Community environmental law because it established the formal responsibility of a member state to people living in the EC but beyond its national frontiers. Article 8 requires the member states to ensure that all people 'liable to be affected' by a major accident be informed 'in an appropriate manner' of the necessary safety measures and emergency procedures. Neighbouring member states also must be provided with this information in the course of bilateral relations.

5. Future Developments

In the past and at present the European Community's environmental policy has focused on the protection of specific sectors of the environment - in particular air and water. The horizontal approach of the 6th Amendment, where information about the impacts of a chemical on all sectors of the environment is required, has not yet been integrated into the framework of specific tools for the control of chemical pollution that were created by many earlier EC environmental laws. One consequence of this previous sectoral approach is that as standards were tightened in one area they may have led to increased emissions in another area. Hence, today's hazardous waste disposal problems may stem, in part, from the ever-tightening controls on water and air pollution that have been enacted over the past decade.

Public interest in a chemical-specific approach is growing. This would involve a multi-sectoral evaluation of the impacts of separate chemicals, including:

Assessing the pathway and exposure of a particular pollutant through the air, water, and soil to target organisms.

Assessing the health and environmental effects of such exposures.

Adopting a strategy to control risks to health and the environment that take into account the entire life cycle of the chemical.

Such an integrated, chemical-specific control strategy could be initially applied to the priority list of existing chemicals, which the Euopean Commission is in the process of developing under the Third Environmental Action Programme. Currently, some 90,000 chemicals are known (*i.e.* listed in EINECS), of which 20,000 are suspected of having hazardous properties. Owing to the ubiquity of chemicals in everyday life, and consequently in the environment, the problems posed by these existing chemicals cannot be ignored or attacked partially, as in the past. The priority list could also be used as a basis for gathering information from industry about hazardous chemicals on the international market and for setting up joint toxicological and ecotoxicological testing programmes with a view to determining which chemicals should be regulated.[16]

Two further priorities for chemicals regulation cover international trade in investment in chemicals production and in chemical products. In 'New Directions', the Commission has called for 'urgent action... to develop at international level adequate control measures and notification and authorization procedures, which will provide a high degree of security without hampering legitimate manufacture and trade in dangerous products'. The EC is interested in world-wide codes of practice to supplement legislation and the Commission has promised to take initiatives to this end. In addition, the Commission is proposing legislation covering the export and import of banned or severely regulated substances and is continuing to work in the OECD and United Nations Organizations on these issues. (The Commission has announced agreement on the draft Council Regulation but has not yet issued the final text.)

A third priority area is biotechnology.[17] Genetically altered organisms and their products can bring about great social benefits, but there is growing concern about the health and environmental hazards that they may pose. In 1985, the European Commission established a Biotechnology Interservice Regulation Committee (BRIC) that is developing proposals for EC legislation intended to establish a broad framework for the health and environmental regulation of biotechnology which would, at the same time, protect the common market in biotechnological processes and products. The Commission has also initiated discussion about a wider international harmonization of biotechnology regulation in the OECD that perhaps could lead to similar success in the international harmonization of test protocols, laboratory practice, and the intepretation of data that was obtained through the OECD Chemicals Programme.

6. References

1.   First Programme of Action on the Environment, *Official Journal of the European Communities*, **16**, No C112, 20.12.73.

2.   Second Programme of Action on the Environment, *Official Journal of the European Communities*, **20**, No C134, 16.6.77.

3.   Third Programme of Action on the Environment (1982-1986), *Official Journal of the European Communities*, **26**, No C46, 13.6.83.

4.   Commission of the European Communities, 'New Directions in Environment Policy', Communication from the Commission to the Council, COM(86) 76 final, 19th February 1986, para. 31.

5.   261 UNTS 140; 298 UNTS 167; 298 UNTS 3.

6.   Ref. 4, para. 3;

7.   See also, N. Haigh, 'EEC Environmental Policy & Britain. An essay and a handbook', Environmental Data Services, London, 1984, Chapter 10, p.206.

8.   Commission of the European Communities, *Guidance Document*, The Testing of New Substances within the Framework of the Directive 79/831/EEC, internal document, 1981, Brussels.

9.   *Official Journal of the European Communities*, Vol. L 251 of 19.9.1984, p.1.

10. Commission of the European Communities, Proposal for a *Council Directive* on the harmonization of the laws, regulations, and administrative provisions relating to the application of the principles of good laboratory practice and the verification of their application for tests on chemical substances, COM(85) 380 final, Brussels, 15th July 1985.

11. Commission of the European Communities, Commission Decision concerning the list of chemical substances notified persuant to Council Directive 67/548/EEC, *Official Journal of the European Communities*, No.L 30, p.33, 2nd February 1985.

12. Commission of the European Communities, *Classification and Labelling of Dangerous Substances*, XI/316/84-EN.

13. Organisation for Economic Co-operation and Development, *The OECD Chemicals Programme*, Paris: 1984.

14. Commission of the European Communities, Council Directive on the approximation of the laws, regulations and administrative provisions of the Member States relating to restrictions on the marketing and use of certain dangerous substances and preparations (76/769/EEC), *Official Journal of the European Communities*, No. L 262 of 27th September 1976.

15. Council Directive on the major accident hazards of certain industrial activities (82/501/EEC), *Official Journal of the European Communities* No. L 230 of 5th August 1982.

16. Jan Henseimans, The Management of Existing Chemicals in the European Community; Current Policies and Action, Study prepared for the Commission of the European Communities, Directorate-General for Environment, Consumer Protection and Nuclear Safety, Utrecht, 1985, pp.32-36.

17. Biotechnology Regulation Interservice Committee, the Commission's Approach to the Regulation of Biotechnology, BRIC/1/85, rev. 25th February 1986; Communication of Commissioners Cockfield, Clinton Davis, and Marjes to the Commission on the Regulation of Biotechnology, 23rd April 1986.

22
Regulatory Affairs: A European View Relating to Legislation, Hazard/Risk, and Chemicals

By B. Broecker

HOECHST AG, POSTFACH 80 03 20, D-6230 FRANKFURT AM MAIN 80, WEST GERMANY

1. Introduction

The fact that chemicals can be hazardous is by no means a recent discovery. That this has been known for some time is backed up by the long history of industrial toxicology and industrial hygiene. Nevertheless, the belief was held for many years that the risks caused by chemical substances could be controlled by relying on the industry's own sense of responsibility and that legal measures had to be taken only in those areas where chemicals were actually intended for human application, for example, with pharmaceutical products, foodstuff additives, cosmetics, or pesticides. Only with the discovery that chemicals can cause injury, which often only becomes apparent after many years, as well as being able to pollute the environment in a harmful manner, has awareness increased of the dangers which can emanate from chemicals.

It was for this reason that the first considerations were already underway in the United States of America at the end of the 1960s about how the hazards from industrial chemicals could be better controlled by relevant legal measures. The plethora of chemical substances produced and their various properties and use patterns made the enactment of legal measures extremely difficult, particularly since negative consequences for the development of new substances, as had occurred in substance areas already regulated, ought to be avoided.

It was thus necessary to find concepts which, on the one hand, would guarantee better control of chemical substances but, on the other, would encroach as little as possible upon the benefits which chemical substances had indisputably afforded society.

The necessity to balance up risk against benefits was emphasized in the clearest manner in Section 2 of the American chemicals law passed by Congress in 1976, the 'Toxic Substances Control Act'. This reads:

'Authority over chemical substances and mixtures should be exercised in such a manner as not to impede unduly or create unnecessary economic barriers to technological innovation while fulfilling the primary purpose of this Act to assure that such innovation and commerce in such chemical substances and mixtures do not present an unreasonable risk of injury to health or the environment'.

2. Definition of the Legislative Problems

<u>2.1. Difference between Old and New Substances</u>. One of the major problems in the control of chemicals lies in the fact that the number of industrially manufactured chemicals is already extraordinarily high. Every country which was looking into appropriate measures came to recognize very quickly that the resources available would be in no way sufficient to enable precise legal measures to be passed, for example, in the form of testing obligations, for all chemical substances already being manufactured.

Firstly, it was therefore necessary to differentiate between those substances which were already being manufactured industrially, the so-called 'Existing substances', and those which had only been newly put into production, the so-called 'New substances'.

Secondly, it was absolutely essential to limit the assessment primarily to individual chemicals, that is to the substances and not, for instance, to try also to pass comprehensive regulations for the several million preparations which can be manufactured by the mixing together of chemical substances. It has to be sufficient to extrapolate the dangerous properties of such mixtures from the properties of the chemical substances contained within them, and to disregard in the first instance, synergistic or antagonistic effects.

In order to be able to differentiate clearly between so-called new and existing substances, the idea, originally developed in the USA, of nicknamed 'Existing substance' inventories gained acceptance worldwide. This implied that all substances already produced must be compended in relevant lists. However, this task proved to be extremely difficult. On the other hand, definitions had to be found for what is understood by a chemical substance but, in so doing, the border between mixtures was frequently not clearly definable. On the other hand the manufacturers of such substances had to be urged to report their substances to the authorities.

Experience gained in the practical implementation of such inventories has led to the number of substances contained in these lists increasing more and more. Whilst only some 25,000 substances are contained in the first existing substance inventory compiled in Japan in 1973, the American inventory worked out in 1979 and 1980 already contains 50,000 substances and the Existing substance inventory of the European Communities, started in 1982 and still not yet completely finished, will contain around 100,000 substances. One factor which doubtless contributed to this development was that the industry concerned recognized increasingly how important it was for them that their substances were included in these lists since the automatic exemption from certain legal conditions is connected with this.

Unfortunately, the three Existing substances lists available are only comparable with each other to a very limited extent. Whilst the Japanese and American inventories emphasize industrial manufacture or import, the European inventory concerns itself only with substances being marketed. In addition, there are considerable differences in the substance definitions and exemption rulings, of which the most important is indeed that the

EC European inventory, contrary to those of America or Japan, does not contain any polymerizates, polycondensates, or polyadducts but only the monomers used in their manufacture.

2.2. Concepts for the Control of New Chemicals. As far as new substances to be produced or marketed for the first time are now concerned, the legislator had two basic problems to face.

Firstly, it had to be decided upon which criteria for tests for the hazardous properties of these substances were to be established. Solutions to this question varied from country to country.

As already indicated, the most difficult problem with legal measures on hazard/risk control is the setting up of norms which, on the one hand, are universally applicable but, on the other hand, however, do justice to the individual case assessment necessary from a scientific point of view in risk estimation. In order to be able to estimate the risk from a chemical substance, its hazardous properties must be related to the expected exposure of humans and the environment from its manufacture, use, and disposal. In the first instance, therefore, tests should be carried out which give information about these hazardous properties. The type and extent of the tests to be carried out here are, however, again dependent upon the exposure to be expected. Far fewer tests are necessary for a substance which is only used in the industrial sector as an intermediate than for another which is contained in household products.

In this respect, one must, in order to be able to establish proper test requirements, actually know for what exact purpose the substance will be later used to be able to undertake a relevant estimate of the exposure. Opposing this, however, is the fact that the legal concepts used for substance control consciously dispense with an authorization procedure for a specific application purpose, but are content with a more informal notification procedure (see below).

The USA have tried to solve the problem by not requiring specific tests for new substance notifications, but to leave it essentially up to the notifier as to which tests he submits (he must, however, submit all test results in his possession or to which he has access). The European Communities, on the other hand, have established with the 6th Amendment to Directive 67/548/EEC a standardized test programme for all new substances (see Section 3 and Chapters 11 and 12). Japan also lays down in its chemicals law, passed as long ago as 1973, certain test requirements which differ, however, from those of the 6th Amendment and lay their main emphasis on ecological assessment.

A much more standardized procedure was adopted, however, for the second problem, namely the way in which these new substances are made known to the authorities. One possible solution would have been to use here too the authorization/permit concept already used for pharmaceutical products and pesticides. No use was made of this possibility, however, principally for two reasons. Firstly, it must be assumed that the number of new substances being produced for the first time or marketed for the first time would be considerably higher than for pharmaceuticals or pesticides. Consequently, it would only have been possible to cope with an authorization procedure with considerable additional bureaucratic expenditure. Secondly, the rate of innovation, that is, the number of newly produced substances per annum, ought, in the sense of the remarks made at the start, to be little impaired or obstructed.

In view of this, most countries which have enacted relevant laws have decided to adopt a notification procedure in place of an authorization procedure. Basically, this means that the new substance with relevant data is only notified to the authorities and it then remains up to them as to whether corresponding measures are introduced or not. Formal authorization is not, however, granted from the part of the authorities. All the same, this pure notification characteristic is still broken in as much as a defined pre-notification period is prescribed and production and marketing may not be commenced until this period has expired. This then does indeed afford the notification procedure authorization-like characteristics (see Section 4).

Japan alone has so far introduced an authorization procedure for industrial chemicals.

3. Possible Concepts for Legal Rulings on Hazard Assessment of Chemicals

3.1. **Historical Development in Europe.** In view of the basic difficulties in establishing generally valid principles for the hazard assessment of chemicals, the first legal rulings in the European Community were primarily aimed at warning the users of chemical products in a general way about certain hazardous substances. In so doing, no real hazard assessment was undertaken but certain substances were classified as hazardous on account of their harmful effects and, indeed, independently of the exposure to be expected in an individual case with their use.

Following this concept, the first EEC Directive on the classification, packaging, and labelling of dangerous substances was passed as early as 1967. The most important elements of this directive were the Annexes in which the hazardous substances were clearly listed and the warnings with which these substances were to be labelled were laid down (see Appendix 1 and Chapters 11, 12, and 23).

With this, it should be guaranteed that warnings are given in a harmonized manner in all countries of the Community about hazards resulting from these substances. This concept, however, had two serious disadvantages. Firstly, it extended by way of definition only to named substances listed in the directive Annex. Certainly, this substance list was and is being continually expanded by the so-called 'Technical Progress Committee' of the EC Commission. This technically costly procedure cannot, however, keep pace with development. Perhaps even more significant is the fact that the data available, on the basis of which the classification was undertaken, were very incomplete for many substances. In this respect, classification often took place as a result of structure-activity relationships or experiences having been gathered with the substances during use. Long-term effects were often not taken into consideration at all. Secondly, the environmental aspect remained to a large extent disregarded.

3.2.   Enactment of the 6th Amendment (79/831/EEC).   In this
respect, the European Commission took a significant change of
direction with the so-called 6th Amendment of the above Directive
for these substances in 1979, not least under the influence of the
Japanese and American chemicals laws which had meanwhile been
passed.

The most important elements of these regulations were:

Test and notification obligation for new substances.   Corresponding to the American model, a testing and notification obligation
was introduced for new substances.   This, however, did not relate
to the commencement of production as did the Japanese and American
laws, but to the marketing.   In order to ensure a unified
application of this test and notification obligation in the Member
States, a testing programme which, in contrast to the American
model, was essentially the same for all new substances was
established and is to be submitted with the notification.

In order to prevent this test programme being too costly and
thereby hindering the marketing of new substances, only a limited
number of tests were required to be carried out before marketing
(see Appendix 2).   Further, more expensive tests can be requested
if two marketing quantity limits laid down in the Directive are
reached (Step sequence plan, see Appendix 3).

Deviations from this test programme are possible; that is to
say, in a justified individual case, the notifier can omit certain
tests and the authorities can demand certain tests in addition
(see Chapter 11).

**Definition principle for the classification of dangerous
substances.**   Also with labelling and classification a new concept
was introduced changing from the so-called list principle in
favour of the definition principle.   This means that, in future,
not only those substances contained in the relevant Annexes will
have to be classified and labelled, but, in principle, all
hazardous substances which can be found on the market.   The
Directive does not stipulate though that relevant data tests have
to be carried out, rather that the classification and labelling
obligation refers only to the data already available.

For classification and labelling, a guide has been published
in which is established in detail which properties lead to
classification in which hazard category and what warning advice
must then be applied (Annex VI Part IID; see Chapter 11 for full
details).

3.3.   Classification of Dangerous Preparations.   Another question
which is similarly not yet resolved is that of the classification
and labelling of preparations.   The Community had indeed passed,
as early as the start of the 1970s, sector directives on the
classification and labelling of dangerous preparations.   These,
however, are limited to certain product classes or certain use
patterns, for example to products containing certain hazardous
solvents or to paints, varnishes, *etc.*   In both cases, the hazard
assessment was made on the basis of the content of dangerous
components, whereby, regrettably, differing concepts are used.
With the so-called Solvents Directive,[1] the classification is made
by means of a calculation process using the percent proportion of
certain solvents and a so-called index value assigned to each

solvent. The Paints *etc*. Directive[2] on the other hand uses so-called limit concentrations above which classification must take place.

Discussion on a general labelling directive for dangerous preparations is at present in the draft stage whereby the Commission proposal envisages an extremely complicated calculation procedure in which the individual injurious effects must be calculated separately. It cannot yet be foreseen which solution will finally win through here.

3.4. OECD Concept for the Hazard Assessment of Chemicals.
International Organizations, in particular the OECD, which has made a lasting contribution to this question with the compiling and passing of harmonized test methods for chemicals, have naturally also been constantly considering the question of a suitable concept for the hazard assessment of chemicals.

In view of the different legislative systems already outlined above, agreement on one harmonized concept, however, proved to be impossible. Admittedly, the OECD was able to agree, after long discussions, on one test programme, the so-called Pre-market Minimum Set of Data (MPD), which ought to be worked out before the marketing of new substances. This corresponds (see Appendix 2) essentially to the Base Set Testing in accordance with the 6th Amendment Annex VII and only exceeds it in some cases.

Since, however, a binding obligation on the use of this programme would not have been compatible with the American and Japanese chemicals laws, the OECD could only come to an agreement about passing the implementation of this concept as a recommendation that could serve as a basis for a first risk assessment. Consequently, this recommendation has not yet changed anything from the differences already existing with regard to the legal requirements. Certainly though, some countries who have not yet issued the relevant legal regulations, *e.g.* Canada and the Scandinavian countries, are adapting themselves primarily to this concept.

4. Discussion of the EC Concept

For the reasons outlined in the preceding sections, the European Communities had decided with the 6th Amendment to prescribe only a limited test programme prior to marketing. Further, more costly tests were only to be requested if the marketing of these substances had reached certain quantity limits.

With this, a harmonized application of the directive in the Member States ought to be ensured. All the same, in order to take the scientific precept of individual case consideration into account, this so-called step sequence plan ought only to be regarded as a framework concept from which both the authorities as well as the notifier can deviate in individual cases.

It is not yet possible to say to what extent this concept will prove to be a success, since the number of fully notified substances is still very small and, as far as individual chemicals are concerned, cannot amount to more than 70 substances at the moment.

As far as is known, only one substance has as yet reached

Testing Stage 1 and is shortly to reach Testing Stage 2. Thus the experiences gained so far are limited almost exclusively to the Base Set Tests. Even here it is difficult to get a fairly comprehensive picture since many of the details are naturally confidential and cannot be disclosed by the authorities.

With current limited knowledge, however, a number of dificulties can already be identified which have been identified in the practical application of this concept.

4.1. Lack of Flexibility of the Concept. The flexible management of Base Testing originally aimed for has obviously not been realized in practice. Rather, it must be assumed from experiences to date that deviations from the test concept called for in the directive have only been permitted in a few cases. Here it is almost exclusively a matter of those in which the implementation of certain tests was not possible for technical reasons, for example if the n-octanol/water distribution coefficient for surface-active substances cannot be determined because the phases do not separate, or, with dyestuffs, when the determination of vapour pressure has to be dispensed with since it cannot be measured.

As far as tests for toxicological and ecological properties are concerned, to all appearances exceptions have only been granted in a very few cases, and then similarly only if for certain technical reasons the tests could not be carried out. However, it has been shown that it is extremely difficult, for scientific reasons, to do without certain tests. One of the main obstacles is the fact that with the 6th Amendment, as well as with similar laws, it is not a case of authorization procedures which permit one substance for one specific application, but that the substance can, in principle, be used for all conceivable end applications.

In this respect, it is extremely difficult to omit specific tests with reference to certain exposure conditions expected in applications. It can neither be excluded with certainty in most cases that people might come into contact repeatedly with this substance even if only in very small concentrations, nor that the substance, even if again possibly only in small quantities, might be released into the environment. Hence, it is hardly scientifically convincing to support the assumption that, for example, the relatively costly tests for sub-acute toxicity are unnecessary in certain cases.

The tendency is therefore emerging to regard the Base Set Testing actually as a minimum requirement which is to be fulfilled in every case. Differences in opinion which obviously originally existed in the EC Member States as regards this question have levelled off by the individual states having the possibility in accordance with the Directive of demanding the submission of any missing tests for notifications which have been made in other Member States. The so-called harmonization pressure resulting from this has added to the development described above.

For the same reasons, on the other hand, it is also obviously difficult for the responsible authorities to justify convincingly that, with individual notifications, other tests that are, in themselves, intended for Test Stages 1 or 2, must be submitted straight away or earlier. In some cases, structure-activity considerations have obviously been used as justification.

4.2. Problems of Definition for Certain Hazard Categories.

As indicated in Section 3.2, the 6th Amendment introduces for labelling and classification the definition principle enabling manufacturers to classify and to label their products (see also Chapter 23). A guide with detailed classification criteria has been elaborated. However this still leaves significant questions open.

One obviously difficult problem in this context is the question of what is to be understood by so-called substances which are harmful to the environment. At the moment, there are two opposing schools of thought of which one is of the opinion that the concept 'Dangerous for the Environment' can only be seen by combination of data on effects and exposure data, that is to say one can only speak of a hazard to the environment once the expected exposure approaches concentrations with which harmful effects have already been observed.

In contrast the other school is of the opinion that the concept 'Dangerous for the Environment', must be defined as is already the norm for human toxicological data, with the exclusive aid of data on efects, for example because the substance exhibits a certain fish toxicity. This would correspond to the practice already pursued with classification whereby, for example, a substance is classified as toxic if an acute oral toxicity test on rats shows LD_{50} of 25 to 200 mg kg^{-1} body weight, totally independent of whether the actual exposure will reach such values or not. The concept assumes that a detailed exposure/effect observation is difficult anyway for the reasons stated above and therefore, in the interest of a preventive environmental protection policy, would have to depend on substances with certain dangerous properties being no longer released at all into the environment.

Considerable difficulties of definition continue to exist with carcinogenic, mutagenic, and teratogenic substances. For these substances, different categories are indeed envisaged in the Annex VI D Guide adopted by the EC Commission in 1981 (see Section 3.2). The criteria given for the classification of substances into these categories are, however, extraordinarily vague. This reflects the fact that, at the moment, a generally accepted opinion does not exist within the scientific world about the question of whether a categorization is at all possible with these substances and, if so, in what form.

The problem is that either precise classification criteria are formulated which cannot do justice to the individual case, as experience has so far shown, or which must by necessity be vague in relation to the criteria, which makes their practical application difficult. The conceivable solution to pass on the decision to a committee of experts has proved itself to be problematic up until now, at least in the European Community, since even within this committee of experts a unanimous opinion has not always been possible.

Nevertheless, it is precisely here that a harmonized policy is essential, since it would be surely more than regrettable from the point of view of health protection as well as environmental protection if the same substance were to be classified as carcinogenic by one country and then as non-carcinogenic by another.

However, industry has taken up this question more and more. A working group from the European Chemical Federation (CEFIC), the American chemical associations (CMA/SOCMA), and the Canadian association (CCPS) passed a first position paper in the Autumn of 1985 in which an internationally harmonized categorization of such substances was proposed.

4.3. Harmonized Concepts on Risk Assessment Do Not Exist.

This problem can be regarded as only a partial aspect of a more fundamental problem, namely that there are as yet no internationally accepted ideas about hazard or risk assessment. In this lies also an essential weakness of the EC Directive: where it is indeed laid down in detail which tests must be carried out but, however, not which conclusions are to be drawn from the individual results. Admittedly, it would be very difficult to establish unified assessment criteria for the multitude of chemical substances having sometimes completely differing end applications.

To sum up, it is not possible at present to assess whether the test concept envisaged in the 6th Amendment really permits a sufficient risk assessment for chemicals put onto the market for the first time. This must inevitably remain incomplete since the test requirements were limited partly, voluntarily and essentially for economic reasons, and also because there is not enough information available about the exposure to be expected.

In attempting to combat this latter deficiency by the development of models which permit, with the aid of a few substance properties (molecular weight, vapour pressure, water solubility, biotic and abiotic degradability, and adsorption/desorption), an assessment of which environmental media certain substances prefer to remain in and - as far as data are available on the quantities emitted in the environment - what approximate concentrations are to be expected.[9]

To what extent such models, however, really permit realistic estimations is questionable. In particular, local conditions are very difficult to predict by this approach.

5. Problems Still to be Resolved

5.1. Existing Chemicals.

As explained in detail in Section 2, all regulations differentiate between so-called existing chemicals already on the market or already manufactured and those which are being manufactured or marketed for the first time. Only the new substances are subjected to notification after testing. This solution is naturally unsatisfactory in so far as damage which could possibly arise for man and his environment from chemicals at the moment is inevitably caused by these existing substances and not by the new substances which are much more insignificant on the basis of their number as well as their tonnage. In this respect, the existing substance problem cannot be disregarded.

The United States have tried to solve the problem in their TSCA by giving to the so-called 'Interagency Testing Committee' specially set up for this purpose the task of sorting out every year from the great number of existing substances a few for which, in the opinion of the Committee, further tests must be carried out for clarification of the risk possibly caused by them to man and his environment.

Provided that the Environmental Protection Agency (EPA) concerned with the implementation of this law follows these suggestions, it has the possibility of demanding tests for these substances from the manufacturers and importers. In its practical implementation, however, this provision has proven itself to be extremely problematic. This is essentially because, for many existing substances, there are simply just not enough data both on exposure as well as on effect. In this respect, one often has to work here with assumptions and analogous considerations. The issuing of formal test requirements presupposes that the authorities can prove to some extent that a so-called 'unreasonable risk' does in fact stem from these substances. More precisely expressed, this means that the results of the tests must actually already be available in order legally to require their being carried out.

The number of substances for which as yet real use has formally been made of this authority is still small. All the same, in many cases, the manufacturers have readily agreed on a voluntary basis to undertake such tests.

In the European 6th Amendment, from its inception, no attempt has been made to adopt a test and notification obligation for old substances. As already discussed in detail in Section 3.2, it is merely prescribed that, the data available for existing substances must be examined to establish whether a classification as 'dangerous' must be made or not. If, however, there are no data available, there is presumably also no classification obligation, never mind a testing obligation.

In taking the 6th Amendment into national law, a number of Member States have exceeded the text of the 6th Amendment in that they have sometimes stipulated testing and notification obligations for existing substances as well. Thus, for example, Denmark and France stipulate in their laws that existing substances must be tested and notified if the quantity marketed or their application changes essentially. As far as is known, these conditions have, however, found scarcely any practical application as yet.

There is also a paragraph contained in the German chemicals law (S4 Par 6) according to which the authorities are authorized under certain pre-conditions to include old substances in the test and notification procedure. Here too, difficulties similar to those which exist in the United States have rendered the practical application of this condition impossible until now. In the last few years, however, some Member States, as well as the OECD, have devoted themselves more and more to this problem. Thus, the Dutch Government had already started in 1982 to compile a first priority list of *ca*. 400 substances and to pick out from this list, using systematic selection criteria, some 20 substances for which a detailed assessment would seem necessary (see ref.4, p.179).

In Germany, two bodies had also been set up by the beginning of the 1980s, on the basis of voluntary agreements between authorities and industry, which concern themselves with the selection, testing, and assessment of existing substances primarily from the point of view of worker protection and

environmental protection.   A first concept of how such selection
can proceed has just been published.[5]   The OECD then further tried
in 1983 and 1984 to develop corresponding ideas of a general
nature.[6]

Common to all concepts is the difficulty that it is impossible
to process the majority of existing substances contained in the
old substance inventory systematically since, for many of these
substances, no documentation on exposure or effect is available.
All known selection schemes attempt to obviate this problem by
trying to select from the great number of existing substances
those which are to be regarded as suspicious on the basis of
either exposure estimations or supposed dangerous effects.

The disadvantage in such selection is that it only covers
those substances for which, for whatever the reason, at least some
data are available.   For this reason most concepts used so far
have resulted in the selection of nearly the same chemicals.   It
remains uncertain as to how to handle the large number of
chemicals with inadequate data.

That progress in the processing of such substances is
relatively low as yet is due to the difficulties already described
in Section 4.   Depending upon the prevailing scientific and
political conditions in certain countries, one and the same
substance can be classified as dangerous and correspondingly
restricted or prohibited completely, whilst in other countries it
can still be marketed.   Examples of this are the fluorocarbons, in
particular their use in aerosol sprays, or phosphate substitutes
such as nitrilotriacetic acid (NTA).   This unfortunately results
in many of the existing substances regarded as suspicious
remaining permanently under discussion without any final
assessment being reached.   It would possibly be more effective to
take the opposite path, namely to separate those substances which
are to be regarded as non-dangerous according to the present state
of knowledge.   This would presuppose though, in view of the
quantities, application, and effect data available, that the
'Existing substance' inventories are analysed more exactly.   Even
then it would probably be difficult in view of our limited state
of knowledge, especially in the area of ecotoxicology, finally to
classify a substance as non-dangerous.

5.2. Risk/Benefit Assessment.   All chemical regulations assume in
principle that the risk caused by a chemical substance to man and
his environment may be related to its benefits.

The practical implementation of this principle is, however,
far more difficult than generally supposed because economic
considerations play a role.

In the case of a substance for which, for whatever reasons,
the economic prospects of success are small, the extent to which
its dangerous properties are assessed will also be limited simply
because the small profit prospects do not justify high
expenditures for tests.   The case is different for substances for
which the profit prospects are good and for which, for example,
for these reasons, considerable financial investments are also
made in order to be able to manufacture them.   Here, one is rather
inclined to carry out more extensive tests for dangerous
properties in order to guarantee that it can also later be
marketed without difficulty and product liability.   However, the

economic prospect of success cannot be any measure for the range of the hazard of chemicals.

This dilemma could only be avoided if it were possible to develop test methods which could be used cost-effectively even in the development of new substances, and with the aid of which fairly certain pointers to dangerous properties would be achieved.

Although it will be inevitably difficult scientifically, for example, to recognize fairly certainly a long-term effect by means of a short-term test, there are initial developments which can point in this direction. In this connection can be mentioned so-called screening tests for mutagenic properties, from which conclusions are drawn about possible carcinogenic properties. The differing predictability of such tests with different chemical substances nevertheless makes the application of such concepts difficult.[7]

To avoid this difficulty if one uses whole batteries of such tests, that which most often occurs is that some tests are positive and some are negative, making a clear decision again difficult. The other alternative of estimating the dangerous properties on the basis of physical/chemical properties like vapour pressure or water solubility or on the basis of the chemical structures has only currently proven itself as to be workable to a very limited extent.

A recent ECETOC report[8] has described how such considerations can essentially only be used successfully if the effect causing a particular result is clearly known, for example if it clearly correlates with a physico-chemical dimension. In all other cases, the setting up of relevant structure-activity relationships is very costly and with this much more expensive than the implementation of the relevant tests.

Whether fundamental improvements will be achieved in the foreseeable future is rather doubtful. Since the expected economic advantage for a substance cannot be determinative for the extent with which the risk resulting from it is assessed, the only alternative would be to classify substances into categories according to their application and to gauge the extent of tests accordingly. To a certain extent, this is already happening today.

Household chemicals, for example detergent raw materials, which due to their use in great quantities come into direct contact with man and his environment, are already currently examined extensively for their dangerous properties before they are brought onto the market. In contrast, it is certainly far less necessary to carry out relevant costly tests on intermediary products, which only remain in the industrial sector, providing it is guaranteed that the quantities released into the environment are only very small.

In this respect, the ideal would be to strive to modify the concept of the 6th Amendment, with its harmonized test regulations for all new substances, in favour of a varying catalogue of measures according to application categories. Certainly the setting up of such categories would have to avoid problems of definition.

The question remains as to what extent with known harmful effects the advantage of a substance outweighs the risk. Naturally, it must first of all be pointed out here that the question of what can be regarded as a 'socially adequate' risk has never yet been quantified. In the USA, attempts are being made to solve this problem statistically and, for example, to classify a risk of less than 1:1,000,000 as harmless since it lies below the normal risk threshold of human life (see also Chapter 21).

This concept, however, only feigns safety since the suppositions which are based on the corresponding mathematical risk extrapolations are highly doubtful and uncertain. In this respect, it remains finally a political decision in the individual case as to whether under certain conditions a risk is regarded as acceptable or not. A harmonization of such considerations is therefore neither feasible not desirable.

6.   References

1.   Labelling of Dangerous Preparations (Solvents), EEC Directive, 73/173/EEC and 80/781/EEC.

2.   Council Directive 77/728/EEC, November 7, 1977, on the approximation of the laws, regulations and administrative provisions of the Member States relating to the classification, packaging and labelling of paints, varnishes, printing inks, adhesives and similar products, as last amended by Directive 82/265/EEC.

3.   'Environmental Modelling for Priority Setting Among Existing Chemicals', Workshop, November 13-15 1985, Munich, Gesellschaft fur Strahlen- und Umweltforschung mbH, Munich, and Umweltbundesamt, Berlin.

4.   'Chemicals on which Data Are Currently Inadequate: Selection Criteria for Health and Environmental Purposes', OECD Report, Vol.I, p.178, 1984.

5.   GDCh-Beratergremium für Umweltrelevante Altstoffe, 'Umwelt-relevante Altstoffe - Kriterien und Stoffliste', GDCh, Frankfurt/M., 1985.

6.   'Chemicals on which Data are Currently Inadequate: Selection Criteria for Health and Environmental Purposes', OECD Report, Vols. I and II, 1984.

7.   'Evaluation of Short-term Tests for Carcinogens', Report of the International Programme on Chemical Safety's Collaborative Study on *in vitro* Assays': 'Progress in Mutation Research', ed. J. Ashby, F. J. De Serres, M. Draper, M. Ishidate Jr., B. H. Margolin, B. E. Matter, and M. D. Shilby, Vol. 5, 752pp., Elsevier, Amsterdam, 1985.

8.   ECETOC monograph No.8, 'Structure-Activity Relationships in Toxicology and Ecotoxicology: An Assessment' February 24, 1986.

Appendix 1

Cas No   95-95-4 [1]
           88-06-2 [2]

No 604-012-00-2

2,4,5-Trichlorphenol (1)
2,4,6-Trichlorphenol (2)
2,4,5-Trichlorphenol (1)
2,4,6-Trichlorphenol (2)
2,4,5-Trichlorophenol (1)
2,4,6-Trichlorophenol (2)
2,4,5-Trichlorophénol (1)
2,4,6-Trichlorophénol (2)
2,4,5-Triclorofenolo (1)
2,4,6-Triclorofenolo (2)
2,4,5-Trichloorfenol (1)
2,4,6-Trichloorfenol (2)

R: 22-36/38

S: 26-28

Cas No 58-90-2

No 604-013-00-8

2,3,4,6-Tetrachlorphenol
2,3,4,6-Tetrachlorphenol
2,3,4,6-Tetrachlorophenol
2,3,4,6-Tétrachlorophénol
2,3,4,6-Tetraclorofenolo
2,3,4,6-Tetrachloorfenol

R: 25-36/38

S: 26-28-37-44

Appendix 2

TEST REQUIREMENT FOR NEW CHEMICALS

EEC -
BASE SET

Chemical designation

Use patterns

Quantity

Precaution measures
  (e.g. labelling)

Physico-chemical parameters (13)

Acute toxicity
    (two routes of administration)

Skin irritation

Eye irritation

Sub-acute toxicity (28 days)

Sensitization

Pre-screening tests on
  carcinogenic effects

Acute fish-toxicity

Acute daphnia toxicity

Degradation

Information on disposal

Disassociation constant

Particle size

Adsorption/desorption

Daphnia reproduction 14 days

Alga growth inhibition

MPD

Appendix 3 (see also Chapter 11)

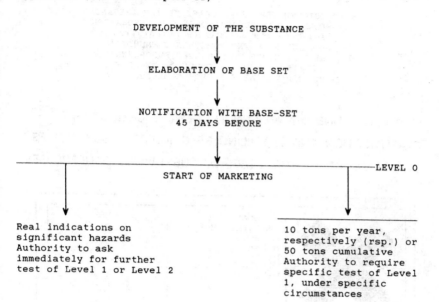

DEVELOPMENT OF THE SUBSTANCE

ELABORATION OF BASE SET

NOTIFICATION WITH BASE-SET
45 DAYS BEFORE

START OF MARKETING ————LEVEL 0

Real indications on
significant hazards
Authority to ask
immediately for further
test of Level 1 or Level 2

10 tons per year,
respectively (rsp.) or
50 tons cumulative
Authority to require
specific test of Level
1, under specific
circumstances

LEVEL 1   LEVEL 1

At 100 tonnes per year resp. 500 tonnes cumulative

authority to require tests listed in Level 1

DIALOGUE ON LEVEL 2   LEVEL 2

At 1000 tonnes per year resp. 5000 tonnes cumulative

authority to require tests listed in LEVEL 2

(in dialogue with the notifier)

Appendix 4

KOMMISSIONEN FOR DE EUROPAEISKE TAELLESSKADLR
KOMMISSION DER EUROPAISCHEN GEMEINSCHAFTEN
ΕΠΙΤΡΟΠΗ ΤΩΝ ΕΥΡΩΠΑΙΚΩΝ ΚΟΙΝΟΤΗΙΤΩΝ
COMMISSION OF THE EUROPEAN COMMUNITIES
COMMISSION DES COMMUNAUTES EUROPEENNES
COMMISSIONE DELLE COMUNITA EUROPEE
COMMISSIE VAN DE EUROPESE GEMEENSCHAPPEN

EINSTUFUNG UND KENNZEICHNUNG GEFÄHRLICHER STOFFE
CLASSIFICATION AND LABELLING OF DANGEROUS SUBSTANCES
CLASSIFICATION ET ETIQUETAGE DE SUBSTANCES DANGEREUSES

Lokalirriterende
Reizend
Irritant
Irritant
Irritante
Irriterend

Giftig
Giftig
Toxic
Toxique
Tossico
Vergiftig

Let antændelig
Leichtentzündlich
Highly flammable
Facilement inflammable
Facilmente infiammabile
Licht ontvlambaar

Appendix 5

For the purpose of classification and labelling, (see also Chapter 23), and having regard to the current state of knowledge, such substances are divided into three categories:

Category 1:

Substances known to be carcinogenic to man. There is sufficient evidence to establish a causal association between human exposure to a substance and the development of cancer.

Category 2:

Substances which should be regarded as if they are carcinogenic to man. There is sufficient evidence to provide a strong presumption that human exposure to a substance may result in the development of cancer, generally on the basis of:

- appropriate long-term animal studies,

- other relevant information

Category 3:

Substances which cause concern for man owing to possible carcinogenic effects but in respect of which the available information is not adequate for making a satisfactory assessment. There is some evidence from appropriate animal studies, but this is insufficient to place the substance in Category 2.

23
Labelling

By S.G. Luxon

MAIDENSGROVE HOUSE, RIVERDALE, BOURNE END, BUCKS. SL8 5EB, UK

1. Historical

As discussed in the Introduction, over the past twenty-five years there has been an increasing awareness of the hazards associated with chemicals of all kinds. Simultaneously, there has been a demand for more open identification of the risks which may arise where such chemicals are manipulated, handled, or stored. One of the outcomes of these pressures has been a demand for the more obvious universal identification of hazardous materials coupled with information on the risks which they pose during use.

Recent surveys have shown that one in every four workers may be exposed to some thousand common hazardous chemicals and almost one in ten to substances which have shown some inferred carcinogenic effect. There are some one hundred thousand trade-named chemicals in use and in almost one half of these uses at least one other trade-named product was also present to form a mixture. In such unidentified products neither the worker nor the employer is in a position to identify possible hazards. Even when the correct chemical nomenclature or accepted trivial name is shown, often neither worker nor management has an adequate knowledge of the hazards involved. The demand, therefore, for a labelling system appears to be more than justified by practical considerations.

In the 1950s when this need for action was first recognized, discussions were taking place on possible general classification systems under the aegis of a number of international organizations, the United Nations, the International Labour Organization, and the Council of Europe being perhaps the more important. The United Nations' groups were primarily concerned with transport, which perhaps has the widest international implications, while the International Labour Organization and the Council of Europe were concerned principally with hazards to the user which were of greater national importance. It was recognized at this early stage that there was a need for unification, but it proved impossible to agree a single system in each of these fields, let alone in both. In 1959, the ILO adopted symbols on a uniform orange coloured square or rectangular background which subsequently became the norm for labelling systems in respect of users, even though these symbols represented a compromise, *e.g.* in respect of corrosive materials incorporating, as it does, two indications, the corroded hand and the corroded bar. On the other hand the UN transport system had adopted a diamond format with differing colours to indicate the type of hazard, but using similar symbols in the upper half to those forming the user system together with key words in the lower half. Attempts have been made over the years to bring together the two systems. Unfortunately the exponents of each could not agree to sufficient flexibility to enable meaningful discussions to take place. They served merely to pin-point the underlying difficulties, such as the different effects in the long- and short-term of certain toxic substances, *e.g.* benzene, where in transport the principal risk is of fire, whereas in use it is the long-term irreversible toxic effects that are much the more serious. EC member states enacted the 1967 Directive (67/548/EEC). Certain countries eventually enacted the 1967 Directive on user labelling systems (see Figure 1) thus enshrining the differences in their own legislation. More recently the classification of individual substances has followed a different pattern so that harmonization is not merely now a question of agreeing universal systems and formats, but must take account of detailed differences, making harmonization the more difficult (see also Chapter 21).

2. User Labelling Systems

In view of the difficulties which had appeared in the earlier deliberations, it had become apparent to all concerned that any attempt to formulate a truly international system of user labelling would be an impossibly protracted task. The Council of Europe accordingly took on board a more limited objective of developing recommendations for a European system for the labelling of pure substances. The basis chosen for this system was the ILO symbols referred to above.

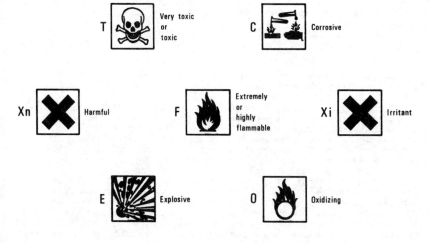

The above symbols are painted in black on an orange-yellow background

Figure 1 *Classification symbols giving indication of risk*

These deliberations resulted in the publication of the earliest formalized system in the so-called Yellow Book entitled 'Dangerous Chemical Substances and Proposals Concerning Their Labelling'. This book sets out the basis on which all the subsequent work has been based. Following its publication, the Commission of the European Communities proposed that this system should be adopted by Member States under Article 100 of the Treaty of Rome to facilitate free trade in such products between Member States. The basic directive of 1967 which gave effect to these proposals was the result.

In order to understand the underlying principles, it is perhaps worth looking in more detail at the background which led to the development of this system. Its basis must be to convey the information set out below to the user by means of a label attached to a container. The information should be based on an evaluation of the physical and toxicological properties of the particular substance.

i)   The correct chemical or accepted trivial name of the substance so that it can be readily identified and, if necessary, further relevant information obtained by literature search.

ii)   The name of the manufacturer or supplier so that advice can be sought as necessary.

iii)   Means of immediately identifying the dangerous classification which has been adopted and which transcends language barriers, *i.e.* a symbol.

iv)   Indication of the special risks which may arise, *i.e.* the Risk or R-phrases (see Chapter 12, Appendix).

v)   Simple advice as to the precautions which should be taken, *i.e.* Safety or S-phrases (see Chapter 12, Appendix).

Where possible such labelling should be supplemented by a leaflet setting out more detailed toxicological information.

Originally, with relatively few substances under consideration, it was possible to decide on labelling in respect of each individual substances in the light of practical experience. It soon became clear, however, that such a course of action would in the longer term result in anomalies, particularly where a severe hazard had been found to be associated with a particular chemical which might have been due to local circumstances rather than the toxic properties of the substance itself. As firm toxicological evidence increasingly became available, and to avoid such pitfalls, efforts were made to rationalize the system as far as was possible at the time. Attention was therefore directed towards the classification of hazardous substances and certain broad parameters have been developed.

For toxic and harmful substances Table 1 shows the parameters which have been adopted.

Table 1 *Classification of toxic substances*

Category	Symbol	LD_{50} Oral Rat mg kg^{-1}	LD_{50} Percutaneous Rat or Rabbit mg kg^{-1}	LD_{50} Inhalation Rat mg l^{-1} 4 h
Very toxic	T	$\leq$ 25	$\leq$ 50	$\leq$ 0.5
Toxic	T	25 - 200	50 - 400	0.5 - 2
Harmful (irritant)	Xn or Xi	200 - 2000	400 - 2000	2 - 20

A simplistic approach such as this based on acute toxicity in animals cannot adequately take account of the many long-term effects which are difficult to quantify. The overall assessment of toxicity must be related to experience gained from epidemiological studies, human experiences in cases of accidental poisoning, possible sensitization, carcinogenic and mutagenic effects. Each substance must therefore be reviewed individually to determine whether or not this routine classification was adequate. It must be emphasized, however, that such a procedure must not be allowed to distort the overall uniformity of the system.

In the table above, it will be noted that harmful substances carry a symbol Xn or Xi. The symbol Xi was originally used for irritant substances applied externally to the body, but it subsequently became used for substances irritant to the respiratory system.

2.1. Corrosive Substances. These are defined as substances and preparations which may on contact with living tissue destroy them. These substances carry a C symbol.

2.2. Irritant Substances. Non-corrosive substances which, through immediately prolonged or repeated contact with the skin or mucous membrane, can cause inflammation. These substances are given a symbol Xi.

2.3. Extremely Flammable Substances. These are substances having a flash point lower than 0°C and a boiling point lower than, or equal to 35°C.

2.4. Highly Flammable Substances.

i) Substances and preparations which may become hot and finally catch fire in contact with air at ambient temperature without any application of energy.

ii) Solid substances and preparations which may readily catch fire after brief contact with the source of ignition and which continue to burn or to be consumed after removal of the source of ignition.

iii) Liquid substances and preparations having a flash point below 21°C.

iv) Gaseous substances and preparations which are flammable in air at normal pressure.

v)   Substances and preparations which in contact with water
or damp air evolve highly flammable gases in dangerous
quantities.

    It should be noted that the above two categories, *i.e.*
extremely flammable and highly flammable, attract a T symbol.

2.5. Flammable Substances.   Liquid substances and preparations
having a flashpoint equal to or greater than 21°C and less than or
equal to 55°C.   In this category the word 'flammable' is used
without a symbol where a substance attracts another symbol.

2.6. Explosive Substances.   Substances and preparations which may
explode under the effect of flame or which are more sensitive to
shocks or friction than dinitrobenzene attract the E symbol.

2.7. Oxidizing Substances.   Substances and preparations which give
rise to highly exothermic reaction when in contact with other
substances, particularly flammable substances, attract the O
symbol.

2.8.   Criteria Used to Determine the Allocation of Symbols.   The
total number of symbols is limited to two and for this purpose a
system of priority has been established.   The order of precedence
is shown in Table 2.

Table 2 *Order of precedence of symbols*

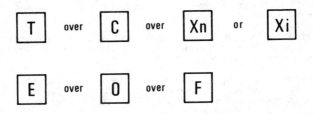

    The Risk-phrases are set out in the Directives and Legislation
referred to in this chapter.   They are allocated on the basis of
the symbol used and the known risks of the substance concerned.
Their number is normally limited to three so as to avoid the
danger of making the label too complex.

    As with Risk-phrases, safety advice phrases are set out in the
relevant Directives and Legislative documents.   They are allocated
on an *ad hoc* basis and in general supplement the Risk-phrases and
sometimes give general guidance on local precautionary measures.
Again a maximum number of three S-phrases is the norm.

    An attempt was made to simplify the system further by adopting
combination phrases so that a set of R- or S-phrases could be
combined into a single sentence with a consequent reduction in the
number of words required to convey the same amount of
information.

3. The Packaging and Labelling Regulations 1978 and Subsequent Amending Regulations

The Regulations should be consulted as to detailed requirements on specific substances listed therein, but the general provisions may be summarized as follows:

3.1. <u>Application</u>. The regulations apply to the supply, which is defined in the Regulations, of some 1000 dangerous substances and require labels to be applied to containers subject to the following exceptions:

i) When the substance is supplied for use as motor fuel munitions or as a pesticide.

ii) Where the substance is a medicinal product under Section 130 of the Medicines Act 1968, or in an Order made under that Act.

iii) When the substance is delivered by pipe, or into a storage tank or vessel provided by the user.

iv) A gas which is compressed, liquified, or dissolved under pressure.

v) When it is supplied for export.

vi) The retail sale of paraffin.

vii) Under customs bond.

3.2. <u>Packaging</u>. There are general requirements regarding the design and construction of containers to ensure safety in handling and storage.

3.3. <u>Labelling</u>. The label must indicate the following:

i) The name of the substance, normally the International Union of Pure and Applied Chemistry (IUPAC) nomenclature, or the recognized trivial name.

ii) Name and address of the manufacturer, importer, wholesaler, or supplier.

iii) The symbol and key word.

iv) The Risk- and Safety-phrases.

Figure 2 shows an example of a label made up in this way.

There are certain exceptions to these requirements, the most important of which is that the Risk- and Safety-phrases need not be shown on the label if the container has the capacity of 125 ml or less, but the Risk-phrases only must be shown if a T, C, or E symbol is necessary. Where the containers are labelled in accordance with transport rules and, in the case of very small quantities which do not constitute a danger, a label is not required except in the case of substances attracting T, C, or E symbols.

Toxic

Aniline

Toxic by inhalation, in contact with skin and if swallowed

Danger of cumulative effects

Wear suitable protective clothing and gloves

After contact with skin, wash immediately with plenty of soap and water

If you feel unwell, seek medical advice (show the label where possible)

XYZ Chemicals Ltd

Figure 2

3.4. Application of Marking. The particulars are required to be indelibly marked on a part of the container reserved for the purpose, or on a label securely fixed with its entire surface in contact with it. Where this is impossible, it must be attached in a suitable manner. The Safety-phrases may be shown on the document accompanying the container. The colour and nature of the marking or label should be such that the orange-yellow symbol stands out. The label sbould be capable of being easily read when the container is in its normal position. There are minimum requirements for the size of labels for differing sizes of container, and, in general, the symbol shall occupy at least one-tenth of the label area and be not less than one square centimetre in size.

3.5. Conveyance or Transport Labelling Systems. Reference has already been made to different systems of labelling for transport or conveyance by road, rail, or air. This labelling system may be summarized in Table 3.

It will be noted that, in this system, parameters are not laid down for toxic substances. In general it is suggested that those set out for this classification in the supply or user system will provide useful guidance.

The particulars shown on the label for transport are, of necessity, somewhat different from those required for user labelling, in that provision must be made for assistance to be obtained in an emergency. It is a requirement that such labels should include the following:

i) The name and address or telephone number of the consignor or other nominated person from whom expert advice on the dangers created by the substance may be obtained.

ii) The name of the substance.

iii) The identification number, if any, of the substance.

iv) The hazard warning sign.

v) Where the quantity is more than 25 l, the nature of the dangers to which the substance may give rise and the emergency action that should be taken, except that this information may be shown on a separate statement accompanying the package.

In the labelling requirements for conveyance there is an exception where labelling is not required for small containers.

It will be seen that the conveyance and user systems are different and two conflicting labels could therefore be required. To avoid this a derogation is provided in the relevant legislative requirements such that where a package consisting of a single receptacle is correctly labelled in accordance with the conveyance system, the label for the user system need not be affixed, provided that the label shows the name and address of the supplier, the name of the substance, the Risk-phrases, and Safety-phrases. Where the package consists of one or more receptacle in outer containers, then the conveyance label only can be applied to that outer package.

Table 3 *Classification and symbols used for substances dangerous for conveyance*

1 Characteristic properties of the substance	2 Classification	3 Hazard Warning Sign
A substance which—		
(a) has a critical temperature below 50°C or which at 50°C has a vapour pressure of more than 3 bar absolute; and (b) is conveyed by road at a pressure of more than 500 millibar above atmospheric pressure or in liquified form, other than a toxic gas or a flammable gas.	Non-flammable compressed gas	 Green background
A substance which has a critical temperature below 50°C or which at 50°C has a vapour pressure of more than 3 bar absolute and which is toxic.	Toxic gas	 White background
A substance which has a critical temperature below 50°C or which at 50°C has a vapour pressure of more than 3 bar absolute and is flammable.	Flammable gas	 Red background

A liquid with a flash point of 55°C or below except a liquid which—

(a) has a flash point equal to or more than 21°C and less than or equal to 55°C; and

(b) when tested at 55°C in the manner described in Schedule 2 to the Highly Flammable Liquids and Liquified Petroleum Gases Regulations 1972 (a) does not support combustion.

Flammable liquid

Red background

A solid which is readily combustible under conditions encountered in conveyance by road or which may cause or contribute to fire through friction.

Flammable solid

Red hatched background

A substance which is liable to spontaneous heating under conditions encountered in conveyance by road or to heating in contact with air being then liable to catch fire.

Spontaneously combustible substance

Red lower half

A substance which in contact with water is liable to become spontaneously combustible or to give off a flammable gas.

Substance which in contact with water emits flammable gas

Blue background

A substance other than an organic peroxide which, although not itself necessarily combustible, may by yielding oxygen or by a similar process cause or contribute to the combustion of other material.

Oxidizing substance

Yellow background

A substance which is—
(a) an organic peroxide; and

(b) an unstable substance which may undergo exothermic self-accelerating decomposition.

Organic peroxide

Yellow background

A substance known to be so toxic to man as to afford a hazard to health during conveyance or which, in the absence of adequate data on human toxicity, is presumed to be toxic to man.

Toxic substance

White background

A substance known to be toxic to man or, in the absence of adequate data on human toxicity, is presumed to be toxic to man but which is unlikely to afford a serious acute hazard to health during conveyance.

Harmful substance

White background

A substance which by chemical action will—

Corrosive sub-stance

White background

(a) cause severe damage when in contact with living tissue;

(b) materially damage other freight or equipment if le=kage occurs.

A substance which is listed in Part 1A of the approved list and which may create a risk to the health or safety of persons in the conditions encountered in conveyance by road, whether or not it has any of the character-properties set out above.

Other dangerous substance

White background

Packages containing two or more dangerous substances which have different characteristic properties.

Mixed hazards

White background

4. Preparations and Mixtures

Pure substances can, by the systems set out above, be labelled with symbols and Risk- and Safety-phrases to indicate their toxicity in a relatively simple way. Preparations, *i.e.* substances which are a mixture prepared by the manufacturer from two or more 'pure' substances, present more difficult problems. Ideally, it would be desirable to evaluate the toxicity of each preparation by the methods set out in this book. This has in fact been applied in one specific area, *i.e.* pesticides, where, because of their *raison d'etre*, toxicity testing is always undertaken. Unfortunately this technique would prove impossibly expensive and time consuming if it were universally applied to the very large number of other preparations on the market, particularly in view of the frequency with which their composition is changed. A simpler first approach was therefore required, but a suitable system has proved very difficult to develop given possible additive and synergistic effects of the component parts.

The Commission of the European Communities has developed what can best be described as a compromise solution and which is perhaps an over-simplification of the problem. It ascribes classes of toxicity to each individual component and attempts to combine these by the use of a summation formula to give an overall indication of hazard by which the classification of the mixture is determined.

5. Solvents

Each solvent is placed in a class as set out in the Approved List published by the Health and Safety Commission on the 25th July 1984 entitled 'Information Approved for the Classification, Packaging and Labelling of Dangerous Substances', which gives effect in the UK and the EEC Directive.[2-6] The classes of toxic and harmful substances are then assigned indices in accordance with Table 4.

Table 4 *Classficiation of solvents*

	Class	I_1	I_2
Very toxic and toxic	I/a	500	500
	I/b	100	100
	I/c	25	25
Harmful	II/a	5	20
	II/b	2	8
	II/c	1	4
	II/d	0.5	2

The preparation is considered to be toxic if the sum of the figures obtained by multiplying the index I_1 for each substance by its concentration in percent by weight in the preparation exceeds 500. The preparation is considered harmful if the sum of the figures obtained by multiplying the Index I_1 for each such substance by its concentration in percent by weight in the preparation is equal to, or less than 500, and if the sum of the figures obtained by multiplying the index I_2 for each such substance by its concentration in percent by weight in the preparation exceeds 100. In addition, where the preparation contains an impurity or additive which, although not a solvent, is a substance which would be classified as very toxic or toxic, it shall be treated as being in Class I/a for the purposes of the above formula. In a similar way, substances classified as harmful shall be deemed to be in Class II/a. However, no account shall be taken of any substance which is at a concentration of less than 0.2% by weight in Class I or 1% by weight in Class II.

Similarly, corrosive and irritant substances present in the solvent must be combined. This is effected by summing the weight percentage of each constituent divided by the concentration specified as corrosive. Where the product is greater than one, the substance is deemed to be corrosive, but no account is taken of any substance present at a concentration of less than 1% by weight.

With irritant substances the situation is more complicated because both irritant and corrosive substances may be present. Here the sum of the weight percentage of each corrosive substance, divided by the lower limit concentration value for irritancy for that substance given in the Approved List, together with the sum of the weight percentage of each irritant substance, divided in that case by the concentration for that substance given in that list exceeds one, then the solvent mixture is deemed to be irritant.

6. Paints, Varnishes, Printing Inks, Adhesives, and Similar Products

These preparations normally contain solvents to which the solvent labelling requirements will apply. They may however contain other toxic ingredients which modify hazards arising in their use. To take such matters into account, such products are considered toxic if they contain solvents classified as toxic as set out above, except that the concentration of each component of the solvent shall be calculated as the percentage of the total weight of the preparation as supplied. In addition, where they contain one or more toxic or very toxic substances, the total concentration of which exceeds 1% of the total weight of the preparation, they are also classified as toxic. No account is taken, however, of any such substance unless its concentration exceeds 0.2% of the total weight of the preparation.

Such products are classified as harmful if they contain solvents classified as harmful by the methods set out above. In addition, where they contain one or more substances which are classified as harmful, and where the total concentration of such substances exceeds 10% of the total weight, then they attract a 'harmful' classification. No account is taken, however, of any such substance unless its concentration exceeds 1% of the total weight.

Such preparations are considered corrosive if the total concentration of corrosive substances exceeds 5% of the total weight; no account is taken of any substance whose concentration is less than 1% of the total. Similarly, they are classified as irritant if they contain irritant substances exceeding 5% of the total weight, but taking no account of those substances present at less than 2% of the total weight of the preparation.

7. Pesticides

As indicated above, pesticides can be treated more logically in that additional toxicological information is normally available. The classification parameters are somewhat different from those for 'pure' substances and are set out in Table 5. The more restrictive requirements are used to determine the classification.

Where the pesticide contains only one active ingredient, the toxicity of the preparation may be calculated in accordance with the formula $(L \times 100)/C$ where L is the LD_{50} of the active ingredient, and C is its concentration as a percentage by weight. This, of course, requires judgement that the toxicity as determined would not differ substantially from the toxicity determined by biological testing. Account should of course be taken of any known factors which affect the risk to human health.

8. Environmental Hazards

No formalized general method of labelling substances hazardous to the environment has been devised. The user system has been extended to include special risks to the general public and in particular to children. Some Risk-phrases have been developed to deal with environmental hazards associated with the disposal of certain chemicals. While it is accepted that global environmental issues are inappropriate to labelling in that they have wider

Table 5 *Classification of pesticides*

Classification	LD50 absorbed orally in rat mg kg⁻¹		LD50 absorbed percutaneously in rat or rabbit mg kg⁻¹		LD50 absorbed by inhalation in rat mg l⁻¹ (4 h)
	Solids other than baits and tablets	Liquids and bait preparations and pesticides in tablet form	Solids other than baits and tablets	Liquids and bait preparations and pesticides in tablet form	Gases, liquified gases, fumigants and aerosols, powders having particle size $\leqslant 50$ μm
Very toxic	$\leqslant 5$	$\leqslant 25$	$\leqslant 10$	$\leqslant 50$	$\leqslant 0.5$
Toxic	> 5 to 50	> 25 to 200	> 10 to 100	> 50 to 400	> 0.5 to 2
Harmful	> 50 to 500	>200 to 2000	>100 to 1000	>400 to 4000	> 2 to 20

implications than those that can be dealt with by the user there is a need to warn universally of the dangers to the environment associated with the uncontrolled escape of certain hazardous substances to air, water, or land.

9. Summary

Labelling has now developed to a stage where labels conforming with accepted rules and parameters can be devised for the great majority of 'pure' substances.

For three categories of preparations or mixtures, composite labelling systems have been proposed. What is really required, however, is an overall system for preparations which would enable any mixture of 'pure' substances to be labelled from a knowledge of the properties of each substance. Purists may argue that such an approach is impossible, but pragmatically it must be realized that labelling at its best is but a coarse warning system, which merely serves to alert those who may come into contact with the substance of the possible dangers. It can never be a substitute for a proper evaluation of the toxicological hazards of each substance as set out in the relevant chapters of this book.

10. References

1. Dangerous Chemical Substances and Proposals Concerning Their Labelling, 4th Edition, 1978, Supplement 1980, Council of Europe, Strasbourg.

2. Classification, Packaging and Labelling of Dangerous Substances, EEC Directive 67/548/EEC, 79/831/EEC, 81/957/EEC, 75/409/EEC, 76/907/EEC, 79/270/EEC, and 83/467/EEC.

3. Labelling of Dangerous Preparations (Solvents), EEC Directive 73/173/EEC, and 80/781/EEC.

4. Classification, Packaging and Labelling of Paints, Varnishes, Printing Inks, Adhesives and Similar Products, EEC Directive 77/228/EEC, and 83/265/EEC.

5. Classification, Packaging and Labelling of Dangerous Preparations (Pesticide), EEC Directive 78/631/EEC, and 81/187/EEC.

6. Classification, Packaging and Labelling of Dangerous Preparations, EEC Directive, 85/C 211/03.

7. Classification, Packaging and Labelling of Dangerous Substances Regs., 1984, SI 1984, No.1244, HMSO.

8. Information Approved for the Classification, Packaging and Labelling of Dangerous Substances, Health and Safety Commission, HMSO.

Appendix A: Glossary of Terms

Abiotic:-Unconnected with living organisms.

Acceptable daily intake *(ADI)*:-The daily intake of a chemical that is considered without appreciable risk on the basis of all the facts known at the time it is defined.

Acute toxicity:-The adverse effect occurring within a short time of [oral] administration of a single dose of a substance or multiple doses given within 24 hours, *cf*. **Chronic toxicity** and **Subchronic toxicity.**

Aetiology:-The science of the investigation of the cause or origin of disease.

Additive effect:-An effect which is the result of two chemicals acting together and which is the simple sum of the effects of the chemicals acting independently. See **Antagonistic effect** and **Synergistic effect.**

Adenoma:-A tumour, usually benign *(q.v.)*, occurring in glandular tissue.

Adenocarcinoma:-A malignant tumour originating in glandular tissue.

Adjuvant:-In immunology, a substance injected with antigens (usually mixed with them but sometimes given prior to or following the antigen) which non-specifically enhances or modifies the immune response to that antigen.

Adverse effect:-An undesirable or harmful effect to an organism, indicated by some result such as mortality, altered food consumption, altered body and organ weights, altered enzyme levels or visible pathological change.

Ambient Standard:-See **Environmental quality standard.**

Aneuploidy:-Deviation from the normal number of chromosomes, excluding exact multiples of the normal haploid *(q.v.)* complement.

Antagonistic effect:-The effect of a chemical in counteracting the effect of another; for example, the situation where exposure to two chemicals together has less effect than the simple sum of their independent effects; such chemicals are said to show antagonism.

331

Antigen:-A substance that elicits a specific immune response when introduced into the tissues of an animal.

Atrophy:-Wasting of a tissue or an organ.

Autophagosome:-A membrane-bound body within a cell, containing degenerating cell organelles (*q.v.*).

Backfile:-Some large databases have had to be divided into segments, usually according to the entry dates of the references, giving a current file together with one or more backfiles, which may need to be searched separately.

Base pairing:-The linking of the complementary pair of polynucleotide chains of nucleic acids by means of hydrogen bonds between the opposite purine and pyrimidine pairs.

Benign:-Relating to a growth which does not invade surrounding tissue   (Not malignant).

Bioaccumulation:-See **Bioconcentration.**

Biochemical mechanism:-A chemical reaction or series of reactions, usually enzyme catalysed, which produces a given physiological effect in a living organism.

Bioconcentration:-The uptake and retention of xenobiotics by organisms from their immediate environment.

Biological half-life ($t_{1/2}$):-The time taken for the concentration of a xenobiotic in a body fluid or tissue to fall by half by a first-order process.

Biological monitoring:-Analysis of the amounts of potentially toxic substances or their metabolites present in body tissues and fluids as a means of assessing exposure to these substances and aiding timely action to prevent adverse effects. The term is also used to mean assessment of the biological status of populations and communities of organisms at risk in order to protect them and to have an early warning of possible hazards to human health.

Biomagnification:-Bioconcentration of xenobiotics up a food chain *e.g.* from prey to predator.

Biotransformation:-The enzyme-mediated transformation of xeno-biotics *via* **Phase 1** (*q.v.*) and **Phase 2** (*q.v.*) **reactions.**

Cancer:-The disease which results from the development of a malignant tumour and its spread into surrounding tissues. See **Tumour.**

Carcinogenesis:-The production of cancer (see above). Any chemical which can cause cancer is said to be carcinogenic.

Carcinoma:-A malignant epithelial tumour (*q.v.*).

Catalase:-A haem-based enzyme which catalyses the decomposition of hydrogen peroxide into oxygen and water. It is ·found *e.g.* in peroxisomes (*q.v.*) located in the liver.

Ceiling value *(CV):-*The airborne concentration of a potentially toxic substance which should never be exceeded in the breathing zone.

Cell Line:-A defined population of cells which has been maintained in a culture for an extended period and which has usually undergone a spontaneous process of transformation conferring an unlimited culture lifespan on the cells.

Chemosis:-A swelling around the eye - a consequence of oedema of the conjunctiva *(q.v.)*

Cholinesterase and pseudocholinesterase inhibitor:-A substance which inhibits the enzyme cholinesterase and thus prevents transmission of nerve impulses from one nerve cell to another or to a muscle.

Chromosomal aberration:-An abnormality of chromosome number or structure.

Chromosome:-The heredity-bearing gene *(q.v.)* carrier situated within the cell nucleus and composed of *DNA (q.v.)* and protein.

Chronic toxicity:-The effect of a chemical (or test substance) in a mammalian species (usually rodent) following prolonged and repeated exposure for the major part of the lifetime of the species used for the test. Chronic exposure studies over two years are often used to assess the carcinogenic potential of chemicals. (*cf.* **Subchronic toxicity** and **Acute toxicity**).

Clastogens:-Agents which cause chromosome breakage.

Cohort:-A group of individuals, identified by a common characteristic, who are studied over a period of time.

Competent authority (in terms of the Sixth Amendment Directive, see below):-An official government organization or group receiving and evaluating notifications of new chemicals. Such notifications are made under the provisions of national legislation implementing the European Communities Directive 79/831/EEC (The Sixth Amendment of Directive 67/548/EEC which relates to the Classification, Packaging and Labelling of Dangerous Substances).

Conjugate:-A water soluble derivative of a chemical formed by its combination with glucuronic acid, glutathione, sulphate, acetate, glycine *etc.*

Conjunctiva:-The mucous membrane that covers the eyeball and the under-surface of the eyelid.

Control limit:-The limiting airborne concentration of potentially toxic substances which are judged to be 'reasonably practicable' for the whole spectrum of work activities and which must not normally be exceeded.

Corrosion (of tissue):-The process of contact damage due to a destructive agent.

Covalent binding:-The irreversible interaction of xenobiotics or their metabolites with macromolecules such as lipids, proteins, nucleic acids.

Cytochrome P-450;-A haemprotein involved, *e.g.* in the liver, with Phase I reactions *(q.v.)* of xenobiotics

Cytogenetics:-The branch of genetics that correlates the structure and number of chromosomes with heredity and variation.

Cytoplasm:-The ground substance of the cell in which are situated the nucleus, endoplasmic reticulum *(q.v.)*, mitochondria and other organelles *(q.v.)*.

Cytotoxic:-Causing disturbance to cellular structure or function often leading to cell death.

Databank:-A databank contains preselected factual information in summary form, with a sophisticated search system to enable the right information to be located.

Database:-Usually online computer-based bibliographic databases which provide references and in some cases abstracts of papers in the more recent literature.

Deoxyribonucleic acid *(DNA)*:-The constituent of chromosomes which stores the hereditary information of an organism in the form of a sequence of nitrogenous bases. Much of this information relates to the synthesis of proteins.

Dermal irritation:-A localized skin reaction resulting from either a single or multiple exposure to a physical or chemical entity at the same site. It is characterized by the presence of irritation (redness), oedema and may or may not result in cell death.

Distribution:-Dispersal of a xenobiotic and its derivatives throughout an organism or environmental matrix, including tissue binding and localization.

Dose-effect curves:-Demonstrate the relation between dose and the magnitude of a graded effect, either in an individual or in a population. Such curves may have a variety of forms. Within a given dose range they may be linear but more often they are not.

Dose-response curves:-Demonstrate the relation between dose and the proportion of individuals responding with a quantal effect *(q.v.)*. In general, dose-response curves are S-shaped (increasing), and they have upper and lower asymptotes, usually but not always 100 and 0%.

Dose-response relationship :-The systematic relationship between the dose (or effective concentration) of a drug or xenobiotic and the magnitude (or intensity) of the response it elicits.

Ecotoxicology:-Is concerned with the toxic effects of chemical and physical agents in living organisms, especially on populations and communities within defined ecosystems; it includes transfer pathways of these agents and their interaction with the environment.

Embryo:-see **Foetus**.

Emission standard:-A quantitative limit on the emission or discharge of a potentially toxic substance from a particular source. The simplest system is uniform emission standard where the same limit is placed on all emissions of a particular contaminant. See **Limit value**.

Endoplasmic reticulum:-A complex pattern of membranes that permeates the cytoplasmic matrix of cells.

Environmental quality objective *(EQO)*:-The quality to be aimed for in a particular aspect of the environment, for example, 'the quality of water in a river such that coarse fish can maintain healthy populations'. Unlike an environmental quality standard, *(q.v.)* the *EQO* is not usually expressed in quantitative terms.

Environmental quality standard *(EQS)*:-The concentration of a potentially toxic substance which can be allowed in an environmental component, usually air (air quality standard - or water, over a defined period. Synonym: ambient standard. See **Limit value**.

Enzymic (or enzymatic) process:-A chemical reaction or series of reactions catalysed by an enzyme or enzymes. An enzyme is a protein which acts as a highly selective catalyst permitting reactions to take place rapidly in living cells under physiological conditions.

Epidemiology:-The statistical study of categories of persons and the patterns of diseases from which they suffer in order to determine the events or circumstances causing these diseases.

Epigenetic changes:-Changes in an organism brought about by alterations in the action of genes. Epigenetic transformation refers to those processes which cause normal cells to become tumour cells without any mutations having occurred. See **Mutation, Transformation** and **Tumour**.

Erythema:-Redness of the skin due to blood vessel distension.

Eschar:-The slough or dry scab that forms for example on an area of skin that has been burnt.

Eukaryote:-An organism *(e.g.* plant and animal) whose cells contain a membrane-bound nucleus and other membranous organelles *(cf.* **Prokaryote**).

European Inventory of Existing Chemical Substances *(EINECS)*:-This is a list of all chemicals either alone or as components in preparations supplied to a person in a Community Member State at any time between 1st January 1971 and 18th September 1981.

Eutrophication:-A complex series of inter-related changes in the chemical and biological status of a water body most often manifested by a depletion of the oxygen content caused by decay of organic matter resulting from a high level of primary productivity and typically caused by enhanced nutrient input.

First order process:-A chemical process where the rate of reaction is directly proportional to the amount of chemical present.

First-pass effect:-Biotransformation of a xenobiotic before it reaches the systemic circulation. The biotransformation of an intestinally absorbed xenobiotic by the liver is referred to as a hepatic first-pass effect.

Foci:-A small group of cells occurring *e.g.* in the liver, distinguishable, in appearance or histochemically, from the surrounding tissue. They are indicative of an early stage of a lesion which may lead to the formation of neoplastic nodules (*q.v.*) or hepatocellular carcinomas (*q.v.*).

Foetus:-The young of mammals when fully developed in the womb. In human beings, this stage is reached after about 3 months of pregnancy. Prior to this, the developing mammal is at the embryo stage.

Frame-shift mutation:-A change in the structure of DNA such that the transcription of genetic information into RNA is completely altered because the start point for reading has been altered: *i.e.* the reading frame has been altered.

Fugacity:-The tendency for a substance to transfer from one environmental medium to another.

Gene:-A part of the DNA (*q.v.*) molecule which directs the synthesis of a specific polypeptide chain.

Genome:-All the genes (*q.v.*) carried by a cell.

Genetic toxicology:-The study of chemicals which can produce harmful heritable changes in the genetic information carried by living organisms in the form of deoxyribonucleic acid (DNA).

Genotoxic:-Able to cause harmful heritable changes in DNA.

Genotype:-The genetic constitution of an organism *cf.* Phenotype.

Guinea pig maximization test:-One of a number of skin tests for screening possible contact allergens. Considered to be a useful model for predicting likely moderate and strong sensitizers in humans.

Haemosiderin:-An iron-protein molecule; *inter alia*, a source of the iron required for haemoglobin synthesis.

Haploid:-The condition in which the cell contains one set of chromosomes.

Hazard (Toxic):-The set of inherent properties of a chemical substance or mixture which makes it capable of causing adverse effects in man or the environment when a particular degree of exposure occurs. (*cf.* Risk).

Hepatocyte:-Liver cell; more specifically a parenchymal (*q.v.*) cell of the liver.

Hepatotoxic:-harmful to the liver.

Histology:-The study of the anatomy of tissues and their cellular structure.

Histopathology:-The study of microscopic changes in tissues.

Homeostasis:-The tendency in an organism toward maintenance of physiological and psychological stability.

Hydrosphere:-Water above, on or in the Earth's crust, including oceans, seas, lakes, groundwater and atmospheric moisture.

Hypocholesterolaemia:-A lowering of the cholesterol content of the blood.

Hypotriglyceridaemia:-A lowering of the triglyceride content of blood.

Immediately dangerous to life or health concentration *(IDLHC)*:-The maximum exposure concentration from which one could escape within 30 minutes without any escape impairing symptoms or any irreversible health effects. This value should be referred to in respirator selection.

Immune response:-The development of specifically altered reactivity following exposure to an antigen. This may take several forms, *e.g.* antibody production, cell-mediated immunity, immunological tolerance.

Immunotoxic:-Harmful to the immune system.

Initiator:-An agent which starts the process of tumour formation, usually by action on the genetic material.

In Vitro:-Biological processes occurring (experimentally) in isolation from the whole organism.

In Vivo:-Within the living organism; *cf. in vitro*.

Ischaemia:-A deficiency of blood supply to a part of the body relative to its localized requirements.

i.v.:-Abbreviation for intravenous (administration).

Lauric acid hydroxylation, Laurate hydroxylase:-Lauric acid 11-(= ω-1) or 12-(= ω) hydroxylation is used as a marker for the measurement of an isoenzyme of cytochrome P-450 *(q.v.)*. This *microsomal* enzyme is induced in rodent liver by several *peroxisomal* *(q.v.)* proliferators.

LC_n:-The concentration of a toxicant *(q.v.)* lethal to n% of a test population.

LD_n:-The dose of a toxicant *(q.v.)* lethal to n% of a test population.

Lesion:-A pathological disturbance such as an injury, an infection or a tumour.

Limit value *(LV)*:-The limit at or below which Member States of the European Community must set their environmental quality standards and emission standards. These limits are set by Community Directives.

Liver nodule:-A small node, or aggregation of cells within the liver.

Lysimeter:-A laboratory column of selected representative soil or a protected monolith of undisturbed field soil with facilities for sampling and monitoring the movement of water and chemicals.

Macrophages:-A large phagocytic *(q.v.)* cell found in connective tissues, especially in areas of inflammation.

Macroscopic (gross) pathology:-The study of tissue changes which are visible to the naked eye.

Malignancy:-A cancerous growth. (A mass of cells showing both uncontrolled growth and the tendency to invade and destroy surrounding tissues).

Malignant:-See **Tumour**.

Maximum allowable concentration *(MAC)*:-Exposure concentration not to be exceeded under any circumstances.

Median effective concentration *(EC$_{50}$)*:-The concentration of toxicant or intensity of other stimulus which produces some selected response in one half of a test population.

Median effective dose *(ED$_{50}$)*:-The statistically derived single dose of a substance that can be expected to cause a defined nonlethal effect in 50% of a given population of organisms under a defined set of experimental conditions.

Median lethal concentration *(LC$_{50}$)*:-The concentration of a toxicant lethal to one half of a test population.

Median lethal dose *(LD$_{50}$)*:-The statistically derived single dose of a chemical that can be expected to cause death in 50% of a given population of organisms under a defined set of experimental conditions. This figure has often been used to classify and compare toxicity among chemicals but its value for this purpose is doubtful. One commonly used classification of this kind is as follows:-

Category	LD$_{50}$ Orally to Rat mg/kg body weight
Very toxic	<25
Toxic	>25 to 200
Harmful	>200 to 2000

Mesothelioma:-A tumour of the mesothelium of the pleura, pericardium or peritoneum, arising as a result of the presence of asbestos bodies. A locally malignant spreading tumour *(q.v.)* diagnostic of exposure to asbestos.

Metabolic activation:-The biotransformation *(q.v.)* of relatively inert chemicals to biologically reactive metabolites.

Mixed function oxidases:-Oxidizing enzymes which are involved in the metabolism of many foreign compounds giving products of different toxicity from the parent compound.

Multigeneration study:-A toxicity test in which at least three generations of the test organism are exposed to the chemical being assessed. Exposure is usually continuous.

Mutagenesis:-The production of mutations. Any chemical which causes mutations is said to be mutagenic. Some mutagenic chemicals are also carcinogenic. See **Carcinogenesis** and **Transformation**.

Mutation:-Any relatively stable heritable change in the genetic material.

Neoplasm:-Any new and morbid formation of tissue *e.g.* a malignancy.

Nephrotoxic:-Harmful to the kidney.

Non-target organisms:-Those organisms which are not the intended specific targets of a particular use of a pesticide.

No observed effect level (*NOEL*):-The maximum dose or ambient concentration which an organism can tolerate over a specific period of time without showing any adverse effect and above which adverse effects are detectable.

Occupational hygiene:-The applied science concerned with the recognition, evaluation and control of chemicals, physical and biological factors arising in or from the workplace which may affect the health or well-being of those at work or in the community.

Ocular:-Relating to the eye.

Organelle:-A structure with a specialized function which forms part of a cell.

Parakeratosis:-Imperfect formation of horn cells of the epidermis.

Parenchyma(-al):-The specific or functional constituent of a gland or organ.

Partition coefficient:-A constant ratio that occurs when a heterogeneous system of two phases is in equilibrium; the ratio of concentrations (or strictly activities) of the same molecular species in the two phases is constant at constant temperature.

Permissible exposure limit (*PEL*):-See **Threshold limit value**.

Peroxisome:-A cytoplasmic organelle (*q.v.*) present in animal and plant cells, and characterized by its content of catalase and other (peroxidase) oxidative enzymes.

Pesticides:-Those chemicals used in agriculture to control the severity and incidence of pests and diseases which reduce agricultural yields; in addition, they have a number of non-agricultural uses.

Phagocytosis:-The ingestion of micro-organisms, cells, and foreign particles by phagocytes; hence phagocytic **Macrophages** (*q.v.*).

Pharmacodynamics:-The study of the way in which xenobiotics exert their effects on living organisms. Synonym: toxicodynamics.

Pharmacokinetics:-The study of the movement of xenobiotics within an organism. Such a study must consider absorption, distribution, biotransformation,storage and excretion. Synonym: toxicokinetics.

Phase 1 reactions:-Enzymic modification of a xenobiotic by oxidation, reduction, hydrolysis, hydration, dehydrochlorination or other reactions.

Phase 2 reactions:-Enzymic modification of a xenobiotic by conjugation. See **Conjugate**.

Phenotype:-The organism itself as opposed to its genetic constitution, the genotype *(q.v.)*

po:-Abbreviation for oral (administration).

Potentiation:-The effect of a chemical which does not itself have an adverse effect but which enhances the toxicity of another chemical.

Predicted environmental concentration:-The concentration in the environment of a chemical calculated from the available information on certain of its properties, its use and discharge patterns and the associated quantities.

Prokaryote:-An organism, *e.g* bacterium, whose cells contain no membrane-bound nucleus or other membranous organelles *(cf.* eukaryote*)*.

Promoter (carcinogenicity):-An agent which increases tumour production by a chemical when applied after exposure to the chemical.

Pulmonary alveoli:-Minute air-filled sacs in a vertebrate lung, thin walled and surrounded by blood-vessels.

Quantal effect:-An effect that either happens or does not happen, *e.g.* death. Synonym: all-or-none response.

Recommended limit:-A maximum concentration of a potentially toxic substance which is suggested to be safe. Such limits often have no statutory implications and in which case a control *(q.v.)* or statutory guide level should not be exceeded.

Renal:-Associated with the kidneys.

Reproductive toxicology (mammalian):-The study of the effects of chemicals on the adult reproductive and neuroendocrine systems, the embryo, foetus, neonate and prepubertal mammal.

Reticuloendothelial system:-A system of cells that have the ability to take up and retain certain dyes and particles ingested into a living animal. This term has generally been replaced by the term Mononuclear phagocyte system.

Ribonucleic acid *(RNA)*:-A generic term for a group of nucleotide molecules, similar in composition to deoxyribonucleic acid *(DNA, q.v.)*, which perform a number of functions in programming the genetic code in cells. There are several types of *RNA* e.g. messenger *RNA*, ribosomal *RNA*, transfer *RNA*.

Risk (Toxic):-The predicted or actual frequency of occurrence of an adverse effect of a chemical substance or mixture from a given exposure to humans or the environment (*cf*. **Hazard)**

Risk assessment:-The process of decision making applied to problems where there are a variety of possible outcomes and it is uncertain which event will happen.

Risk evaluation:-The determination of the significance of risk to those affected.

Risk management:-Judgements concerning the acceptability of risk.

Safety (Toxicological):-Can be defined as the high probability that injury will not result from use of a substance under specific conditions of quantity and manner of use.

Short term exposure limit *(STEL)*:-The time weighted average *(TWA)* airborne concentration to which workers may be exposed for periods up to 15 minutes, with no more than 4 such excursions per day and at least 60 minutes between them.

Sister chromatid exchange:-A reciprocal exchange of DNA *(q.v.)* between the two DNA molecules of a replicating chromosome. *(q.v.)*

Stochastic:-Obeying the laws of probability.

Structure-activity relationship *(SAR)*:-The correlation between molecular structure and biological activity. It is usually applied to observing the effect that the systematic structural modification of a particular chemical entity has on a defined biological end-point.

Subchronic toxicity:-The adverse effects occurring as a result of the repeated daily [oral] dosing of a chemical to experimental animals for part (not exceeding 10%) of the life span. (Usually 1-3 months) *cf*. Acute toxicity.

Suggested no adverse response level *(SNARL)*:-The maximum dose or concentration which on the basis of current knowledge is likely to be tolerated by an organism without producing any adverse effect.

Synergistic effect:-An effect of two chemicals acting together which is greater than the simple sum of their effects when acting alone. *cf*. **Additive effect, Antagonistic effect, Potentiation.**

Temporary safe reference action level *TSRAL*:-Inhalational exposure level which is safe for a short time but which should be reduced as soon as possible or respiratory protection employed.

Teratogenesis:-Defects in embryonic and foetal development caused by a substance.

Threshold limit value *(TLV)*:-The airborne concentration of a potentially toxic substance to which it is believed healthy working adults may be exposed safely through a 40 hour working week and a full working life. This concentration is measured as a time weighted average concentration. Synonym: permissible exposure limit *(PEL)*.

Time weighted average concentration *(TWA)*:-The concentration of a substance to which a person is exposed in the ambient air, averaged over a period, usually 8 hours. For example, if a person is exposed to 0.1 mg m^{-3} for 6 hours and 0.2 mg m^{-3} for 2 hours, the 8 hour *TWA* will be $(0.1 \times 6 + 0.2 \times 2) \div 8 = 0.125$ mg m^{-3}.

Toxicant:-Any substance which is potentially toxic.

Toxicity:-Any harmful effect of a chemical or a drug on a target organism. See also **Acute, Chronic** and **Subchronic toxicity**.

Toxicodynamics:-See **Pharmacodynamics**.

Toxicokinetics:-See **Pharmacokinetics**.

Toxin:-A toxic organic substance produced by a living organism.

Transformation (neoplastic):-The conversion of normal cells into tumour cells (see below). Frequently this is the result of a genetic change and the same term is used to describe the genetic modification of bacteria for biotechnological purposes. See **Epigenetic changes, Genetic toxicology, Genotoxic** and **Mutation**.

Trans-science:-Questions which can be asked of science and yet which cannot be answered by science. They are thus 'transcended science', although epistemologically they are questions of fact.

Trohoc:-An epidemiological study which starts with the outcome and looks backwards for the causes.

Tumour (neoplasm):-A growth of tissue forming an abnormal mass. Cells of a **benign** tumour will not spread and cause cancer. Cells of a **malignant** tumour can spread through the body and cause cancer.

Tumourigenic:-Causing tumour formation.

Urate oxidase:-A hepatic peroxisomal *(q.v.)* enzyme which catalyses the oxygen mediated conversion of uric acid into allantoin.

Xenobiotic:-A chemical which is not a natural component of the living organism exposed to it. Synonyms: drug, foreign substance or compound, exogenous material.

Xenobiotic metabolism:-The chemical transformation of compounds foreign to an organism by various enzymes present in that organism. See also **Biotransformation** and **Xenobiotic**.

Appendix B: Useful Addresses

Sources of Information include:

<u>COMPUTER BASED INFORMATION SOURCE</u>

Blaise-Link. The British Library, BLAISE, 2 Sheraton Street, London WlV 4BH, England.

BRS. Bibliographic Retrieval Service, 1200 Route 7, Latham NY 12110, USA.

Datacentralen. DC Host Centre, I/S Datacentralen, Retortvej 6-8, DK-2500 Valby, Copenhagen, Denmark.

Data-star. Radio Suisse Ltd., 61 Schwartztorstrasse, CH-3000 Berne, Switzerland.

Data-star UK Office. Plaza Suite,114 Jermyn Street, London SW1Y 6HJ, England.

Derwent SDC Search Service, Stuart House, 47 Crown Street, Reading, Berkshire RG1 2SG, England.

Dialog. Dialog Information Services Inc., 3460 Hillview Avenue, Palo Alto, CA 94304, USA.

Dialog Information Retrieval Service, P.O. Box 8, Abingdon, Oxfordshire OX13 6EG, England.

DIMDI. Deutsches Institut fur Medizinische Dokumentation und Information, Weisshausstrasse 27, Postfach 42 05 80, D-5000 Koln 41, West Germany.

ESA-IRS. Via Galileo Galilei, CP64, I-00044 Frascati (Rome), Italy.

IRS Dialtech, Dept. of Trade & Industry, Room 392, Ashdown House, 123 Victoria Street, London SW1E 6RB, England.

NLM. National Library of Medicine, 8600 Rockville Place, Bethesda MD 20209, USA.

Pergamon InfoLine. Pergamon InfoLine Ltd., 12 Vandy Street, London EC2A 2DE, England.

Pergamon InfoLine Inc., 1340 Old Chain Bridge Road, McLean, VA 22101, USA.

Questel. Telesystèmes, Direction Diffusion de L'information, 83-85 Boulevard Vincent-Auriol, 75013 Paris, France.

343

Questel Inc., 1625 I Street NW, Suite 719, Washington DC 20006, USA.

SDC.   SDC Information Services, 2500 Colorado Avenue, Santa Monica, CA 90406, USA.

STN International.   STN-Columbus, c/o Chemical Abstracts Service, American Chemical Society, 2540 Olentangy River Road, PO Box 2228, Columbus, Ohio 43202, USA.

STN-Karlsruhe,   c/o Fachinformationszentrum, Energie, Physik, Mathematik GmbH, Postfach 2465, D-7500 Karlsruhe 1, West Germany.

MISCELLANEOUS SOURCES

Royal Society of Chemistry, Burlington House, Piccadilly, London W1V 0BN, England. (From whom a list of consultants can also be obtained).

American Conference of Government Industrial Hygienists, 6500 Glenway, Building D-5, Cincinnati, Ohio 45211, USA.

Association of the British Pharmaceutical Industry, 12 Whitehall, London SW1A 2DY, England.

Association for Information Management, Information House, 26/27 Boswell Street, London WC1N 3JZ, England.

British Agrochemical Association, Alembic House, 93 Albert Embankment, London SE1 7TU, England.

British Food Manufacturing Industries Research Association, Randalls Road, Leatherhead, Surrey KT22 7RY

British Industrial Biological Research Association, Woodmansterne Road, Carshalton, Surrey SM5 4DS, England.

British Library, Lending Division, Boston Spa, Wetherby, West Yorkshire LS23 7BQ, England.

British Telecom, PSS National Sales Office, 1 Swan Lane, London EC4B 4TS, England.

British Toxicology Society (BTS), P.O. Box 10, Faversham, Kent ME13 7HL, England.

Chemical Industries Association Ltd., Alembic House, 93 Albert Embankment, London SE1 7TU, England.

Commission of the European Communities, Joint Research Centre, Ispra Establishment, 21020 Ispra (Varese), Italy.

Commonwealth Agricultural Bureaux, Farnham House, Farnham Royal, Slough SL2 3BN, England.

Department of the Environment (DOE), 2 Marsham Street, London SW1P 3EB, England.

Derwent Publications Ltd., 128 Theobalds Road, London WC1X 8RP, England.

ECDIN Group, Commission of the European Communities, Joint Research Centre, I-21020 Ispra (Varesse), Italy.

ECETOC, Avenue Loise 250, B63, Brussels 1050, Belgium.

ECHO Customer Service, 177, Route d'Each, L-1471, Luxembourg.

Environmental Mutagen Information and Environmental Teratology Information Centre, Oak Ridge National Laboratory, Oak Ridge, Tennessee, TN 37830, USA.

Environment Protection Agency (EPA), Washington DC.20460, USA.

Health & Safety Executive (HSE), Baynards House, 1 Chepstow Place, London W2 4TF, England.

Health & Safety Executive Library and Information Services, Broad Lane, Sheffield S3 7HQ, England.

Industry and Environment Office, United Nations Environment Programme, 17 Rue Margueritte, 75017 Paris, France.

Infoterra/PAC, United Nations Environment Programme, PO Box 30552, Nairobi, Kenya.

Institute of Information Scientists (IIS), 44 Museum Street, London WC1A 1LY, England.

International Agency for Research on Cancer, 150 Cours Albert Thomas, F-69372 Lyon, Cedex 2, France.

International Labour Office, CH-1211 Geneva 22, Switzerland.

International Programme on Chemical Safety (IPCS), World Health Organization, CH-1211 Geneva 27, Switzerland.

International Register of Potentially Toxic Chemicals, United Nations, Palais des Nations, CH-1211 Geneva 10, Switzerland.

Inter-Research Council Committee on Pollution Research, c/o NERC Headquarters, Polaris House, North Star Avenue, Swindon, Wiltshire SN2 1EU, England.

Laboratory of the Government Chemist, Department of Trade and Industry, Cornwall House, Waterloo Road, London SE1 8XY, England.

Marine Biological Association of the United Kingdom, The Laboratory, Citadel Hill, Plymouth, Devon PL1 2PB, England.

National Institute for Occupational Safety & Health (NIOSH), Robert A. Taft Laboratories, 4676 Columbia Parkway, Cincinnati, Ohio 45226, USA.

OECD Switchboard, Chemicals Division, Environment Directorate, Organisation for Economic Co-operation and Development, 2 Rue André-Pascal, 75775 Paris, Cedex 16, France.

Paper Industry Research Association, Randalls Road, Leatherhead, Surey KT22 7RV, England.

Pascal, Centre Nationale de la Recherche Scientifique, Centre de Documentation Scientifique et Technique, 26 Rue Bayer, 75971 Paris, Cedex 20, France.

Waste Management Information Bureau, Building 7.12, Harwell Laboratory, Oxfordshire OX11 ORA, England.

Water Research centre, WRc Environment, PO Box 16, Henley Rd., Medmenham, Marlow, Buckinghamshire SL7 2HD, England.

World Health Organization (WHO), 1211 Geneva 27, Switzerland.

Subject Index

Abstract journals, 25, 27
Abstracts, information retrieval, 20
Abstracts on Hygiene and Communicable Diseases, 25
Acceptable Daily Intake, 95, 213, 219
Accidents
  causation, 7
  death rate, 212
  EC regulations, 291
  frequency rate, 6
  gassing, 8
  incident rates, 6
  size of organization, 7
Accumulation, pesticides, 248
Acetic acid, nitrilotri-, legislation, 306
Acetylation, dogs, 94
ACGIH - - - see American Conference of Governmental Industrial Hygienists
Acid dyes, biodegradation, 179
Acid Red 14, carcinogenicity, 173
Acid Red 18, carcinogenicity, 173
Acronyms, 54
Acrylics, allergy, 223
Acrylonitrile, cancer, 161
ACTS - - - Advisory Committee on Toxic Substances
Actuarial tables, 211
Acute toxicity
  assessment, 93
  colorants, 170
  testing, 135, 137
    aquatic organisms, 70
Adhesives, labelling, 327
ADI - - - Acceptable Daily Intake
Adipic acid, di-(2-ethylhexyl) ester, peroxisome proliferation, 240
Administration, route and mode, toxicology, 55
ADRS - - - see British Library Document Supply Centre's

Automatic Document Request Service
Adsorption
  environmental transport, 103
  toxicity tests, aquatic organisms, 67
Advisors, standards, control, 11
Advisory Committee on Asbestos, 267
Advisory Committee on Pesticides, 264
Advisory Committee on Toxic Substances, 73, 264
Aflatoxin B_1, carcinogenicity, 172
Aflatoxins, cancer, 160, 164
Agricultural chemicals, birds, 251
Agricultural Materials Analysis Information Service, 37
Agriculture
  colorants, 178
  occupational diseases, 224
  sewage sludge, 123
AGRIS, 19
Air
  partition between soil and, 105
  partition between water and, 104
Air pollutants
  ecosystem tests, 101
  reviews, 31
Air pollution
  exposure, 197
  sampling, 197
ALARA - - - see As Low as Reasonably Achievable
Alcohol, exposure, 229
Aldicarb, invertebrates, 256
Algae
  colorants, toxicity, 178
  ecotoxicity testing, 100
Alkanes, halo-, threshold limit values, 95
Alkylating mustards, laboratory animals, 95
Allergy
  colorants, 170, 183
  polymers, 223
Allometric law, 94
Aluminium production, cancer, 161
AMAIS - - - see Agricultural Materials Analysis Information Service
American chemical associations, 304

American Conference of Governmental
   Industrial Hygienists, 11, 73,
   172, 265
Ames test, 145
Amines, colorants, exposure, 176
Ammonia
   gassing accidents, statistics, 8
   ionization, toxicity tests, 67
Amosite, risk assessment, 268
Analytical methods
   accuracy, 69
   human exposure, 79
Aneuploidy, testing, 139
Animal testing, 226
   di-(2-ethylhexyl) phthalate, 233
   welfare, 136
Animals, bioaccumulation, 108
Anthrax, legislation, 3
Anti-hypertensive drugs, cadmium
   accumulation, 190
AQUALINE, 19, 37
Aqualine Abstracts, 38
Aquatic organisms, toxicity tests,
   errors, 65-71
Aquatic Sciences and Fisheries
   Information System, 37
Aromatic compounds
   amino, poisoning, 8
   chloro, poisoning, 8
   nitro, poisoning, 8
Arsenic compounds, cancer, 160, 162
Arsenic poisoning
   chemical form, 55
   legislation, 3
   reviews, 31
Arsenobetaine, 55
As Low as Reasonably Achievable, 218
Asbestos
   cancer, 160, 162
   control policies, 271
   exposure, 270
   lung damage, 229
   regulations, 266
   reviews, 31
   risk assessment, 264
   Soviet reviews, 31
Asbestosis, 269
ASFA, 19
ASFIS - - - see Aquatic Sciences
   and Fisheries Information System,
   37
ASLIB, 20
Assassination, 215
Asthma
   occupational, colorants, 170
   toluene di-isocyanate, 226
Auramine, manufacture, cancer, 160
Azathioprine, cancer, 160, 162
Azo compounds, colorants, carcinogenicity, 171
Azo dyes, biodegradation, 179

B_2-microglobulin, cadmium poisoning, 60
Bacteria, wastewater, colorants, 178
Bald eagles, mortality, 252
Basic dyes, biodegradation, 179
Bees, pesticides, 255
Benign tumours, 155
Benzafibrate, peroxisome proliferation,
   240
Benzene
   cancer, 160, 162
   labelling, 315
   poisoning, statistics, 8
   regulations, 264
Benzene, nitro-, toxicokinetics, 90
Benzenesulphonic acid, linear alkyl,
   disposal, 123
Benzenesulphonic acid, 3-amino-5-
   chloro-4-hydroxy-, carcinogenicity,
   173
Benzenesulphonic acid, 2-amino-5-
   chloro-4-methyl-, carcinogenicity,
   173
Benzenesulphonic acid, 3-amino-4-
   hydroxy-, carcinogenicity, 173
Benzidine
   cancer, 160, 162
   colorants, carcinogenicity, 171
   dye powders, exposure, 176
Benzo[a]pyrene, cancer, 160
Beryllium compounds, cancer, 161
Bias, epidemiological studies, 199
Bibliographies, governmental services,
   39
Bioaccumulation, 98, 108
   colorants, 180
Biochemical oxygen demand, 107, 121
Bioconcentration, 108
Biodegradation (see also Degradation),
   103, 107
   colorants, 179, 183
Biomagnification, 98, 108
BIOSIS, 18, 20
Biotechnology, EC regulations, 293
Biotechnology Interservice Regulation
   Committee, 293
Biphenyl, 4-amino-, cancer, 160, 162,
   164
Biphenyl ether, peroxisome proliferation,
   240
Birds
   biochemistry, pesticides, 254
   mortalities, 250
   pesticides, 250-4
      indirect effect, 254
      sub-lethal effects, 254
   physiology, pesticides, 254
Birds of prey
   agricultural chemicals, 251
   insecticides, 251
Blood pressure
   cadmium, 190-1

control, 89
measurement, 199
BOD - - - see Biochemical oxygen demand
Books, 24
Boolean logic, search strategy, 17
Boot and shoe manufacture, cancer, 160
Bottle-feeding, 199
Breast cancer, incidence, variation, 150
Breast-feeding, 199
British Library, 27
British Library Document Supply Centre's Automatic Document Request Service, 35
British Occupational Hygiene Society, 271
British Reports, Translations and Theses, 35-6
Brown pelicans, insecticides, 251
BRTT - - - see British Reports, Translations and Theses
Butane-1,4-diol, dimethanesulphonate cancer, 160, 162
Butterflies, pesticides, 255
Byssinosis, 194

CAB Abstracts, 18
Cadmium
    blood pressure, 190-1
    poisoning, B_2-microglobulin, 60
        protein loss, 89
    reviews, 31
    Soviet reviews, 31
    statistics, 8
Cadmium oxide, absorption, 229
Cadmium sulphide, absorption, 229
Calibration, monitors, human exposure, 79
Campaign for Lead-free Air, 274
Cancer
    colorants, 183
    incidence, variation, 150-1
    risks, variation, 150
    rubber antioxidants, 226
    threshold limit values, 95
Cancerlit, 19-20
Car-driving, risk, 212
Carbon disulphide
    exposure, 196
    heart disease, 191
    ischaemic heart disease, 188
    textile workers, 192
Carbon monoxide, gassing accidents, statistics, 8
Carbon tetrachloride
    regulations, 291
    toxicokinetics, 90
Carcinogenicity, 3, 60
    animal data, 155

assessment, 93
classification, 140
colorants, 171
dose-response curve, 95
humans, 150-64
testing, 134, 139
definition, 303
labelling, 290
regulations, 264
risk assessment, 262
safe doses, 214
threshold, 269
CAS-Online, substructure searching, 20
Case-control studies
    epidemiology, 189
    representative samples, 189
Cell transformation, mutagenicity tests, 157
CHAIS - - - see Consumer Hazards Analytical Information Service
Chelation, cadmium, anti-hypertensive drugs, 190
Chemical Abstracts, 18-20, 25
    registry number, 54
Chemical exposure (see also Exposure)
    nature, 76
    sources, 76
Chemical industry, occupational diseases 224
Chemical nomenclature, 54
    synonyms, 33
Chemical Nomenclature Advisory Service, 37
Chemical oxygen demand, 107, 121
Chemical reactivity, environmental models, 103
Chemical spillage, toxicology information, 53
Chemical structures, 33
Chemicals Information Switchboard, 35
Chemicals
    production, information services, 33
    uses, information services, 33
Chemline, 18
Chlorambucil, cancer, 160, 162
Cholesterol, prospective studies, 190
Chlorine, gassing accidents, statistics, 8
Chlornaphazine - - - see 2-Naphthyl-amine, N,N-bis(2-chloroethyl)-
Chloroform, toxicokinetics, 90
Chromosomes
    damage, hair dyes, 193
    mutagenicity tests, 156
Chromatography, internal standardization 69
Chromium compounds, cancer, 160, 162
Chronic toxicity
    assessment, 93
    colorants, 171
    testing, 134

Chrysotile
   control, 271
   mesothelioma, 269
   risk assessment, 268
C.I. 14720, carcinogenicity, 173
C.I. 16255, carcinogenicity, 173
C.I. 40300, 169
C.I. Acid Blue 1, 182
CICLOPS, 38
Cigarette smoking, 95
   risk, 209
Ciprofibrate, peroxisome
   proliferation, 240
CIS Abstracts, 25
CIS-ILO, 19
Classification
   dangerous preparations, 300
   dangerous substances, 300
   pesticides, 328
   solvents, 326
   toxic substances, 317
   transport, labelling, 322
Clastogenicity, testing, 139
Clean Air Act (1970), 273
CLEAR - - - see Campaign for Lead-
   free Air
Clofibrate, peroxisome proliferation
   239-40
Clorocyclizine, peroxisome
   proliferation, 240
CMEA - - - see Council for Mutual
   Economic Assistance
CNAS - - - see Chemical Nomenclature
   Advisory Service
Coal
   gasification, cancer, 160
   tar, cancer, 160, 162, 164
Coastal marine systems, laboratory
   models, 100
COD - - - see Chemical oxygen demand
Coffee, ischaemic heart disease,
   190
Coke production, cancer, 160
Colorants
   applications, 167
   classification, 167
   identification, 167
   low-dusting powders, 176
   manufacture, exposure, 176
   non-dusting granules, 176
   processing, exposure, 176
   purity, 169
   structure-activity, 166
   toxicity, 166-84
Cohort studies, 190
Colour Index, 167
   Constitution No., 168
   Generic Name, 168
Commission of the European
   Communities, 11
   ecotoxicity testing, 99

Concentration
   exposure, 102
   solid material, toxicology tests, 61
   toxicity tests, aquatic organisms,
   65, 66
Confounding, epidemiological studies,
   200
Congenital malformation, trace elements,
   193
Consumer Hazards Analytical Information
   Service, 37
Control, human exposure data, 75
Control Limits, 95
Control of Pollution Act 1974, 51, 118
Control of Substances Hazardous to
   Health, 230
Conveyance (see also Transport)
   classification, labelling, 322
   labelling, 321
Coosa River Basin, colorants, 182
Copper
   chelation, toxicity tests, 67
   seawater, 52
   Shayer Run, fish, 101
Corrosion, labelling, 290
Corrosive substances, labelling,
   314, 317, 326
Corvids, insecticides, 251
Cosmetics, colorants, 174-5
Cost-benefit analysis, 216
Council for Mutual Economic Assistance,
   44
Council of Europe, 44
   labelling, 314-15
CRIB - - - see Current Research in Britain
Critical assessment, experimental
   data, 70
Crocidolite
   control, 271
   mesothelioma, 269
   risk assessment, 268
Cross-sectional surveys, epidemiology,
   189
Cultural variation, risk assessment,
   259-78
Current Research in Britain, 35, 39
Cyclodienes, plants, surface uptake, 106
Cyclones, risk, 209
Cyclophosphamide, cancer, 160, 162
Cycling, risk, 212

2,4-D, reviews, 32
DAGAS - - - see Dangerous Goods Advisory
   Service
Dangerous Chemical Substances and
   Proposals Concerning Their Labelling,
   316
Dangerous Goods Advisory Service, 37
Dangerous preparations, classification,
   300
Dangerous substances

classification, 300
labelling, 133
Daphnia, ecotoxicity testing, 99
Data
  analysis, toxicological tests, 59
  collection, human exposure data,
    74
  consistency, toxicological tests
    60
  evluation, 44-5
  quality, IRPTC, 41-7
    toxicology, 55
  reliability, toxicology, 56
  retrieval, computer, 13-22
    manual, 24-7
  selection, IRPTC, 41-7
    toxicology, 51-63
  toxicology, interpretation,
    87-96
  validation, IRPTC, 45
    toxicology, 51
Databanks, 16-17, 41
  governmental services, 38
Databases, 17-19
DDE, egg-shell thinning, 254
DDT
  data validation, 52
  mammals, 255
  plants, surface uptake, 106
Deafness, 224
Degradation (see also Biodegradation)
  anaerobic tests, 121
  exposure assessment, notified
    data, 121
Delaney Amendment, US Food and Drug
    Act, 218
Denmark, legislation, 'old'
    chemicals, 305
Department of Health and Social
    Security, 211, 274
Department of Transport, 211
Dermatitis, allergic contact,
    dyed textiles, 177
Detection, information services, 33
Detergents
  disposal, 123
  testing, 307
Dibenzo-p-dioxin, 2,3,7,8-tetra-
    chloro-, toxicology, animal
    tests, 61
Dieldrin
  foxes, 255
  uptake from air, 105
Diet, high-fat, peroxisome
    proliferation, 240
Diethyl sulphate, cancer, 161
Dimethyl sulphate, cancer, 161
Dimethyl sulphoxide, solvent
    toxicological tests, 58
Diquat, reviews, 32

Direct-reading instruments, calibration
  human exposure, 79
Direct Yellow 106, 169
Directory of Environmental Research
  Projects in the European Communities,
    34,39
Disperse dyes, allergic contact
  dermatitis, 177
Dissolved oxygen content, toxicity
  tests, aquatic organisms, 67
Distribution, chemicals, environment,
    103
DNA, damage, mutagenicity tests, 155
Document retrieval, 53
Dogs
  acetylation, 94
  di-(2-ethylhexyl) phthalate toxicity,
    234
Dose
  estimation, toxicological tests, 58
  laboratory animals, 94
Dose-response curve
  carcinogenicity, 155
  di-(2-ethylhexyl) phthalate, 238
  extrapolation, 94
  toxicological tests, 59
Dosing, toxicity testing, 142
Doves, insecticides, 251
Drinking water
  colorants, 182
  fluorides, 188
Drugs, colorants, 174
Ducks, insecticides, 251
Dust, colorants, exposure, 176
Dyes.
  carcinogenicity, 172
  definition, 167
  detection, wastewater, 182
  disposal, 123
  toxicity, fish, 177

Earthworms, aldicarb, 256
EC_{50}, 62
EC_{50}s, 66
ECDIN – – – see Environmental Chemicals
  Data and Information Network
ECE – – – see Economic Commission for
  Europe
Ecological properties, testing, 302
Economic Commission for Europe, 44
Ecosystems
  field tests, 101
  laboratory models, 100
  processes, 248
Ecotoxicity
  colorants, 166-84
  interpretation, 98-109
  microcosm testing, 99
  multispecies testing, 99
  single species tests, 99

testing, 99, 135
Ecotoxicology, notification
  requirements, 131
EEC - - - see European Economic
  Community
Egg-shell thinning, pesticides,
  254
EINECS - - - see European
  Inventory of Existing Chemical
  Substances
Elimination, colorants, 179
Emulsions, toxicity tests,
  aquatic organisms, 67
ENREP - - - see Directory of
  Environmental Research Projects
  in the European Communities
Enviroline, 19
Environment
  carcinogenicity, 150
  chemical distribution, 103
  chemicals, transport, 98
  colorants, 177, 180
  dangerous substances, 303
  effects data, 125
  European Community Regulations,
    285-95
  hazards, labelling, 327
  information services, 29, 34
Environmental Chemicals Data and
  Information Network, 16, 30,
  33, 38
Environmental Hazard Assessment,
  117-32
Environmental Health Criteria, 31
Environmental Mutagen Information
  Center, 152
Environmental Protection Agency,
  265, 305
  carcinogens, 261
  human exposure model, 231
Environmental Teratology
  Information Center, 152
Enzymes, tissue content, aquatic
  organisms, 69
EPA - - - see United States
  Environmental Protection Agency
Epidemiological data, 188-201
Epidemiological studies
  design, 188-92
    errors, 192-201
  prior hypothesis, 192-3
  risk, 211, 212
  size, 197-9
Epidemiology
  carcinogenicity, 154
  human exposure data, 75
Epoxy resins, allergy, 223
Errors, toxicity tests, aquatic
  organisms, 65-71
Ethers, bis(chloromethyl),
  cancer, 160, 162

Ethers, chloromethyl methyl, cancer,
  160, 162
Ethylene, dichloro-, gassing accidents,
  statistics, 8
Ethylene, tetrachloro-, regulations, 291
Ethylene, trichloro-
  engineering works, 4
  gassing accidents, statistics, 8
  laboratory animals, 96
  regulations, 291
  reviews, 32
Ethylene óxide, mutagenicity, 157
Europe
  existing substances lists, 297
  hazard assessment, legislation, 299
European Chemical Federation, 304
European Common Command Language, 15
European Communities Directive on the
  Classification, Packaging and Labelling
  of Dangerous Substances, 117
European Community
  chemicals control, 286
  information services, 33-5
European Community Chemical Notification
  Scheme, 117-32, 133-49
European Community Regulations
  chemical products, 291
  'new' chemicals, 286-95
  'old' chemicals, 285-95
European Economic Community, 44
European Inland Fisheries Advisory
  Commission, 71
European Inventory of Existing Chemical
  Substances, 118, 287
Evaluation, data, carcinogenicity, 154-8
Evaluation, databanks, 16
Event-tree analysis, 211, 213
Excerpta Medica, 18, 20
Experimental data, critical assessment,
  70
Experimental Lakes Area, 101
Existing substances, lists, 297-8
Explosions, notification regulations, 6
Explosive substances, labelling, 318
Exposure
  asbestos, 270
  assessment, notified data, 120
  colorants, 166, 174
    dust, 176
  concentration, 102
  data (see also Chemical exposure)
    evaluation, 74
    humans, 72-82
    interpretation, 81
    measurement, 77
    validation, 81
  dyed textile, 177
  effects versus, 125
  estimation, 120
  measurement, 90
  period, toxicity tests, aquatic

organisms, 65
realistic estimates, 196
risk, 228
route, toxicological tests, 59
testing, 298
time, toxicity tests, aquatic
    organisms, 69
Eye irritancy, testing, 135, 138

Famphur, wildlife, 253
FAO - - - see Food and Agriculture
    Organization
Fault-tree analysis, 211, 213
Fenofibrate, peroxisome proliferation,
    239, 240
Fertility, chemicals testing, 134
Finger paints, colorants, exposure,
    177
Fires, notification regulations, 6
Fish
    bioaccumulation, testing, 108
    colorants, accumulation, 180
        toxicity, 177
    ecosystem field tests, 101
    ecotoxicity testing, 100
Fishing, occupational diseases, 224
Flammable substances, labelling, 317
Flax mills
    dust, bias, 200
    exposure, 197
Fluorocarbons, legislation, 306
Fluorescent whitening agents, 182
Fluorides, domestic water supply,
    188
Food
    additives, risk, 217
    colorants, 174-5
Food and Agriculture Organization, 44
Forest birds, insecticides, 251
Forestry, occupational diseases, 224
Formaldehyde
    gassing accidents, statistics, 8
    laboratory animals, 95
    Soviet reviews, 31
Foundries, risk assessment, 227
Foxes, pesticides, 247, 255
France, legislation, 'old' chemicals,
    305
Fungi, pesticides, 256
Fungicides, plants, 256
Furniture manufacture, cancer, 160

Game birds
    agricultural chemicals, 251
    insecticides, 251
Game conservancy, cereals and game
    birds project, 254
Gassing, accidents, 8
Gastrointestinal cancers, asbestos,
    269
Geese, insecticides, 251

Gemfibrozil, peroxisome proliferation, 240
Genetic activity, mutagenicity tests, 157
Genetic toxicity, assessment, 93
Genotoxicity
    colorants, 166-84
    testing, 134, 139, 145
Germany, legislation, 'old' chemicals, 305
GESAMP - - - see Inter-Agency Group of
    Experts on the Scientific Aspects of
    Marine Pollution
GIFAP - - - see International Group of
    National Associations of Manufacturers
    of Agrochemical Products
Glutamic acid imide, $\underline{N}$-phthalyl- - - - $\underline{see}$
    Thalidomide
Glyphosate, 256
Good Laboratory Practice, 136
    EC environmental legislation, 289
    Regulations, 57
Goose pimples, 60
Grebes, insecticides, 251
Grey literature, 36, 55
    document retrieval, 53
    information services, 34
Groundwater, radionucleides, 101
Gulls, insecticides, 251

Haematite, mining, cancer, 160
Hair dyes, chromosomal damage, 193
Harwell Environmental Safety Group, 37
Hat manufacture, mercury poisoning, 3
Hazardous Substances Data Bank, 16
Hazards
    aquatic organisms, 65
    assessment, 117, 233
    assessment, actions arising from, 126
        legislation, 299
        notified data, 120
        OECD, 301
    definition, 43, 91, 208, 303
    EC environmental legislation, 287
    ecotoxicological data, critical
        assessment, 70
    environment, labelling, 327
    identification, 225
    regulations, Europe, 296-308
    toxicology information, 53
Health and Safety at Work Act 1974, 5,
    51, 117, 218
Health and Safety Commission, 264, 271
    ecotoxicity testing, 99
Health and Safety Executive, 45, 211, 267
    sampling, 78
Health, European Community Chemicals
    Notification Scheme, 133-49
Health, Safety and Environment Committee
    (HSEC), 5
    role, 9
Health statistics, 3
Heart disease, carbon disulphide, 191
Heavy metals, solubility, toxicity tests,
    66

Henry's Law, 104
Hepatocarcinogenicity, di-
  (2-ethylhexyl) phthalate, 241
Heptachlor, foxes, 255
Herbicides
  non-target organisms, 247
  wildlife, 252
Herons, insecticides, 251
Hexanoic acid, 2-ethyl-,
  peroxisome proliferation, 240
1-Hexanol, 2-ethyl-, peroxisome
  proliferation, 240
2-Hexanone, toxicity, 89
History, chemical health hazards,
  3
Homophones, 54
Hospital patients, sampling, 194
HSE - - - see Health and Safety
  Executive
HSE-Line, 19
Hubbard Brook Experimental Forest,
  101
Humans, exposure data, 72-82
Hydrocarbons
  chlorinated, gassing accidents,
    statistics, 8
  polychlorinated aliphatic,
    peroxisome proliferation, 240
Hydrogen chloride, gassing
  accidents, statistics, 8
Hydrogen sulphide, gassing accidents,
  statistics, 8
Hydrolysis, toxicity tests,
  aquatic organisms, 67
Hypolipidaemic drugs, peroxisome
  proliferation, 239-40

IARC - - - see International
  Agency for Research on Cancer
Ill health, occupational disease,
  75
ILO - - - see International
  Labour Office
Industrial diseases, statistics, 8
Industry, risk, 223
Industry and Environment, 32
Industry and Environment File, 32
Industry and Environment Office, 32
Infertility, threshold limit
  values, 95
Information Approved for the
  Classification, Packaging and
  Labelling of Dangerous Substances,
  326
Information retrieval - - - see
  Data retrieval
Information services
  computer-based, 15
  governmental, 29-39
INFOTERRA - - - see International
  Environmental Information System

Inhalation, toxicity testing, 142
Insecticides
  birds, 251
  wildlife, 252
INSTAB, 38
Inter-Agency Group of Experts on the
  Scientific Aspects of Marine Pollution,
  44
Interlaboratory testing, human
  exposure, 81
Internal standardization, analytical
  methods, 69
International Agency for Research
  on Cancer, 31, 44, 151, 171
  monographs, 151
  supplements, 158
International Committee on Radiological
  Protection, 218
International Directory, 32
International Environmental Information
  System, 31, 32, 38
International Group of National
  Associations of Manufacturers of
  Agrochemical Products, 30
International Labour Office, 44
International Labour Organization,
  labelling, 314
International Programme on Chemical
  Safety, 31
  data evaluations, 44
International Register of Potentially
  Toxic Chemicals, 29, 38, 41-7
  ambitions and limitations, 42
  data quality, 41-7
  data selection, 41-7
  information files, 42
  reliability of data, 44
  sources of information, 44-5
  toxicity data, 45
Interpretation
  ecotoxicity, 98-109
  toxicology data, 87-96
Inter-Research Council Committee on
  Pollution Research, 36
Intervention studies, epidemiology, 192
Invertebrates, pesticides, 255
Ionization, ammonia, toxicity tests, 67
IPCS - - - see International Programme
  on Chemical Saftey
IQ, lead, 274
IRCOPR - - - see Inter-Research Council
  on Pollution Research
Iron and steel founding, cancer, 161
IRPTC - - - see International Register
  of Potentially Toxic Chemicals
IRPTC Bulletin, 30-1
IRPTC Legal File, 30
Irritant substances, labelling, 317, 326
Irritation, labelling, 289
ISI, 19
Ischaemic heart disease

beverages, 190
carbon disulphide, 188
Isopropyl alcohol, manufacture,
  cancer, 160

Japan
  existing substances lists, 297
  industrial chemicals, authorization
    procedure, 299
  testing, new substances, 298

Knee jerk reflex, organophosphates,
  89

Labelling, 314-29
  carcinogens, 174
  dangerous substances, 133, 300
  EC environmental legislation, 290
  hazardous substances, 4
  regulations, 319
  toxic substances, 136
  transport, 321, 322
  user systems, 315
Laboratory animals, 92
Laboratory of the Government
  Chemist, 36
Lakes, laboratory models, 100
LAS - - - see 'linear alkyl'
  under Benzenesulphonic acid
Lawther Report, 274
LC$_{50}$, 62
LC$_{50}$s, 66
LD$_{50}$, 62
Lead
  air pollution, sampling, 197
  epidemiological studies,
    confounding, 200
  exposure, 196-7
  paint, 185
  petrol, 198, 272
  regulations, 266
  poisoning, 190
    intervention studies, 192
    legislation, 3
    red blood cells, 89
    statistics, 8
  risk assessment, 264
Leather, colorants, 175
Legislation, 10
  chemical health hazards, 3
  Europe, 296-308
  hazard assessment, 299
  history, 5
  human exposure data, 75
  information services, 33
  'old' and 'new' sybstances, 297
  risk assessment, 263
Legumes, exotoxicity testing,
  100
Lethality, testing, 137
Libraries, 27

Lightning, risk, 209
Lindane, mammals, 255
Lipids, solubility, bioaccumulation
  testing, 108
Lysimeters, 104
Liver, nitro aromatic compounds, 89
Lung cancer
  incidence, variation, 150
  smoking, risk perception, 225
Lung damage, exposure risk, 229
Lymphomas, therapy, cancer, 160-2

MAC - - - see Maximum Allowable
  Concentrations
Magenta, manufacture, cancer, 161
Magpies, insecticides, 253
Malignant tumours, 155
Mammals
  colorants, toxicity, 166-84
  pesticides, 255
  poison, wildlife, 252
Manufacturer's data sheets, 26
Marfans syndrome, 194
Marine Biological Association, 37,
  39
Marine Pollution Research Titles, 37
Marketing
  EC environmental legislation, 287
  new substances, legislation, 300
Marking, toxic substances, 321
Martindale Online, 16
Matches manufacture, phossy jaw, 3
Matching of controls, 194
Mathematical methods, risk assessment
  220
Maximum Allowable Concentrations,
  USSR, 73
MBA - - - see Marine Biological Association
Medical procedures, death rate, 212
Medical Research Directory, 36
Medline, 18, 20
Melphalan, cancer, 160, 162, 164
Mercury
  poisoning, hat manufacture, 3
    reviews, 31
    Soviet reviews, 31
    statistics, 8
Mesothelioma, asbestos, 268-9
Methoxalan, cancer, 160, 162
MICIS - - - see Microbial Culture
  Information Service
Mice, di-(2-ethylhexyl) phthalate
  toxicity, 234
Microbial Culture Information Service,
  37
Microcosms, ecotoxicity testing, 101
Micro-organisms
  biodegradation, 107
  soil, pesticides, 256
Microwaves, reviews, 32
Migrants, cancer, variation, 150

Migration, soils, testing, 103
Mineral oils, cancer, 160, 162
Mining
    occupational diseases, 224
    physiology, 230
    risk assessment, 227
Ministry of Agriculture, Fisheries
    and Food, 264
Mites, ecotoxicity testing, 100
Mixtures, labelling, 325
Models, environment, chemical
    distribution, 103
Molluscicides, wildlife, 252
Monetary value of life, 216
Monitoring
    biological, human exposure data,
    78
    continuous systems, calibration,
    79
    human exposure data, 77
Monkeys, di-(2-ethylhexyl) phthalate,
    toxicity, 234
MOPP, cancer, 160, 162
Morbidity, 60
Mortality, 60
    rates, 210
    statistics, 211
    toxicity tests, aquatic
    organisms, 68
Mortality Statistics Cause, 228
Motor cycling, risk, 212
Mustard gas, cancer, 160, 162
Mutagenicity, 60
    classification, 140
    di-(2-ethylhexyl) phthalate, 235
    2-ethylhexyl phthalate, 236
    short-term testing, 174
    testing, 135, 139, 156, 226
Mutagens
    definition, 303
    labelling, 290
    testing, 307
Mutations, mutagenicity tests, 156
Myleran - - - see 'dimethane-
    sulphonate' under Butane-
    1,4-diol

Nafenopin, peroxisome proliferation,
    240
Naphthalene, 2-amino-, carcino-
    genicity, 173
Naphthalene-1,3-disulphonic acid,
    8-amino-7-hydroxy-, carcino-
    genicity, 173
Naphthalene-1-sulphonic acid, 2-amino-
    carcinogenicity, 173
    toxicity, 172-3
Naphthalene-1-sulphonic acid, 4-amino-,
    carcinogenicity, 173
Naphthalene-1-sulphonic acid, 3-amino-
    4-hydroxy-, carcinogenicity, 173

2-Naphthylamine
    cancer, 160, 162, 164
    toxicokinetics, 90
2-Naphthylamine, N,N-bis(2-chloro-
    ethyl)-, cancer, 160, 162
Nasopharynx cancer, incidence,
    variation, 150
National Academy of Science, 262
National Cancer Institute, 171, 235
National Institute for Occupational
    Safety and Health, 265
    sampling, 78
National Institute of Health, 265
National Library of Medicine, 18
National Radiation Protection Board,
    220
National Research Council of Canada,
    45
Negative results, carcinogenicity, 154
Nematicides, wildlife, 252
Nephroblastoma, incidence, variation,
    150
Nerve conduction velocity, 89
Netherlands, legislation, 'old'
    chemicals, 305
Neutron activation, risk assessment,
    231
Nickel compounds, cancer, 161
Nickel refining, cancer, 160
NIOSH - - - see United States National
    Institute of Occupational Safety and
    Health
Nitrates
    Soviet reviews, 31
    vasodilation, 88
Nitro compounds, aromatic liver, 89
Nitrogen mustard, cancer, 161, 162
Nitrogen oxides, reviews, 31
Nitrous fumes, gassing accidents,
    statistics, 8
NOEL - - - see No observable effect
    level
Noise, reviews, 32
No observable effect level, 95, 213
Notification
    chemicals marketing, 118
        data submitted, 119
        degradation, 129
        ecotoxicology, 129
        identity, 127
        Level 1 requirements, 130
        physico-chemical properties, 128
        toxicology, 128
    EC environmental legislation, 287
    new substances, Europe, 300
        marketing, 299
Notification of New Substances
    Regulations, 133
NRCC - - - see National Research
    Council of Canada
Nuclear power, risk perception, 225

Occupational diseases, 224
  investigations, 75
Occupational exposure limits, 73
Occupational health, information
  retrieval, 15
Occupational health and safety,
  information services, 33
Occupational hygiene, 72-82
Occupational Safety and Health
  Administration, carcinogens, 261
OECD - - - see Organisation for
  Economic Co-operation and
  Development
Oestrogens, cancer, 160, 162
Office of Management and Budget, 266
Office of Population, Census and
  Statistics, 211
Office of Population Censuses and
  Surveys, 191
Oils, toxicity tests, aquatic
  organisms, 67
Oilseed rape, pesticides, 255
Ontario Royal Commission on
  Matters of Health and Safety
  Arising from the Use of Asbestos,
  270
OPCS - - - see Office of Population
  Censuses and Surveys
Oral contraceptives, cancer, 161
Organization for Economic Co-
  operation and Development, 44
  chemical group, 30
  Chemicals Information Switchboard
    35
  Chemicals Programme, 287
  ecotoxicity testing, 99
  Environment Committee, 287
  hazard assessment, 301
  Switchboard, 38
Organophosphates
  knee jerk reflex, 89
  toxicity, 89
Organophosphorus compounds,
  reviews, 32
Osteoacryolysis syndrome, vinyl
  chloride, 89
Otters, pesticides, 255
Oxidising substances, labelling, 318
Oxymethalone, cancer, 161

Packaging, regulations, 319
Packaging and Labelling Regulations
  1978, 319
Packaging materials, colorants, 177
Paints
  labelling, 327
  lead, 188
Paints etc Directive, 301
Palacability, toxicity testing, 142
Paper, colorants, 175
Paraquat, 256

reviews, 32
Partridges, pesticides, 254
Pasture grasses, ecotoxicity testing
  100
Patents, 26
PCA - - - see Pollution Control
  Authority
PEC - - - see Predicted environmental
  concentration
Peregrine falcons, pesticides, 247
Peroxisomes, proliferation, rats,
  239, 240
Persistence times, pesticides, 248
Pesticides
  classification, 328
  ecosystem tests, 101
  labelling, 325, 327
  non-target organisms, 247-57
  organisms affected by 249
  reviews, 31
  solubility, toxicity tests, 66
  Soviet reviews, 31
Petrol, lead, 198, 272
Pharmaceuticals, books, 25
Pharmacokinetics, 90
Phenacetin, cancer, 160-2
Phosgene, gassing, accidents,
  statistics, 8
Phosphorus
  phossy jaw, 223
  poisoning, legislation, 3
    statistics, 8
  yellow, phossy jaw, 3
Phossy jaw
  matches manufacture, 3
  phosphorus, 223
Photodegradation, tests, 121
Photolysis, toxicity tests, aquatic
  organisms, 67
Phthalic acid
  di-(2-ethylhexyl) ester, mechanism
    of action, 240
    metabolism, 236
    rats, 237
    mode of action, 238
    peroxisome proliferation, 240
    toxicity, 233-43
    lifetime studies, 235
  di-(n-hexyl) ester, peroxisome
    proliferation, 239
  di-(n-octyl) ester, peroxisome
    proliferation, 239
  2-ethylhexyl ester, metabolism, 236
    toxicity, 236
Pharmacokinetics, di-(2-ethylhexyl)
  phthalate, 236
Physico-chemical properties
  exposure assessment, notified data,
    120
  information services, 33
Pigeons, insecticides, 251

Pigments, organic, definition, 167
Plasticizers, peroxisome proliferation, 240
Plastics, colorants, 175
Plant integrity, human exposure data, 75
Plants
  bioaccumulation, 108
  chemical transport, 105-6
  pesticides, 256
Pneumoconiosis, slate workers, 197
Pollution Abstracts, 19
Pollution Control Authority, 126
Pollution Research, 36
Polyesters, allergy, 223
Polychlorinated biphenyls, regulations, 291
Polychlorinated triphenyls, regulations, 291
Polyvinyl chloride, di-(2-ethylhexyl) phthalate, 234
Ponds, laboratory models, 100
Pottery industry, respiratory diseases 3
Predators, ecotoxicity testing, 100
Predicted environmental concentration, 125
Prednisolone, cancer, 162
Pre-market Minimum Set of Data, 301
Preparations, labelling, 325
Printing inks, labelling, 327
Probability, 209
Procarbazine, cancer, 161-2
Production, exposure assessment, notified data, 120
Professional bodies, legislation, 4
Prospective studies, epidemiology, 190
Protein loss, cadmium poisoning, 89
Pulmonary sensitization, colorants 170, 176
Pumps, calibration, human exposure, 79
Purity, toxicological tests, 58
Pyrethroids
  peroxisome proliferation, 240
  toxicokinetics, 90

Quantitative Structure Activity Relationship, 172
Quarrying, occupational diseases 224

Radioactive chemicals, exposure data, measurement, 78
Radiofrequency, reviews, 32
Radionucleides, groundwater, 101
Raptors
  agricultural chemicals, 251
  egg-shell thinning, 254
Rats

di-(2-ethylhexyl) phthalate,
  dose-response studies, 238
  metabolism, 237
  toxicity, 234
ecotoxicity testing, 99
Random sampling, epidemiology, 194
Rayon industry, carbon disulphide, 196
Reactive dyes, pulmonary sensitization, 177
Record keeping, human exposure, 81
Red blood cells, lead poisoning, 89
Referees, publishing, toxicity tests, 69
Referral Systems, governmental services, 38
Register of Research and Surveys, 34
Registry of Toxic Effects of Chemical Substances, 16, 30
Regulations
  Europe, 296-308
  labelling, 319
  'old' chemicals, 304
  risk assessment, 263
  toxicology information, 53
Regulatory science, 261
Relevant outcome measures, epidemiology, 195
Repeated Dose Toxicity, testing, 139
Respiratory diseases, pottery industry, 3
Research in British Universities, Polytechnics and Colleges, 35
Research Registers, governmental services, 39
Respiratory cancer, asbestos, 269
Response, toxicity tests, aquatic organisms, 65, 68
Representative population sample, epidemiology, 193
Reproduction, assessment, 93
Retrospective follow-up studies, epidemiology, 191
Rhine, sediments, colorants, 182
Risk
  acceptable, 215
  accepted versus acceptable, 216
  assessment, 207-221, 227
    cultural variation, 259-78
    definition, 262
    divergencies, 260
    harmonized concepts, 304
    legislation, 263
    regulations, 263
  carcinogenicity, 150
  colorants, 167
  definition, 43, 91, 208
  estimation, 210, 231, 262
  evaluation, 214, 262
  exposure, 228

management, 217
definition, 262
practice, 219
monitoring, 230
perception, 215, 225, 262
regulations, Europe, 296–308
voluntary versus involuntary, 216
work force, 223–32
Risk-benefit analysis, 209, 216,
    296, 306
Risk Phrases, 147–8
River Pollution Survey, 122
Rivers
    colorants, 181
    sediments, colorants, 182
    waste concentration, 121
Rodenticides, wildlife, 252
Rooks, agricultural chemicals, 251
Root concentration factor, 106
Royal Society of Chemistry,
    role, 4
Rubber, antioxidants, cancer, 226
Rubber industry, cancer, 160
Rules, information services, 33

Saccharin, 218
Safety
    absolute, 214, 218
    information retrieval, 15
    standards, 219
    training, 9
Safety factor, 213
Safety Phrases, 148–9
Salinity, toxicity tests, aquatic
    organisms, 67
Sampling
    air pollution, 197
    cross-sectional surveys, 190
    human exposure, 78–9
Sampling devices, 197
    human exposure data, 78
Scientific Reviews of the Soviet
    Literature on Toxicity and Hazards
    of Chemicals, 31
Search strategy, 17
Seawater, copper, 52
Sebacic acid, di-(2-ethylhexyl)
    ester, peroxisome proliferation,
    240
Sediments, partition between water
    and, 103
Selenastrum capricornutum, colorants,
    toxicity, 178
Sensitivity, epidemiological studies,
    198
Sensitivity analysis, risk assessment,
    220
Sensitization, assessment, 93
Seveso Directive, 291
Sewage sludge
    colorants, 178–9

disposal, 122
    agriculture, 123
Sewage treatment plants, 121
    waste disposal, 122
Shale oils, cancer, 160, 163
SIGLE – – – see System for Information
    on Grey Literature in Europe
Silent Spring, 247
Shayer Run, fish, copper, 101
Skin contamination, estimation, 78
Skin corrosion, testing, 138
Skin irritancy, testing, 135, 138
Skin sensitization, testing, 135, 139
Slate workers, pneumoconiosis, 197
Smoking
    exposure, 196
    lung cancer, 229
        risk perception, 225
Snail kites, agricultural chemicals,
    251
Soil
    colorants, 182
    micro-organisms, pesticides, 256
    partition, between air and, 105
        between water and, 103
Soil cores, ecotoxicity testing, 100
Solubility
    colorants, 180
    toxicity tests, aquatic organisms, 66
Solubilizers, toxicity tests, aquatic
    organisms, 66
Solvents
    labelling, 326
    toxicity tests, aquatic organisms, 66
    toxicological tests, 58
Solvents Directive, 300
Song birds, insecticides, 251
Soot, cancer, 160, 163, 164
Sorghum, colorants, 178
Soya, colorants, 178
Sponsorship, toxicological research, 56
Spray drift, herbicides, 256
Standards
    legislation, 11
    safety, 219
Standing Technical Committee on
    Synthetic Detergents, 123
Statistics
    health and safety, 5–9
    toxicological tests, 59
Stilboestrol, diethyl-, cancer, 160,
    162, 164
Stomach cancer, incidence, variation, 151
Streams, laboratory models, 100
Stress, 230
Structure, exposure hazards, 120
Structure-activity relationships, 21
    colorants toxicity, 172
    peroxisome proliferation, 239
Strychnine, wildlife, 252
Sub-acute toxicity

assessment, 93
colorants, 170
data, assessment and interpretation,
   143
di-(2-ethylhexyl) phthalate, 234
no-effect level, 144
testing, 135
Sub-chronic toxicity, testing, 134
Sub-lethal effects, toxicity effect
   aquatic organisms, 68, 70
Substructure searching, 20
Sucrose octa-acetate, 177
Sulphur oxides
   air pollution, sampling, 197
   gassing accidents, statistics, 8
   reviews, 31
Sunflowers, colorants, 178
Sweating, 60
Symbols
   allocation, 318
   labelling, 315
   transport, 322
System for Information on Grey
   Literature in Europe, 34, 39

TCDD - - - see Dibenzo-p-dioxin,
   2,3,7,8-tetrachloro-
Tea, ischaemic heart disease, 190
Temperature, toxicity tests,
   aquatic organisms, 67
Teratogenicity
   classification, 141
   testing, 139
   threshold limit values, 95
Teratogens
   definition, 303
   labelling, 290
Teratology, testing, 134
Test animals, toxicology, 58
Test concentration, toxicology, 61
Test duration, toxicology, 59
Test materials, toxicology, 58
Testing (see also Animal Testing)
   chemicals marketing, 134
   EC environmental legislation, 287
   ecotoxicity, 98-9
   new chemicals, Europe, 298
   new substances, flexibility, 302
      legislation, 300
   toxicity, dosage, 142
Textile workers
   carbon disulphide, 192
   ischaemic heart disease, 188
Textiles
   colorants, 175
   home-dying, 177
   occupational diseases, 224
Thalidomide, toxicity, 52
Threshold Limit Values, 73, 95, 219
Tibric acid, peroxisome proliferation,
   240

TLV - - - see Threshold Limit Values
Tobacco, cancer, 160, 162
Tobias acid - - - see Naphthalene-
   1-sulphonic acid, 2-amino-
Toluene, reviews, 32
Toluene di-isocyanate, asthma, 226
o-Toluidine, cancer, 161
Topical toxicity, assessment, 93
Toxic substances, classification, 317
Toxic Substances Control Act, 172,
   272, 296
Toxicity
   assessment, 93
   classification, 130
   clinical, 88-9
      classification, 88
   colorants, 166-84
      fish, 177
   data, IRPTC, 45
   definition, 43, 87
   di-(2-ethylhexyl) phthalate, 233-43
   information services, 33
   physical, 89-90
      functional lesions, 89
      structural lesions, 89-90
   tests, aquatic organisms,
      concentration, 66
         errors, 65-71
Toxicokinetics, 90
Toxicological data, biochemistry, 90
Toxicology
   books, 24
   data selection, 51-63
   human exposure data, 75
   information retrieval, 15
   notification requirements, 130
   physical and chemical form, 54
   procedures, 92
   publications, scientific merit, 57
   responsibilities, 96
   testing, 302
Toxicology Abstracts, 25
Toxicology Data Bank, 16
Toxicometric Parameters of Industrial
   Toxic Chemicals under Single Exposure
   30
Toxline, 18-20
Toys, colorants, 177
Trace elements, congenital malformation,
   193
Train travel, risk, 215
Training
   information retrieval, 21
   safety, 9
Translations, 27, 54
Transpiration stream concentration factory,
   106
Transport
   chemicals, environment, 98
   labelling, 321, 322
   risk, 212
Trans-science, 261

Trees, ecosystem studies, 101
Treosulphan, cancer, 160, 163
Tris-(2,3-dibromopropyl) phosphate,
  regulations, 291
Trophic relations, ecotoxicity
  testing, 102

Uncontrolled release, notification
  regulations, 6
United Kingdom
  Health & Safety Executive, 73
  risk assessment, 260-78
United Nations
  Environment Programme, Industry
    and Environment File, 38
    Systems, 29-33
  labelling, 314
United States of America
  Environmental Protection Agency
    267
    ecotoxicity testing, 100
  existing substances lists, 297
  Food and Drug Act, Delaney
    Amendment, 218
  National Cancer Institute, 151-2
  National Institute of Occupational
    Safety and Health, 30, 44, 267
  Public Health Service Publication
    No. 149, 152
  risk assessment, 260-78
  testing, new substances, 298
Units, toxicology, 61
Uses
  colorants, 174
  exposure assessment, 120
USSR
  Maximum Allowable Concentrations,
    73
  State Committee for Science and
    Technology, 30

Vaccinations
  risk, 215
  whooping cough, risk, 217

Varnishes
  colorants, 175
  labelling, 327
Vasodilation, nitrates, 88
Veterinary Bulletin, 25
Vincristine, cancer, 162
Vinyl chloride
  cancer, 160, 163-4
  osteoacrylysis syndrome, 89
Volatile substances, toxicity
  tests, aquatic organisms, 67
Volatility, determination, 105
Volatilization, from soil, 105

Warblers, insecticides, 251
Waste Management Information Bureau,
  37, 39
Wastewater
  bacteria, colorants, 178
  colorants, 180, 183
  dyes, detection, 182
Water
  partition, between air and, 104
    between sediment and, 103
    between soil and, 103
  sampling, 197
Water Quality Criteria for European
  Freshwater Fish, 71
Water quality standards, toxicity
  tests, 65
Water Research Centre, 37, 39
Well water, colorants, 182
Whooping cough, vaccination, risk,
  217
Wildlife, mortality, 253
Wiswesser Line Notation, substructure
  searching, 21
WMIB - - - see Waste Management
  Information Bureau
Work force, risk, 223-32
WRc - - - see Water Research Centre
WRC Information, 38

Xenobiotic chemicals, disposal, 123